AF785477

Monographs on Endocrinology

Volume 23

Edited by

F. Gross, Heidelberg · M. M. Grumbach, San Francisco
A. Labhart, Zürich · M. B. Lipsett, Bethesda
T. Mann, Cambridge · L. T. Samuels (†), Salt Lake City
J. Zander, München

E. Flückiger
E. del Pozo
K. von Werder

Prolactin

Physiology, Pharmacology and Clinical Findings

With 54 Figures and 14 Tables

Springer-Verlag
Berlin Heidelberg New York 1982

Edward W. Flückiger

Pharmaceutical Division, Preclinical Research, Sandoz Ltd., Basle/Switzerland

Emilio del Pozo

Pharmaceutical Division, Experimental Therapeutics Sandoz Ltd., Basle/Switzerland

Klaus von Werder

Department of Medicine Innenstadt, University of Munich, Munich/Federal Republic of Germany

ISBN-13: 978-3-642-81723-6 e-ISBN-13: 978-3-642-81721-2
DOI: 10.1007/978-3-642-81721-2

Library of Congress Cataloging in Publication Data.

Flückiger, E. (Edward), 1923– . Prolactin: physiology, pharmacology, and clinical findings.
(Monographs on endocrinology; v. 23)
Bibliography: p. Includes index. 1. Prolactin. 2. Prolactin – Metabolism – Disorders.
I. Del Pozo, E. (Emilio), 1932– . II. Werder Klaus von. III. Title. IV. Series.
[DNLM: 1. Prolactin. WI MO57 v. 23 / WK 515 F646p]
QP572.P74F58 616.4'7 81-18314

This work is subject to copyright. All rights are reserved, whether the whole or part of the material is concerned, specifically those of translation, reprinting, re-use of illustrations, broadcasting, reproduction by photocopying machine or similar means, and storage in data banks. Under § 54 of the German Copyright Law where copies are made for other than private use, a fee is payable to 'Verwertungsgesellschaft Wort', Munich.

© Springer-Verlag Berlin, Heidelberg 1982

Softcover reprint of the hardcover 1st edition 1982

The use of registered names, trademarks, etc. in this publication does not imply, even in the absence of a specific statement, that such names are exempt from the relevant protective laws and regulations and therefore free for general use.

2125/3020–543210

Foreword

Lactogenic hormone activity was first observed in bovine pituitary extracts by Stricker and Grüter in 1928, working in Bouin's laboratory in Strasbourg. Since that time prolactin has been shown to exist in anterior pituitary extracts of almost all vertebrate species investigated. Although its biology was extensively studied in many mammalian species, the existence of prolactin in the human was generally doubted, despite the positive evidence produced by such researchers as Pasteels. This can partly be explained by the fact that human growth hormone isolated in 1961, is itself a potent lactogen, in contrast to nonprimate growth hormones, and is present in the normal human pituitary in much greater amounts than prolactin. As a result there was a lag of nearly 10 years until prolactin was unanimously accepted as a hormone of the human pituitary, separate from human growth hormone. In 1970 new bioassay techniques permitted the demonstration of prolactin bioactivity in the serum of postpartum women and galactorrhea patients, and chromatographic methods led to the isolation and purification of human prolactin allowing the establishment of a specific radioimmunoassay for this hormone. This opened the road to the understanding of prolactin physiology and pathophysiology in the human, which has revolutionized clinical neuroendocrinology and reproductive endocrinology. Particularly hyperprolactinemia has turned out to be one of the most common endocrine syndromes. Simultaneously with the discovery of human prolactin, bromocriptine, a drug which had been developed in the late sixties, specifically to inhibit prolactin secretion, was introduced to the clinic as an investigational drug for the treatment of hyperprolactinemic states. The stories of prolactin and bromocriptine are therefore inextricably interwoven. In the past 10 years an explosion of knowledge in endocrinology, pharmacology and clinical therapeutics around prolactin has occurred, which makes it difficult for many physiologists and clinicians who are not directly engaged in prolactin research to remain well informed.

The purpose of this book, therefore, is to review the relevant aspects of prolactin physiology and pharmacology in mammals and particularly in man and to present a review of the clinical aspects of prolactin pathophysiology, i.e., diagnosis and treatment of hyperprolactinemic states. This book has been written mainly for the clinician who feels the need to catch up with the rapidly evolving knowledge about prolactin, but we also hope that the book will serve as a useful review for endocrinologists outside the clinic who are interested in prolactin physiology or pharmacology.

When we accepted the invitation by Springer-Verlag to write this book in the series "Monographs on Endocrinology" we did not anticipate how great a task lay before us. We wish to thank Springer-Verlag for their great patience and for their meticulous help in bringing the manuscript to press.

Basle and Munich, August 1981

E. Flückiger
E. del Pozo
K. von Werder

Contents

1 Prolactin: Synthesis, Fate and Actions

1.1 Molecular Structure

Prolactin from the sheep had already been prepared in a highly purified form in the late 1930's but it was only in 1970/1972 that human prolactin (hPRL) was isolated and differentiated from human growth hormone (hGH) (Frantz and Kleinberg 1970; Lewis et al. 1971; Hwang et al. 1971, 1972). The identification of hPRL as an entity separate from hGH occurred with about 10 years delay, partly because of the high concentration in human pituitaries of GH ($\sim$10 mg/gland) and low content of PRL ($\sim$0.1 mg/gland) and partly because hGH, in contrast to GH from other species, has pronounced prolactin-like activity which makes bioassay-supported fractionation of pituitary extracts to obtain hPRL nearly impossible (Friesen and Hwang 1973). The linear amino acid sequence of hPRL has recently been elucidated (Fig. 1.1) (Shome and Parlow 1977). The molecule is made up of 198 amino acid residues including, typical of prolactin molecules, six half-cystines. The sequence identity of hPRL with PRL of sheep (oPRL), pig (pPRL) and rat (rPRL) is 73%, 77% and 60% respectively. Sequence identity with hGH, however, was found to be only 16%, and with human placental lactogen (hPL) only 13%. This low level of sequence identity between hPRL and hGH or hPL is highly surprising in view of the potent prolactin-like activity of the latter two peptides. Studies of the secondary and tertiary structure of prolactin from various species have been performed using the technique of circular dichroism measurement (Bewley 1979).

Human prolactin has been found to occur in different molecular sizes, as "little" or "small", "big" and "big-big" prolactins (Kruse 1973; Rogol and Rosen 1974; Suh and Frantz 1974; von Werder and Clemm 1974). The majority of immunoreactive prolactin elutes as little prolactin corresponding to a molecular weight of about 22000. Eight percent to 24 percent of total prolactin immunoreactivity elutes as big prolactin with an apparent molecular weight of 40–50000 daltons (Rogol and Rosen 1974; von Werder and Clemm 1974). A minute fraction of immunoreactive prolactin is found in the void volume of the column with an apparent molecular weight greater than 60000 daltons (von Werder and Clemm 1974). The meaning and the source of the different prolactin moieties in the serum is still not clear. Since big prolactin is also found in pituitary extracts (Suh and Frantz 1974) it is likely that both little and big prolactin are secreted from the pituitary. However, the fact that in patients with pituitary tumors and suprasellar extension big and little prolactin was found in the serum whereas only small prolactin was found in the cerebrospinal fluid, suggests that big prolactin may result from the binding of little prolactin to plasma proteins or its aggregations in the presence of serum after release from the pituitary (Jordan and Kendall 1978).

Freezing and thawing of serum leads to conversion of big into little prolactin and the appearance of larger fractions of immunoreactive prolactin eluting in the void

NH_2-Leu-Pro-Ile-Cys-Pro-Gly-Gly-Ala-Ala-Arg-Cys-Gln-Val-Thr-Leu-Arg-Asp-Leu-Phe-

Asp-**Arg-Ala**-Val-Val-**Leu**-Ser-His-Tyr-Ile-His-Asn-Leu-Ser-Ser-**Glu**-Met-Phe-Ser-Glu-

Phe-Asp-Lys-Arg-Tyr-Thr-His-Gly-Arg-Gly-Phe-Ile-Thr-Lys-Ala-Ile-Asn-Ser-**Cys**-His-

Thr-**Ser**-Ser-Leu-Ala-**Thr-Pro**-Glu-Asp-Lys-**Glu**-Gln-Ala-**Gln-Gln**-Met-Asn-Gln-Lys-Asp-

Phe-**Leu**-Val-Ser-Ile-**Leu**-Ile-**Leu**-Arg-Ser-Trp-Asn-Glu-Pro-Leu-Tyr-His-Leu-Val-Thr-

Glu-**Val**-Arg-Gly-(Asx)-Gln-Glu-Ala-Pro-Glu-**Ala**-Ile-Leu-**Ser**-Lys-Ala-Val-Glu-Ile-Glu-

Glu-Gln-Thr-Lys-Arg-Leu-Leu-Glu-Gly-Met-Glu-Leu-Ile-Val-Ser-Gln-Val-His-**Pro**-Glu-

Thr-Lys-Glu-Asp-Glu-Ile-Tyr-Pro-Val-Trp-**Ser**-Gly-Leu-Pro-Ser-Leu-Gln-Met-Ala-**Asp**-

Glu-Ser-Glu-Arg-Leu-Ser-Ala-Tyr-**Tyr**-Asn-**Leu-Leu**-His-**Cys**-Leu-**Arg**-Arg-**Asp**-Ser-His-

Lys-Ile-Asp-Asn-Tyr-**Leu**-Lys-Leu-Leu-Lys-**Cys-Arg**-Ile-Ile-His-Asn-Asn-Asn-**Cys**-OH

Fig. 1.1. The linear amino acid sequence of human prolactin. Residues identical with the human growth hormone sequence are indicated in bold letters. (Shome and Parlow 1977)

volume after column chromatography (Suh and Frantz 1974). This suggests that big prolactin may be a dimer which is in part noncovalently bound and that the prolactin with a molecular weight of more than 60 000 represents a high molecular weight aggregate.

The highest amounts of big prolactin have been found in pregnant females (Suh and Frantz 1974). Though in some patients an extremely high proportion of big prolactin (up to 70% of total immunoreactivity) has been observed (Rogol and Rosen 1974), in most patients with prolactinomas 75% of total prolactin immunoreactivity elutes always as little or monomeric prolactin. Partial pituitary adenomectomy leading to a decrease in prolactin levels to half of the initial value does not lead to a significant change in the distribution of prolactin immunoreactivity.

Recently it was observed that the proportion of big-big prolactin in normal adults was greatly reduced after the injection of TRH with a corresponding increase in little hPRL (Farkouh et al. 1979).

All fractions have shown identical immunoreactivity (von Werder and Clemm 1974; Suh and Frantz 1974). However, a decreased activity in a radioreceptor assay of big prolactin has been reported (rabbit mammary gland membranes) (Garnier et al. 1978; Farkouh et al. 1979). This had not been observed by others (Guyda 1975; Rogol 1975) using the same radioreceptor assay system. Dattatreyamurty and

Sheth (1977), using the rat ventral prostate, found higher receptor activity of the less retarded prolactin fractions compared to monomeric prolactin. The diverging results may be due to the different sources of prolactin and the differing stability of the big prolactin fraction. Thus, Dattatreyamurty and Sheth (1977) used prolactin containing human seminal fluid which showed a different elution pattern compared to serum prolactin. Since freezing and thawing of serum prolactin never results in complete conversion of big into little prolactin (Suh and Frantz 1974; Garnier et al. 1978) these results suggest, in addition, that we are dealing with two different forms of big prolactin. One is bound by noncovalent bonds to another peptide, e.g., to another prolactin molecule, whereas the stable big prolactin is bound by an interpolypeptide chain disulfide bond (Garnier et al. 1978). The latter situation has been found to be true also for hGH (Lewis et al. 1977) and placental lactogen (Schneider et al. 1975). The radioreceptor activity of the two different big prolactins, i.e., convertible and stable big-hPRL has not been investigated.

The polymorphism of hPRL is shared with prolactins of other species (Asawaroengchai et al. 1978). It is not known for certain whether the different froms can be secreted independently and how interconvertible (in vivo) they are. Recently it was found that big-big prolactin may be produced by Sephadex column chromatography, i.e., to represent an artefact (Fang et al. 1979) of prolactin bound to polysaccharides. The consequences of such findings for the study of prolactin biology are not known (Nicoll 1972, 1975; Nicoll et al. 1976). The problems involved have not yet brought concern to the average consumer of RIA data. Still, a recent study in a strain of mice with a high incidence of mammary tumors clearly showed that the interconversion between polymorphic forms of prolactin may be of great importance in pathophysiologic studies (Sinha and Baxter 1979), primarily, but not only, because of the analytical insecurity involved.

It can be assumed that in the future the study of peptide polymorphism and its bearing on the various assay systems, as well as the study of its biologic consequences, will become very important to the progress in endocrinology.

1.2 Synthesis, Storage and Secretion by the Pituitary

Prolactin is produced by cells of the anterior pituitary lobe (pars distalis hypophysis), which belong to the so-called acidophilic cell-type. Histologic staining techniques enable the differentiation of prolactin- and growth hormone-producing cells (Friedgood and Dawson 1940; Herlant 1960; Holmes and Ball 1974). In recent years, immunofluorescence techniques have often been used for the identification of prolactin cells (see, e.g., Duello and Halmi 1979). Recently, application of this technique revealed the presence, in the pituitary of the golden hamster, of cells reacting with human placental lactogen antibody (Campbell et al. 1979). A detailed discussion of the morphology and tinctorial aspects of prolactin cells will be given in Sect. 5.1, in connection with pathologic aspects.

Prolactin synthesis takes place predominantly on membrane-bound polysomes, as studies with GH_3 cells have revealed (Biswas and Tashjian 1974). Using rat pituitaries or the rat pituitary tumor MtTW10 mRNA was isolated, which in a wheat germ cell-free system induced the synthesis of rat preprolactin (Maurer et al. 1976), with a molecular weight of about 28 000 as compared to 22 500 for rPRL. The amino acid sequence of this precursor segment of about 6000 dalton has been partially elucidated (Maurer et al. 1977) and found to resemble precursor segments of other secretory proteins. The chemical nature of the precursor segment with its accumulation of leucine residues would favor the interaction of the protein with membranes. It is not known at which point in the sequence of events in the production of prolactin this precursor segment is cleaved from the N-terminal of prolactin. Messenger RNA for oPRL has been isolated from sheep pituitaries and characterized recently by Shupnik and Gorski (1980). Prolactin is finally sequestered in cytoplasmic membrane-bound ovoid to polymorphous granules of 600–900 nm diameter. Hormone extrusion is obtained when the membrane of a granule fuses with the cell membrane to release the granular hormone to the outside (exocytosis or "reverse pinocytosis") (Trifaro 1977). Hormone granules which are not released from the cell may fuse with lysosomes and be digested (crinophagy, Farquhar 1977). Isolated hormone granules were found to have dopamine associated with them (Gudelsky et al. 1979). The granules were observed to react to TRH, stimulatory hypothalamic extracts, and calcium by release of hormone (Dular and LaBella 1978). It is unclear whether in the intact prolactin cell hormone release depends on cyclic (c) AMP, although dibutyryl-cAMP or theophyllin added to pituitary cells in vitro increase prolactin release (Nagasawa et al. 1973; Sundberg et al. 1976). On the other hand, TRH was found to stimulate prolactin release without inducing an increase in cAMP, and cholera enterotoxin augmented cAMP without stimulating prolactin secretion (Sundberg et al. 1976). The study by Thorner et al. (1980 b) also led to the conclusion that calcium rather than cAMP is the intracellular regulator of prolactin release. The role of calmodulin should also be considered (Cheung 1980). The release process of prolactin is not yet thoroughly understood. On the one hand, analytical problems have been found to become prominent when assessing quantitatively changes of prolactin content in the pituitary of lactating rats (Nicoll et al. 1976; Häusler et al. 1978). On the other hand, Nicoll (1972) and Grosvenor et al. (1979 a, b) have produced evidence that prolactin moves in at least two steps from the prolactin cell to the circulating blood in suckling-stimulated, lactating rats. After the first step, depletion from the prolactin cell, a phase of hormone transformation follows before the hormone is released into the circulation.

The number of prolactin cells is not constant but varies with the physiologic state of the animal (Crane and Loomes, 1967). Also, in women, prolactin cells are more numerous in pregnancy and in the postpartum (Erdheim and Stumme 1909; Pasteels 1972). It is also known that treatment of rats with estrogens which stimulate prolactin secretion may induce pituitary tumor formation. This suggests that prolactin cell proliferation and secretory activity are connected.

Stimulation of prolactin secretion by estrogens increased DNA synthesis (Kalbermann et al. 1979 a, b; MacLeod, 1976) and mitotic activity (Davies et al. 1974; Pawlikowski and Stepien 1979). Continued stimulation of the prolactin cells by estrogens may lead to tumor formation in the pituitary gland (Stevens and Helfenstein 1966) but spontaneously developing prolactin-secreting adenomas may be observed in ageing rats. An early but valuable discussion of prolactinomas in experimental animals and man was given by El Etreby and Günzel (1973) (see also El Etreby 1980). Interestingly, subchronic administration of psychotropic stimulants of prolactin secretion, in contrast to what was observed after estrogen, failed to produce evidence of stimulated synthesis of newly formed glycolytic enzymes (Nagy et al. 1979 a). Inhibition of prolactin secretion led to decreased cell proliferation and DNA synthesis (Davies et al. 1974; Lloyd et al. 1975; Stepien et al. 1978, Pawlikowski and Stepien 1979).

The prolactin-producing pituitary cells of teleosts and mammals including man, show spontaneous electrical activity (Kidokoro 1975; Biales et al. 1977; Douglas and Taraskevich 1977; Ozawa and Sand 1978; Taraskevich and Douglas 1978; Dufy et al. 1979 a, b). Action potentials of low frequency were observed. Both sodium and calcium are involved in the electrical activity. Increase of spike activity in the presence of the prolactin release stimulating TRF or 17β-estradiol and attenuation of spike activity in the presence of prolactin release inhibiting catecholamines have been described. It may be assumed that electrical activity and hormone release are interrelated, calcium being a very likely candidate for the coupling of membrane excitation and hormone release. Calcium was found to be essential for the stimulated prolactin release of GH_3 cells in the presence of TRH or high potassium concentrations (Tashjian et al. 1978) and for the release of prolactin from isolated secretory granules (Dular and LaBella 1978). Nifedipine, a Ca^{2+} blocker, reduced spontaneous prolactin release from rat pituitary halves, but only very high concentrations were able to also reduce TRH-stimulated prolactin secretion (Takahara et al. 1979).

1.3 Extrapituitary Sites of Prolactin Synthesis

Remnants of the fetal pituitary (Rathke pouch) may be left behind and become embedded in the mucoperiosteum of the roof of the nasopharynx. Prolactin and growth hormone have been demonstrated in pharyngeal hypophyses (McGrath 1968) but the relevance of this tissue as a source of prolactin in physiologic or pathologic conditions is not known.

The use of the RIA technique has led to the detection of prolactin in the amniotic fluid in the human and in nonhuman primate species (Hwang et al. 1971). This finding attracts special interest as the prolactin concentration in the amniotic fluid of the human exceeds maternal or fetal serum levels 10- to 20-fold at midgestation (Tyson et al. 1972 a; Clements et al. 1977). In the rhesus monkey, a species where no significant rise in prolactin occurs in the pregnant female nor in the fetus, the amniotic fluid prolactin concentrations were found to be very similar to those in the human species (Friesen et al. 1972). Early studies in man assumed that the prolactin in the amniotic fluid was derived from maternal or fetal pituitary prolactin, but Friesen et al. (1972) concluded tentatively that it was produced by the chorion. Healy et al. (1978) demonstrated that the decidua-chorion over an incubation period of 3 days released more immunoreactive prolactin into the culture medium than it originally contained. Production of prolactin could be reduced by adding puromycin or cycloheximide (Riddick et al. 1978; Healy et al. 1979). In a further study (Bigazzi et al. 1979) only culturing of the decidua produced hPRL by immunologic criteria, while hPL and hCG were synthesized during incubation of placenta but not of amnion, chorion or decidua capsularis tissue. Incubation of human decidual tissue together with ^{3}H-leucine produced radioactively labeled peptides, part of which were specifically precipitated by antiserum to human pituitary prolactin and when chromatographed on Sephadex G-100 eluted in the same fractions as human pituitary prolactin (Riddick et al. 1978). Thus, human decidual tissue synthesizes a peptide similar to human pituitary prolactin. This led to the study of the human endometrium for its capacity to synthesize such peptides. Maslar et al. (1980) observed that the human endometrium is competent in the final days of a normal cycle to produce prolactin-like immunoreactive material and that after conception this synthetic capacity increases rapidly in early pregnancy. Trophoblastic tissue is not necessary for this. The release of prolactin from human decidual tissue, in contrast to pituitary tissue, was unresponsive to dopamine, TRH (Golander et al. 1979) or to bromocriptine (Bigazzi et al. 1979).

Recently, immunohistochemical studies of various brain areas of the rat revealed prolactin-antibody positive neurons in the hypothalamus and preoptic area. The neuronal content of this prolactin-like material was not decreased when the rats were hypophysectomized (Fuxe et al. 1977; Toubeau et al. 1979), suggesting a nonpituitary origin. The perikarya of such neurons were localized in the arcuate, ventromedial, premammillary, supraoptic and paraventricular nuclei of the hypothalamus (Toubeau et al. 1979).

Prolactin may also be produced by cells of the intestinal tract. Prolactin-immunoreactive cells have recently been identified in the jejunal mucosa of patients with celiac disease (McFarland 1979).

Carcinoma cells in the lung and the kidney have also been found to produce prolactin (Turkington 1971; Rees et al. 1974). For further details see Chap. 4.

1.4 Prolactin in the Body Fluids; Metabolic Clearance

Prolactin after its release and secretion from the pituitary enters the blood stream by which it is carried to the target organs and to the eliminating systems.

Prolactin was found in human and rat milk and its concentration correlated well with the plasma levels (Gala et al. 1975; Grosvenor and Whitworth 1976). In the rat, prolactin seems to be retained in the milk either by physical or chemical means. Little back-diffusion was observed. Grosvenor and Whitworth (1976) concluded that the mammary gland in the lactating rat may serve as an organ of prolactin extraction and excretion, being responsible for the increased metabolic clearance rate for prolactin in lactating rats and sheep (Davis and Borger 1973). In postpartum women hPRL was found in high concentrations (ca. 140 ng/ml) during the first 3 days followed by a rapid fall, reaching about 30 ng/ml by the 11th day after delivery (Healy et al. 1980). Unfortunately, no data were given of serum prolactin levels in these women. The authors found that the drop in milk prolactin concentration was preceded by an increase in lactose and a decrease in total milk protein concentrations. Prolactin obtained by neonates in the milk may pass through the intestinal tract wall into the blood, at least in rats up to 13 days of age (Whitworth and Grosvenor 1978).

Prolactin was reported to occur also in human and murine urine and the content seemed to correlate with values in the serum (Sinha et al. 1973), but dialyzation of human urine resulted in complete loss of radioimmunoactivity, while added intact hPRL could be recovered by more than half (Gala et al. 1975). This indicates that urine may contain small sequences of degraded prolactin, with preserved immunoreactivity.

Prolactin is also found in the cerebrospinal fluid (CSF) (Clemens and Sawyer 1974) of rabbits and rats. In rats, CSF levels of prolactin also rose in the serum in the afternoon of proestrus, or after haloperidol (Login and MacLeod 1977), although basal and peak levels were much lower than in blood. It could be assumed that CSF prolactin contributes to the feedback control of prolactin secretion. In man, prolactin was recently found to be a normal constituent of CSF (Login and MacLeod 1977) and the levels showed a good correlation with plasma levels (Assies et al. 1978). In the CSF, the prolactin levels were lower than in the plasma by a factor of > 3 in the majority of nontumor patients, while in pituitary tumor patients, about half had a factor of < 3. Compared to the relative concentrations in the CSF of other proteins (Felgenhauser 1974) the relative prolactin value is astonishingly high and could indicate that prolactin in the CSF is not in simple diffusion equilibrium with the serum prolactin. It has been suggested (Assies et al. 1978) that the surplus of prolactin is due to retrograde transport of prolactin from the pituitary up through the hypothalamus. For further discussion see Kendall and Orwoll (1980).

Basal serum prolactin levels were found to be elevated in patients with impaired renal function (Chirito et al. 1972; Nagel et al. 1973; Czernichow et al. 1976). The reason for this was not clear as renal patients commonly take drugs which may increase prolactin secretion. A recent study has now clarified the issue. Cowden et al. (1978) have shown that in their renal patients, a significant correlation between serum prolactin and creatinine concentrations existed. Normalization of prolactin levels occurred after successful renal transplantation. In patients with nonrenal disease, a 16% arteriovenous prolactin concentration difference could be demonstrated across the kidney. In patients with liver disease, renal extraction was estimated to be about 90–18 μg/h while at the same time hepatic extraction, which was not markedly affected by the severity of the disease, amounted to only about 10% of the renal elimination rate (Bauer et al. 1980). This indicates that, in man, the kidney has an important role in the metabolic clearance of prolactin. In experimental animals, the process of metabolic clearance of prolactin has not been studied in detail. Donatsch and Richardson (1975) have demonstrated in the rat the filtration of injected prolactin in the kidney and also the rapid reuptake of immunoreactant material by the epithelial cells of the proximal tubule. Van der Gugten and Kwa (1970) concluded from their measurements of the disappearance rate of injected bovine prolactin in female rats that at least two processes of elimination exist. In consequence, disappearance rate was not a constant. Metabolic clearance rate was found to be comparable in conscious and anesthetized lactating rats by Grosvenor and Whitworth (1979).

Reported half-life times for circulating prolactin vary greatly, probably depending on the experimental situation. Using a bioassay method, Grosvenor (1967) reported a half-life of 3 min for porcine and bovine prolactin in the serum of adult, castrated female rats. Ovine prolactin showed a half-life of 10–12 min in such animals. Rat prolactin disappeared with a half-life period of 10–12 min from adult castrated female rats and of 11 min from intact rats with corpora lutea, while in lactating rats the half-life was reduced to 3 min. Using ether as a stimulus to release endogenous prolactin in intact and castrated male rats, Chi and Shin (1978), found a disappearance rate (by RIA) with a half-life of 7 min. After prolonged treatment of rats with oPRL Diamond et al. (1980) observed a decrease of the heterologous prolactin clearance rate with the initial short half-life of 6–12 min being maintained and a new slow half-life of 40–>200 min becoming prominent. This agrees with the conclusions of van der Gugten and Kwa (1970) that prolactin clearance is not a constant single process, but this phenomenon may not be relevant for endogenous prolactin levels. In female rabbits radio-labeled prolactin disappeared with a half-life of 16 min (Birkinshaw and Falconer 1972), while in the lactating cow 29 min was measured as the half-life for the disappearance of bovine prolactin (Johke 1969). It was suggested (Bennett and McMartin 1979) that these results may indicate a correlation between the half-life of prolactin and the size of the animal studied.

1.5 Prolactin Receptors

Interactions between a hormone and its target cells are mediated by specific receptors. With polypeptide hormones these receptors are usually situated on the cell membrane, but intracellular binding sites for prolactin have also been described (Nolin and Witorsch 1976; Nolin 1978). Birkinshaw and Falconer (1972) first observed that radioactively labeled prolactin sticks to the serosal side of mammary epithelial cells and Turkington et al. (1973) demonstrated specific binding sites for prolactin on cell membranes of various rodent tissues: mammary gland, liver, kidney, brain (without cortex and cerebellum), adrenals, ovary and seminal vesicles. These findings have been confirmed by others and extended to several species (Kelly et al. 1974 a, b; Posner et al. 1974). Also, new organs were added to the list: the uterus of rabbit and sheep (Posner et al. 1974), the ciliary body of the rabbit eye (Niederer et al. 1975), the prostate and the epididymis (Aragona and Friesen 1975; Barkey et al. 1977), the placenta, the fetal heart and the fetal lung (Josimovich et al. 1977), the ependyma of the choroid plexus in the rat brain (Walsh et al. 1978). A review on prolactin receptor studies has recently appeared (Nagasawa et al. 1979). The prolactin receptor of the rabbit *mammary gland* has been first studied in great detail by Shiu and Friesen (1974 a, b, 1976 a, b), who solubilized and partially purified it and who raised an antiserum to it (Bohnet et al. 1978). The solubilized receptors showed a somewhat higher binding affinity for prolactin than the particulate state, but in other respects, the two forms were identical. The receptor appeared to be a protein molecule of 220000 dalton on the basis of gel filtration experiments on Sepharose 6B. The rabbit mammary gland prolactin receptor is not specific for prolactin but also binds both hGH and hPL, which are known for their prolactin-like (lactogenic) action. The receptor is, therefore, also called lactogen receptor, or lactogenic receptor. In contrast to the high prolactin binding capacity of rabbit gland membranes, it appears that the binding capacity of such material from rats, mice and ruminants is relatively small (Forsyth and Hayden 1977). Mammary prolactin receptors are regulated by various endocrine factors (Nagasawa et al. 1979). Thus, the binding of prolactin to rat mammary glands was found to be enhanced by estrogen, which effect was reduced by progesterone or by hypothyroidism or by hypophysectomy. The effect of hypophysectomy could be partially corrected by prolactin (Hayden et al. 1979). In neither the rat nor the rabbit is sufficient information available to understand clearly the interactions of the mammotropic peptide hormones with estrogens, progestogens, corticoids and thyroid hormones at the receptor level (Djiane and Durand 1977; Waters et al. 1978; Bhattacharya and Vonderhaar 1979; Sakai et al. 1979). Prolactin receptors of the mammary gland like those of the liver were found to be induced and maintained by prolactin (Djiane and Durand 1977; Djiane et al. 1979 a, b) but there is also evidence for a rapid and reversible suppression of such receptors by prolactin (Djiane et al. 1979 a). This down-regulation of binding capacity may be indicative of an enhanced phase of receptor internalization (Davies et al. 1980). Intracellular prolactin receptors in mammary glands have been found (Houdebine and Djiane 1980) and interpreted as a step in signal transmission from the membrane to the nucleus to induce specific protein synthesis. For further details see Cowie et al. (1980).

Receptors from the *ovaries* of rats, cows and human were similar in several respects to the material isolated from mammary glands (Saito and Saxena 1975). The binding capacity of rat ovaries for prolactin was found to vary during the estrus cycle, being lowest in metestrus and highest in proestrus, thus correlating with estrogen levels, as was also found to be true for the mammary prolactin receptor (Friesen 1973; de Sombre et al. 1976).

In the ovary of pseudopregnant rats, prolactin was found to bind to the corpora lutea (Midgley 1973), and this was also the case in the pig (Rolland et al. 1976). The development of the prolactin binding capacity of the rat corpus luteum was independent of whether prolactin was being suppressed by bromocriptine from the time of luteinization onward or not (Holt et al. 1976). Lactogen receptors in luteinized rat ovaries were found to be reduced temporarily after an LH stimulus (Davies et al. 1980). Besides being bound to ovarian cellular membranes, prolactin has also been found within luteal cells, bound to cytoplasmic structures (Nolin 1978).

The *testes* of rodents also have the capacity to bind prolactin. Binding is exclusive to Leydig cells (Charreau et al. 1977; Costlow and McGuire 1977). This binding capacity was found to be greatly reduced in Snell's dwarf mice (DW/J), which are lacking both growth hormone and prolactin and are infertile despite a normal production of gonadotropins (Bartke 1965, 1966 a, b). Prolactin but not GH treatment restored male fertility (Bartke 1966 a) and increased the binding capacity of tested homogenates for prolactin and also for LH to near normal (Bohnet and Friesen 1976). In the rat, exogenous testosterone depressed the prolactin-binding capacity of the testis, as did elevated prolactin serum levels (Barkey et al. 1979). The lactogen-binding capacity of the testis was transiently found reduced after in vivo stimulation by HCG (Davies et al. 1980).

In the *accessory sex organs* of the rat, seminal vesicles and prostate, the prolactin-binding capacity was depressed by castration and increased by exogenous testosterone in intact and castrated animals, whereas suppression of prolactin secretion had no influence on the binding capacity of these organs (Barkey et al. 1979). In the prostate of the normal rat, prolactin was found to bind to the secretory epithelium (Costlow and McGuire 1977). In such cells, intracellular binding sites for prolactin, which were found to be androgen dependent, have also been described (Witorsch and Smith 1977). In the epididymis of rabbits the lactogen binding has been characterized by Orgebin-Crist and Djiane (1979). Binding capacity was found to be quite stable with age while in the testis, prostate and seminal vesicles it was highest at 3 months of age.

The *adrenal cortex* of rats also binds prolactin specifically and this capacity was increased in male rats by dehydration or salt loading (Marshall et al. 1975). Neither aldosterone nor spironolacetone nor renin altered prolactin binding but hydrocortisone and dexamethasone lowered the binding capacity, as did hypophysectomy (Marshall et al. 1978). Adrenal lactogen-binding capacity was only slightly sensitive to changes in thyroid hormone status (Marshall et al. 1979). Immunoreactive prolactin does not only bind to cellular membranes of rat adrenals, especially of the zona reticularis (Costlow and McGuire 1977) but it was also detected within zona fasciculata cells in the cytoplasm and occasionally in the cell nucleus (Nolin 1978). Stimulation of the rat adrenal in vivo by ACTH temporarily reduced the lactogen-binding capacity (Davies et al. 1980).

The *liver* of rodents also contains prolactin binding sites. These have been demonstrated in membrane material isolated from liver homogenates (Turkington et al. 1973; Posner et al. 1974) and also on intact, isolated hepatocytes (Ranke et al. 1976; Herington and Veith 1977). Using liver homogenates, it was found that in male rats castration greatly increased prolactin binding, whereas testosterone replacement restored this to the low level of intact male rats (Barkey et al. 1979). Suppression of prolactin secretion did not influence prolactin binding in intact male rats, but by increasing prolactin levels augmented binding was observed, indicating induction. In Snell's dwarf mice, which lack both growth hormone and prolactin, prolactin treatment increased binding of this hormone to liver membranes in females but not in males (Bohnet and Friesen 1976). Normal male rats could be induced to increase prolactin binding by treatment with estrogens (Kelly et al. 1975; Posner et al. 1975), but this did not occur in hypophysectomized males. Increasing circulating prolactin levels by a pituitary implant under the kidney capsule attenuated the decrease of prolactin-binding capacity in livers of hypophysectomized males. Studies using cycloheximide, an inhibitor of RNA and protein synthesis, indicate that the population of lactogen-binding sites is turned over quite rapidly (Kelly et al. 1975). Prolactin probably induced its own receptors in the liver (Costlow et al. 1975; Posner et al. 1975). In rats of both sexes, bearing GH_3 tumors, secreting both rGH and rPRL, changes in lactogen-binding capacity correlated better with rGH serum levels than with rPRL (Furuhashi and Fang 1979).

Liver prolactin-binding sites are not specific for this hormone but also show affinity to other lactogens, like hGH but not (bovine) bGH. Since the liver is regarded as an important target organ for GH in conjunction with somatomedin formation, the question of whether prolactin-binding sites and GH-binding sites can be differentiated is of importance. In a study using isolated and functionally intact rabbit hepatocytes, Herington and Veith (1977) found only specific binding for prolactin of sheep and man as well as hGH, but no evidence for GH receptors, as evidenced by the lack of specific binding of ovine, bovine or procine GH. In a study conducted with Snell's dwarf mice, which lack GH and prolactin and showed greatly reduced somatomedin serum levels, Holder and Wallis (1977) were able to stimulate growth and increase serum somatomedin by bovine GH as well as with bovine prolactin, although the latter was less potent.

Using human liver slices, Carr and Friesen (1976) have demonstrated the presence of GH receptors to which placental lactogens of sheep, cattle and goats also bind. Lesniak et al. (1977), based on an analysis of published studies and their own findings, have proposed a classification of lactogen and growth hormone receptors according to binding characteristics of hormone (see Table 1.1). From this

Table 1.1 Classification of hormones and receptors based on observed affinities. (Lesniak et al. 1977)

Hormones		Receptors		
		Type 1 for lactogenic hormones	Type 2 for nonprimate GH	Type 3 for primate GH
Type I	Primate GH	100	100	100
	Bovine PL	100	100	100
Type II	Nonprimate GH	0	100	0.03
Type III	Human PL	100	0	0.03
	Ovine PRL	100	0	0.03
	Human PRL	100	0	0

ensues that, at least in the subprimate liver, there exist besides lactogenic hormone receptors, receptors with high affinity to growth hormones but little or no affinity for prolactin. Intracellular lactogen-binding sites have been described in the liver of female rats, using radioiodine labeled hGH: in Golgi and plasmalemma fractions labeling of the Golgi fraction was much higher than of the latter, and lactogen-binding sites seemed more numerous on the cisternal face of the Golgi secretory vesicles than on the outer face (Bergeron et al. 1978).

Davies et al. (1980) recently demonstrated that in the steroidogenic organs of the rat, corpus luteum, testis and adrenal, the capacity to bind lactogen is rapidly but transiently reduced after trophic stimulation of the organ. This is indicative of increased prolactin receptor turnover and was considered an important component of target cell response. The authors suggested that prolactin receptors might be involved in the transport of lipoprotein precursors of steroidogenesis through the cell membrane. This function could then also explain the internalization of prolactin as seen in the mammary gland epithelial cells as well as in ovarian luteal cells, in adrenal cortical cells and also in hepatocytes, where lipoprotein synthesis occurs.

The *kidney* also binds prolactin. According to an autoradiographic study (Costlow and McGuire 1977) this is restricted to the cortical tubules. When ovine prolactin was injected into rats it could be demonstrated by immunofluorescence to appear in the lumen of proximal tubules and then on the brush-border of the epithelium and finally to appear in the cytoplasm of the epithelial cells (Donatsch and Richardson 1975). Kidney membrane preparations of rats bound less prolactin when the animals were thirsting than when normally hydrated (Marshall et al. 1975). In chronic progressive nephropathy of the rat, decreased specific binding was noted (Cowden et al. 1981). Adrenalectomy greatly increased prolactin-binding capacity of rat kidneys while dexamethasone decreased the binding capacity in intact and in hypophysectomized animals (Marshall et al. 1978). Hypophysectomy also decreased prolactin binding by kidney membrane preparations, and so did thyroidectomy (Marshall et al. 1979). This effect could be counteracted by treating the animals with thyroxin. As this hormone had no effect on prolactin binding by adrenal membrane preparations, it was suggested that the stimulation of prolactin binding to renal membranes by thyroxin represents an organ-specific rather than a general metabolic effect (Marshall et al. 1979).

Cell membrane fractions of fetal *lung* of the rhesus monkey were shown to specifically bind hPRL at 69 days of gestation and later (Josimovich et al. 1977), but the total binding capacity was much lower than that in the liver or the placenta. In contrast to these observations, the use of adult lung cell membranes of four species, including monkeys, did not produce evidence for significant binding of oPRL (Posner et al. 1974) indicating possibly a physiologic role of prolactin in the fetal lung but not in the lung after birth.

Cultured human *lymphocytes* (IM-9) were shown by Lesniak et al. (1977) not only to bind hGH, but also oPL with a similar affinity, while binding bGH and oGH as well as oPRL and hPL with only 0.03% of the affinity of hGH. Recently, a lymphoma cell line (rat) was described which responded specifically to lactogenic hormones with an increased rate of mitosis, this effect being in proportion with the activity of such hormones in receptor assays (Tanaka et al. 1980).

1.6 The Spectrum of Actions

Prolactin as a hormone shows a most outstanding spectrum of biologic actions throughout the vertebrate system. In contrast to the other classic pituitary hormones, it is not specialized to regulate one or at most a few functions, but it is involved in a wide variety of physiologic processes of diverse organ systems. These include morphogenetic effects (from metamorphosis in amphibia to breast development and mammary lobuloalveolar differentiation in mammals), effects on specific cell functions (from sodium or water permeability in the fish gill and the amnion of the human fetus to steroidogenesis and the synthesis of specific proteins, like casein, in mammals), as well as behavioral effects in conjunction with reproduction. These examples of the fundamental role of prolactin in "vertebrate function and organization" (Riddle 1963) are so diverse that at present they escape any unifying theory to understand them. Nicoll (1974) has categorized these actions and authoritatively discussed them from a comparative point of view. Important points have emerged from this: many of prolactin's actions, from electrolyte metabolism to behavioral effects, involve some sort of interaction with steroid hormones. The effects of prolactin may depend on the temporal relationship of its release and the peak secretion of other hormones, e. g., steroids (Nicoll 1973, 1974), an aspect which is usually neglected in experimental endocrinology. In the framework of this book, the actions on the mammary gland, the gonads and the male accessory glands, the adrenal cortex, and the kidney are of major interest, but some other aspects will also be mentioned.

1.6.1 Effects on the Mammary Gland (Lactation)

Lactation is usually differentiated into three steps. Mammary growth and differentiation, initiation of milk secretion (lactogenesis), and maintenance of lactation (galactopoiesis) (Cowie and Tindal 1971).

Prolactin is first involved in the development of the mammary duct system from the postpubertal stage to the lactational state. In the rat this consists mainly in lobuloalveolar growth which is promoted by prolactin and progesterone in synergism with estrogens, corticoids, insulin, growth hormone, and thyroid hormones (Nicoll 1974; Vonderhaar and Greco 1979). It is thought that thyroid hormones regulate prolactin-binding capacity on the alveolar cells, while prolactin stimulates cell proliferation (Bhattacharya and Vonderhaar 1979; Hayden et al. 1979). Physiologically, in the rat, the development of the mammary duct system takes place in the course of pregnancy and placental lactogen could then also contribute (Josimovich 1974), but its role is poorly understood. As human newborns quite often produce milk (Hiba et al. 1977), although placental lactogen is only found in trace amounts in fetal blood (Kaplan and Grumbach 1965) it may be concluded that this hormone is not critical in the preparation of the mammary gland for lactation. In the human, the role of progesterone in preparing the mammary gland is probably small, as lactation can be induced in estrogen-primed nonpregnant women by the aid of TRH to stimulate prolactin secretion (Tyson et al. 1976). An observation (Wyss and Del Pozo unpublished) in a male transsexual showed that induction of lactation can be similarly achieved in the human male. The morphogenetic effect of prolactin in the mammary gland duct system is of particular interest in conjunction with the mammary cancer problem (see below).

The next step where prolactin is involved is lactogenesis, which consists of the initiation of milk formation and secretion. This second step usually does not take place until pregnancy is terminated. It is generally accepted that progesterone blocks this process, so that the onset of lactogenesis is timed by the fall of the serum progesterone concentration at the end of pregnancy. In sheep, this role of progesterone has recently been clearly demonstrated (Kann et al. 1978), but in this species prolactin serum levels are also low as long as progesterone levels are high. Prevention of the prepartum rise of prolactin by bromocriptine inhibited lactogenesis in sheep (Kann et al. 1978), in cows (Schams 1976) and in goats (Hart and Morant 1980). In women, estrogens increase together with prolactin throughout pregnancy and their role as blockers of lactogenesis besides progesterone should be considered. Estrogen preparations are in therapeutic use to prevent postpartal lactogenesis (Brun del Re et al. 1973).

The details of lactogenesis have been studied in vitro. The effect of prolactin on the conditioned lobuloalveolar cell consists of the induction (probably after internalization of the receptor-hormone complex) of synthesis of specific ribosomal, messenger and transport RNA, increased RNA polymerase activity, followed by the induction of enzymes which synthesize lactose and casein (Turkington 1972; Delouis and Denamur 1972; Rosen et al. 1979). Recently it was shown by Houdebine and Djiane (1980) that the induction of casein synthesis in rabbit mammary glands in vitro by prolactin could be blocked by cytochalasin B, but the drug did not prevent the accumulation of casein mRNA. The stimulatory effect of prolactin on mRNA and actual casein synthesis was blocked, however, by colchicine. Lysosomal agents (e.g., chloroquine) had no effect. These results suggested to the investigators that degradation of the internalized prolactin-receptor complex by lysosomes is not an obligatory step, and that the integrity of microfilaments is unnecessary in the activation of the casein gene. On the other hand, the microtubule apparatus is involved in the transmission of the prolactin signal to the nucleus.

Once lactogenesis has occurred and milk secretion is established, the continuation of this process, galactopoiesis, depends on a regular emptying of the mammary gland of its product. Prolactin is necessary in certain species for galactopoiesis, but not in all. Thus, in the postpartum woman, in the rat, rabbit, sow, sheep and dog, suppression of prolactin secretion inhibited milk production

(Flückiger 1978a; Kann et al. 1978), but in dairy breeds of goats (Hart 1973) and cows (Schams 1976), prolactin could be suppressed without influence on milk yield. In all species studied, the physiologic stimulation of the teats or nipples by the suckling young, or the mechanical stimulation for experimental purposes induces a short-lasting prolactin release (Grosvenor and Turner 1958; Karg and Schams 1974; Tyson et al. 1972a, b). This effect is coupled to the release of oxytocin necessary for milk ejection (Cowie and Tindal 1971), but the two events are otherwise independent. Prolactin release through nipple stimulation can, to a smaller extent, also be provoked in the male of some species (Karg and Schams 1974). It is therefore not dependent on previous pregnancy or on established lactation. The greater sensitivity of the female to such tactile stimulation may be explained by the observation that estrogens increase the sensory field of certain nerves, as found in rats (Komisaruk et al. 1972, 1973). The amount of prolactin released in women and cows by nipple stimulation depended directly on the intensity and the duration of the stimulus (Schams 1976; Aono et al. 1977). The neural afferent pathway for the stimulation of prolactin release has been analyzed in the rabbit (Tindal 1978; Tindal and Knaggs 1972). Apart from the tactile stimulus, visual and olfactory stimuli from the young may be very effective in releasing prolactin in lactating females, at least in rats (Mena and Grosvenor 1971). In the postpartum woman, basal prolactin serum concentrations are maximal at delivery and then decline to normal levels with an attenuation which depends decisively on the frequency of suckling (Delvoye et al. 1978). This physiologic hyperprolactinemia is accompanied by amenorrhea and has therefore been praised as the natural and efficient spacer of pregnancies (Short 1976; Tyson et al. 1976; Delvoye et al. 1978). For an extended discussion of this endocrine situation see Chap. 3.

Human growth hormone but not nonprimate GH also has lactogenic activity, the potency varying with the test system used. Recently, in a comparative study (Kleinberg and Todd 1979) hGH was found to be of similar lactogenic potency to hPRL itself, if assayed on subhuman primate mammary tissue. This indicates that the mammotropic potential of hGH should not be neglected in the discussion of the pathology of the human breast. An authoritative presentation of the "Hormonal Control of Lactation" has recently been published in the form of a monograph (Cowie et al. 1980).

1.6.2 Prolactin and the Mammary Cancer Problem

It has been shown that long-term treatment of mice with prolactin, or the implantation of ectopic pituitaries in mice, as well as other methods to continuously increase prolactin levels, will lead to an increased incidence of mammary tumors (Mühlbock and Boot 1967; Boot 1969; Welsch et al. 1970; Röpcke 1975). Suppression of serum prolactin by pharmacologic means depresses the development and progression of early neoplastic mammary gland lesions in strains of mice with a high incidence of mammary carcinoma (Welsch 1978), or in mice and rats with carcinogen-induced mammary neoplasias (Heuson et al. 1970; Stähelin et al. 1971; Nagasawa 1977). Such mammary tumors are considered by many as a model of mammary cancer in women, and have therefore been extensively studied either in rodent strains with a high spontaneous frequency of tumor formation or in rodents treated with a carcinogen (mostly polycyclic aromatic hydrocarbons). An interesting aspect of these studies is that the effectiveness of the carcinogen has been found to depend on the mammary gland DNA synthesis at the time of application. If DNA synthesis is high, e. g., in proestrus, the effectiveness of the carcinogen is greater than if DNA synthesis is low (e. g., in diestrus). DNA synthesis is controlled by prolactin serum levels. Also, in strains with a high incidence of mammary tumors this dependence of mammary DNA synthesis on prolactin levels and the close relationship between tumorigenesis and DNA synthesis was observed (Nagasawa et al. 1976; Nagasawa 1977; Nagasawa and Yanai 1978). In rats treated with carcinogen, a dependence of the ex vivo protein-synthesizing activity of mammary tumor ribosomes on in vivo prolactin levels was also demonstrated (Hartmeyer et al. 1975). In women, the question regarding how far prolactin is involved in mammary cancer remains unanswered (Short and Drife 1977). In vitro studies on biopsies from patients have revealed a very small incidence of prolactin dependence in mammary cancer tissue (Hobbs et al. 1973), and clinical studies using pharmacologic suppression of prolactin levels by L-DOPA or bromocriptine have, on the whole, not really been encouraging (Frantz et al. 1973; Schultz et al. 1973; European Breast Cancer Group 1972). At present, the therapeutically relevant question of prolactin involvement in cancer of the breast receives little attention (Dogliotti et al. 1979; Blossey et al. 1979). In a recent study of the distribution of estrogen- and prolactin-binding sites in human breast tumors, di Carlo et al. (1980) found specific hPRL binding in 19 of 53 tumors, of which eight were associated with estrogen-binding sites. In contrast to the distribution of estrogen "receptors" which were relatively more numerous in tissue of postmenopausal patients, prolactin-binding sites were equally frequent in pre- and postmenopausal women. No attempt at therapeutic conclusions was made.

1.6.3 Effects on the Ovary

The second important function that prolactin was found to have in mammals was that of maintaining the corpus luteum in the rat (Astwood 1941; Evans et al. 1941). This led to its second name, luteotrop(h)ic hormone (LTH). In the female rat, when increased prolactin serum levels are induced by implanting an anterior pituitary under the kidney capsule (Zellmarker and Carlsen 1962) or by giving certain drugs which suppress the inhibitory hypothalamic control of prolactin secretion (van Maanen and Smelik 1968; Flückiger et al. 1972), then pseudopregnancy follows. This situation can be interrupted by suppressing prolactin secretion (Zeilmarker and Carlsen 1962; Flückiger et al. 1972, 1976) which is followed by new ovarian cycles. In the inseminated rat, prolactin suppression before ovum implantation interrupts pregnancy and induces a new ovarian cycle. This can be prevented by injecting prolactin (Flückiger et al. 1976) which maintains luteal function and pregnancy. After ovum implantation, prolactin seems to be no longer critical for the functioning of the rat corpus luteum (Morishige and Rothchild 1974).

Luteinization of the rat ovary induces an increased production of progesterone. Hypophysectomy of pseudopregnant animals was found to reduce progesterone synthesis in the ovary. Acetate incorporation into cholesterol was suppressed and at the same time 20-α-hydroxysteroid-dehydrogenase-dependent inactivation of progesterone to 20-α-hydroxyprogesterone increased (Armstrong et al. 1970). Prolactin reversed these effects. Pharmacologic suppression of prolactin secretion in intact rats produced similar effects to hypophysectomy (Lindner and Shelesnyak 1967; Lamprecht et al. 1969; Rodway and Kuhn 1975). In the rat, complete suppression of corpus luteum function in early pregnancy necessitates the suppression of prolactin secretion for about 24 h (Wuttke and Döhler 1973; Flückiger and del Pozo 1978). Interestingly, prolactin was also found to have similar biochemical effects in vitro in the interstitial ovarian tissue of rabbits (Hilliard et al. 1969) and of cows (Bartosik et al. 1967; Bartosik and Romanoff 1969), species in which prolactin is not the critical luteotrope hormone. If cows, pigs and sheep are hysteroectomized, a luteotrope action of prolactin is revealed also in these species (Denamur et al. 1973). Granulosa cells isolated from follicles of the pig revealed a differential response to coincubated prolactin (Veldhuis et al. 1980). Prolactin produces a concentration-dependent inhibition of progesterone synthesis in granulosa cells isolated from small follicles, whereas large follicles produced cells which responded to prolactin with increased progesterone synthesis. In cynomolgus monkeys (M. fascicularis) suppression of prolactin together with a low dose of long-acting estrogen during the luteal phase shortened the life span of the corpus luteum (Castracane and Shaik 1979).

In human follicular granulosa cells in situ, McNatty et al. (1974) observed that high concentrations of prolactin depressed progesterone production, which seems to agree with the inverse relationship between the follicular fluid concentrations of progesterone and prolactin measured at various stages of the menstrual cycle. On the other hand, addition of antiserum to neutralize endogenous prolactin also led to a reduced progesterone production by the granulosa cells. In cultures of mouse ovaries, prolactin was found to enhance progesterone synthesis if corpora lutea were present but to inhibit progesterone synthesis by developing follicles (McNatty et al. 1976). Thus, it seems that in the human ovary prolactin is also involved in steroidogenesis besides LH, possibly in a biphasic function of concentration. It is possible that these observations provide the basis for understanding the occurrence of luteal insufficiency (reduced progesterone production) in women with pharmacologically elevated prolactin levels (Delvoye et al. 1974; Robyn et al. 1976; del Pozo et al. 1979) on the one hand, as well as in women in whom serum prolactin levels were suppressed by supramaximally acting doses of bromocriptine (Schulz et al. 1976, 1978).

Prolactin can influence corpus luteum function in still another way, at least in rodents: in hormonally induced pseudopregnancy in immature rats prolactin was found to induce and maintain LH receptors in the corpus luteum (Grinwich et al. 1976; Holt et al., 1976; Gibori and Richards 1978), i.e., to predetermine the luteotropic action of LH. Prolactin also seems to maintain estrogen receptors (both in the cytosol and in the nuclei) in granulosa cells of the rat (Gibori et al. 1979).

Interestingly, prolactin is not only luteotropic in rodents but it also has a luteolytic action. In hypophysectomized rats regression of corpora lutea was induced by prolactin (Malven et al. 1969) injected 80 h after removal of the pituitary. Suppression of prolactin secretion in cyclic rats induced an accumulation of nonfunctional corpora lutea (Heuson et al. 1970; Billeter and Flückiger 1971) which could be induced to regress by injecting prolactin (Billeter and Flückiger 1971; Wuttke and Meites 1971). In mice, similar observations were made (Grandison and Meites 1972). It was suggested that in rodents prolactin released during the proestrus surge serves to induce lysis of the corpora lutea of the previous ovarian cycle. The basis of this action is still unknown.

1.6.4 Effects on the Testis

Prolactin also has gonadotropic actions in the male, as in the female. Bartke (1964, 1965, 1966 a) was the first to point out that infertile males of dwarf mice (Snell's dwarf = dw/dw and Ames dwarf = dt/df) which lack prolactin and GH, but not gonadotropins, responded positively to treatment with prolactin. Both prolactin injections or implantation of pituitaries from heterozygous animals resulted in marked improvement of spermatogenesis (Bartke and Lloyd 1970 b). As the seminal vesicles in the treated animals became heavier, it was concluded that prolactin probably acted on Leydig cells to increase androgen production (Bartke and Lloyd 1970 a, Bartke et al. 1977). Prolactin has a direct stimulatory effect on the number of LH receptors on rat Leydig cells (Purvis et al. 1979). In hypophysectomized random-bred mice or rats LH, in the presence of prolactin, induced a greater production of androgens than in the absence of prolactin (Bartke 1971; Hafiez et al. 1972; Bartke and Dalterio 1976) in dwarf male mice, the amount of esterified cholesterol in the testes being augmented by prolactin (Bartke 1969) and the activity of 3β-hydroxysteroid dehydrogenase increased (Hafiez et al. 1971). In golden hamsters a short-day regime lowered testicular weight and activity. This could be counteracted by prolactin which augmented the reduced LH-binding capacity to that found with 14 h of daylight (Bex and Bartke 1977). Reverse experiments, i. e., suppressing prolactin secretion in genetically normal adult rodents has brought conflicting results, some authors describing a significant lowering of plasma testosterone and increased LH levels in bromocriptine treated rats (Boyns et al. 1972) and a decrease in testicular LH-binding sites (Belanger et al. 1979), while others found only insignificant changes in testosterone production (Bartke 1973, Harper et al. 1976). In male rats which were desensitized to LH stimulation by treatment with an LHRH analogue, high prolactin levels accentuated the ensuing block in the steroidogenic pathway between progesterone and testosterone, while suppression of prolactin secretion prevented this (Belanger et al. 1979). In male lambs, suppression of circulating prolactin from the tenth to the 21st week of life did not influence LH nor testosterone levels. The establishment of spermatogenesis was not delayed (Ravault et al. 1977). In the human male the role of prolactin in testicular function is not clear. Prolactin has been demonstrated by RIA to occur in human semen (Sheth et al. 1975). Prolactin at physiologic concentrations stimulated c-AMP formation, rate of fructose utilization and glucose oxidation. Prolactin levels in some infertile and impotent men were found to be lower than normal (Deutsch and Sherman 1979; Spitz et al. 1977) but whether this finding points to an essential role of prolactin in human testicular function is unknown.

If circulating prolactin levels in rats were elevated far above normal, e. g., by implanting a prolactin-secreting tumor, then hypogonadism ensued with low levels of testosterone, atrophic testes and accessory sex glands, and slightly elevated levels of LH (Fang et al. 1974). In man, experimentally increased prolactin levels produced ambiguous results. Healthy young men after several days of sulpiride treatment produced a higher testosterone response to HCG than before (Ambrosi et al. 1976). With low doses of haloperidol this positive prolactin effect was also found (Rubin et al. 1976). The testosterone response was not found to be augmented but the dihydrotestosterone response was much reduced, indicating a reduction in 5α-reductase activity. Suppression of prolactin secretion produced a slightly increased dihydrotestosterone response (Magrini et al. 1976). The authors of this study pointed out that clinical observations of hypogonadism with hyperprolactinemia but relatively moderate changes of testosterone might be due to its decreased conversion to dihydrotestosterone. Prolonged hyperprolactinemia is often accompanied by loss of libido and impotence (Besser and Thorner 1976; Fossati et al. 1976 a, b; Thorner 1977). For further discussion of this aspect see below and Chap. 4 of this book.

1.6.5 Prolactin and Hypogonadism

In rodents, in sheep, in monkeys, and in humans of both sexes increased circulating prolactin levels are associated with an attenuation of gonadal function or with sexual dysfunction. Such a state may be obtained physiologically in the female (lactational anestrus or amenorrhea) (Konner and Worthman 1980), by continued mental or physical stress (Bowman et al. 1978; Brisson et al. 1980; Tyson et al. 1975), by pharmacologic stimulation of prolactin release or by the development of pituitary tumors (for details see Chaps. 4 and 5). Several hypotheses have been advanced to explain this phenomenon, which phylogenetically is an old one (Blüm 1977): Prolactin is proposed to act by

a) suppression of gonadotropin secretion,
b) abolition of positive feedback of estrogen on LH secretion in the female,
c) changes in adrenal androgen secretion, and
d) blockade of the effects of gonadotropins at the gonadal level.

These hypotheses are discussed in detail by Thorner et al. (1980 a, b) and in Chap. 4 of this book.

1.6.6. Accessory Glands

In agreement with the demonstration of specific prolactin binding to rodents' secondary sex organs like the prostate and the seminal vesicles, prolactin has also been found to have pronounced stimulatory effects in synergism with androgens (Grayhack 1963; Bartke 1974; Thomas and Manandhar 1974, 1975). In lambs, suppression of prolactin attenuated growth of seminal vesicles and fructose accumulation (Ravault et al. 1977). For a detailed discussion of these aspects of prolactin physiology, see Aumüller (1979). It is not known whether in man prolactin is involved in the functions of these accessory sex organs.

1.6.7 Effects on the Adrenal Cortex

In rats, adrenal progesterone production was found to be stimulated by hLH, oGH and oPRL (Piva et al. 1973). Because of the magnitude of the prolactin effect, a physiologic role of this hormone at the adrenal cortex was thought possible. In cortisone-treated rats, adrenal 5α-reductase activity increased severalfold after hypophysectomy and this was prevented by either ACTH, GH or prolactin injections. In gonadectomized animals prolactin was the only effective hormone (Witorsch and Kitay 1972; Witorsch and Edwards 1976). Prolactin inhibited the formation of reduced steroid metabolites and thereby increased corticosterone output. According to a recent study (Colby 1979), prolactin did not enhance corticosterone synthesis but only suppressed intra-adrenal degradation by inhibition of 5α-reductase.

Fifteen years before human prolactin was isolated Forbes et al. (1954) had already pointed out a connection between the condition responsible for galactorrhea and adrenal function. But even today the role of prolactin in adrenal cortical function in man is not very clear. A discussion of this clinically important aspect will be found in Chap. 4. In the human fetus, prolactin was thought to stimulate adrenal growth (Winters et al. 1975), but suppression of prolactin secretion throughout pregnancy did not modify plasma DHEAS and estrogen levels in either mother or newborn (del Pozo et al. 1978, 1980) which speaks against prolactin as the physiologic corticotroph during pregnancy.

Aldosterone secretion has been found by Edwards et al. (1975) and by Edwards and Jeffcoate (1976) to be reduced in subjects treated with bromocriptine. The authors considered the possibility that this effect was due to a direct action of the drug on the adrenal cortex and not due to a lowering of prolactin levels. Recent studies have produced evidence that serotonin elevated and serotonin antagonists (bromocriptine is a weak 5HT antagonist also) inhibited aldosterone secretion (Glaz et al. 1979; Mantero et al. 1979). The results of another, recent human pharmacology study (degli Uberti et al. 1979) support the idea that the effect of bromocriptine on aldosterone is independent of its effect on prolactin and that prolactin does not play a significant role in aldosterone secretion. Others were unable to find changes in aldosterone serum levels with bromocriptine (Marek and Horky 1976; Ølgaard et al. 1976; del Pozo et al. 1977 a), nor a change in sensitivity to angiotensin II infusion (Carey et al. 1979). On the other hand, DA receptor blockade by metoclopramide increased serum aldosterone concentration in normal men, but did not change the increment after increasing doses of angiotensin II, suggesting that aldosterone in healthy men is maximally inhibited by a dopaminergic mechanism (Carey et al. 1979). The aldosterone response to metoclopramide was recently found to be blunted by simultaneous infusion of DA (dopamine) (Noth et al. 1980), lending further support to the notion of a tonic dopaminergic suppression of aldosterone (for reviews see Ontjes 1980). In conjunction with this it is interesting to note that serum potassium has recently also be observed to be under dopaminergic control (Bevilacqua et al. 1980), reacting (more quickly than aldosterone) to metoclopramide with a fall.

1.6.8 Effects on the Excretory Kidney

Although specific prolactin binding has been demonstrated in the kidney of rodents (see Sect. 1.5) it is still not clear whether prolactin in mammalian species should be considered a regulator hormone of water and electrolyte handling by the kidney. In man, recent studies led to the conclusion that prolactin was acutely not involved in renal water and electrolyte metabolism in any relevant way (del Pozo and Ohnhaus 1976; Baumann et al. 1977; Carey et al. 1977; Vorherr 1979). Prolactin has been found to be involved in the chronic progressive nephropathy of the rat (Richardson and Luginbühl 1976) as it seems to facilitate the formation of protein casts in the collecting ducts. The relevance of these findings to the understanding of degenerative renal lesions in man is not clear.

1.6.9 Effects on Fluid Transport in Nonrenal Organs

Besides the kidney there is the intestine to be considered in the water and electrolyte balance of an organism. Experimental evidence indicates that prolactin may be involved in intestinal fluid transfer. Ramsay and Bern (1972) demonstrated that treatment of rats with oPRL subsequently increased the fluid transfer capacity of the small intestine in vitro. This was confirmed by later studies by Mainoya (1975), who was also able to induce such an effect by stimulation of endogenous prolactin release. It could be hypothesized that in suckling animals prolactin taken up with the milk could thus accelerate the resorption of milk from the intestine. A recent study using 2–3 week-old pigs in which *E. coli* inocculation of ligated jejunal segments was used to induce fluid secretion into the segment did not produce evidence that oPRL could reduce the stimulated fluid secretion (Mulloy and Malven 1979), but the duration of prolactin pretreatment may not have been long enough: in the rat (Mainoya 1975), the maximal effect of prolactin on intestinal fluid transport occurred with a latency of 24–48 h. The participation of prolactin in the regulation of intestinal fluid transfer in the human is unknown.

In the fetus, the regulation of the volume and the composition of amniotic fluid is still incompletely understood. In view of the high hPRL levels in amniotic fluid, especially in the midtrimester, a role of this hormone in amniotic fluid balance must be considered. Using the isolated amnion of the guinea-pig Perks et al. (1978) found oPRL to reduce outward water movement and to increase the outward transport of sodium. Thus, prolactin may be viewed as possibly being involved in the regulation of amniotic fluid volume and composition.

1.6.10 Effects on the Liver

Although the hepatocytes of rats bind lactogenic hormones avidly (see above, Sect. 1.5), suggesting that prolactin has a physiologic role in hepatic function, nothing positive is known about this presumed relationship. In rodents, hepatic steroid metabolism was found to show marked sex differences, which seemed to depend on the presence of pituitary hormone (Colby et al. 1973). Hypophysectomy of rats led to a masculinization of liver enzyme activity, and ectopic implantation of a pituitary in such animals induced feminization of the liver enzyme activity pattern, but this effect showed poor correlation with serum prolactin levels (Eneroth et al. 1977). Pharmacologic suppression of prolactin secretion in female or male rats did not change the hepatic steroid metabolism pattern, and in hypophysectomized rats "feminized" by an ectopic pituitary, suppression of prolactin did not lead to a reversion of this pattern of steroid metabolism (Skett et al. 1978).

In a study of somatomedin serum levels in hyperprolactinemic patients (50–20 000 ng/ml) no correlation between prolactin and somatomedin levels was found, indicating that prolactin is not a relevant regulator of somatomedin in man (Spencer 1980).

1.6.11 Prolactin and Vitamin D_3 Metabolism

During lactation adaptive changes in Ca^{2+} metabolism are necessary. Recent research revealed that prolactin infusions in rats increased plasma calcium concentration and calciuria (Mahajan et al. 1974). Elevation of calcemia in parathyroid hormone- and calcitonin-deprived animals occurred to an even greater extent. It was suggested that this action of prolactin might involve an interaction with vitamin D (Horrobin 1974). In the chicken kidney 25-hydroxycholecalciferol-l-hydroxylase was found to respond to single or multiple injections of ovine prolactin with increased activity (Spanos et al. 1976 a). The activation of the 1-hydroxylase was maximal after 1 h and nearly vanished at 3 h. In hens, 1,25-dihydroxycholecalciferol (1,25 $(OH)_2D$) blood levels were found to increase during egg laying and this increase could also be achieved by injecting prolactin (Spanos et al. 1976 b). In the serum of lactating rats the active metabolite of vitamin D_3 was also found to be elevated severalfold (Boass et al. 1977). Suppression of prolactin secretion in lactating but not in nonlactating rats by pharmacologic means led to a 50% reduction in circulating 1,25 $(OH)_2D$. This effect in lactating rats could be alleviated by concomitant injections of sheep prolactin (MacIntyre et al. 1977).

In the human, elevated circulating levels of 1,25 $(OH)_2D$ were measured in late pregnancy with a gradual return to an upper limit of "normal" range 30 days postpartum (Lund and Selnes 1979), similar to the trend in serum prolactin levels. No significant changes in PTH were observed in this same period. Other researchers, studying nonpregnant normal subjects and patients with hyperprolactinemia, patients with primary hyperparathyroidism and patients with Ca nephrolithiasis, could not find a correlation of 1,25 $(OH)_2D$ with prolactin levels (Adams et al. 1979 b): 1,25 $(OH)_2D$ was found to be elevated in patients with high PTH and normal prolactin (primary hyperparathyroidism) and in patients with Ca nephrolithiasis who showed normal PTH and normal prolactin levels, but it was not elevated in patients with hyperprolactinemia.

Studying acromegalic patients Eskildsen et al. (1979) found high levels of 1,25 $(OH)_2D$ which correlated well with the urinary excretion of growth hormone. Treating these patients with bromocriptine decreased 1,25 $(OH)_2D$ slowly to the normal range. In the untreated of two other metabolites of vitamin D_3, 24,25 $(OH)_2D$ was found to be elevated, while the serum concentrations of 25 (OH)D were in the normal range. Treatment with bromocriptine lowered the levels of the former significantly but not to normal, and did not change the levels of the latter. Urinary GH decreased significantly during therapy. In most patients serum prolactin was elevated before therapy but did not correlate with serum 1,25 $(OH)_2D$ concentrations. During therapy, prolactin levels were depressed.

Steichen et al. (1980) measured 1,25 $(OH)_2D$ in the blood of pregnant women at term, in blood from placental veins of their infants and in neonates 24 h old. Maternal concentrations were elevated above normal adult range, while in the placental vein at birth the concentrations were below the values found in normal adult blood, indicative of low values also in the fetus. No correlation was found with maternal values. In the 1-day-old infants, normal adult values were reached, while at the same time there was an important lowering of the ionized calcium concentration and a trend to higher PTH levels.

From the above findings it follows that in man the role of prolactin in the regulation of vitamin D metabolism is still far from clear.

1.6.12 Effects on Carbohydrate and Fat Metabolism

Lactation, as well as pregnancy, requires adjustments in the carbohydrate and fat metabolism of the adult organism. Therefore, it could be surmised that prolactin, either directly or indirectly, also plays a regulatory role in carbohydrate and/or fat metabolism. Beck et al. (1964) have described that oPRL given to hypopituitary dwarf patients and to hypophysectomized diabetics, induced similar metabolic changes to the injection of hGH, e. g., a diabetogenic effect. In the dog, oPRL was also clearly diabetogenic by several criteria (Rathgeb et al. 1971). This observation raises the question of whether homologous prolactin also has such an action. This was negated for oPRL injected into sheep (Manns and Boda 1965). On the other hand, in man it was observed that hyperprolactinemic patients (without signs of acromegaly) revealed, after a glucose load, an impaired glucose tolerance and an increased insulin response (Tourniaire et al. 1974; Landgraf et al. 1975, 1977). Another group (Gnudi et al. 1977), after pharmacologic suppression of prolactin secretion, noted in insulin-dependent diabetics and in healthy subjects an improvement in glucose tolerance and in the healthy subjects a reduced insulinemia after the glucose load. This metabolic derangement was improved after pharmacologic suppression of prolactin secretion. These observations led to the conclusion that in man hPRL acts as a diabetogenic hormone (Landgraf et al. 1977). A lowering of fasting glucose levels and an improved glucose tolerance was also observed in maturity onset diabetics and an initial improvement of glucose tolerance in healthy controls, after lowering of the (normal) prolactin serum levels by bromocriptine (Barnett et al. 1980). Metoclopramide used to augment circulating prolactin levels, produced a slight impairment of glucose tolerance in the healthy persons but no significant alteration in the diabetics. In these experiments it cannot be decided whether the effects observed are due to altered levels of hPRL or secondary to other actions of bromocriptine and metaclopramide. The question of hPRL in glucose homoeostasis could possibly be better settled by studies in patients before and after successful ablation of a prolactinoma.

Beck et al. (1964) and McGarry and Beck (1972), first reported that in hypophysectomized diabetic patients oPRL induced an increase in serum free fatty acids and triglycerides indicating a fat mobilizing property for prolactin. Recently this aspect of prolactin physiology has not received much attention. In lactating rats an augmented lipoprotein lipase activity in the mammary gland together with a depressed activity in adipose tissue was found (Zinder et al. 1974). Hypophysectomy induced a reversal of these activities and this could be prevented specifically by exogenous prolactin.

1.6.13 Prolactin and Fetal Lung Maturation

Based on theoretical considerations, Hamosh and Hamosh (1977) investigated the effect of oPRL on phospholipid levels in the fetal rabbit lung and found an augmentation of lung lecithin content, which suggested to the authors hat prolactin might be the physiologic trigger of lung surfactant synthesis. In the human neonate, Hauth et al. (1979) observed that the chance of respiratory distress syndrome (RDS) developing was higher in babies if cord-plasma prolactin amounted to less than 200 ng/ml, while no correlation was found between RDS and cord-plasma cortisol, the generally assumed inducer of lung maturation (Giannopoulos 1980). These authors, therefore, also considered the possibility that prolactin has a role in fetal lung maturation. In contrast, Gluckman et al. (1978) could not find a stimulatory action of oPRL on surfactant production in the rabbit and sheep and concluded that RDS and low cord-blood prolactin levels are independent consequences of low estrogen levels.

2 Control of Proclactin Secretion

2.1 Physiologic Aspects

2.1.1 Patterns of Prolactin Secretion

In the developing human prolactin was first detected by Aubert and colleagues in the pituitary and serum of fetuses about 80 days old. With increasing gestational age, pituitary prolactin content increased with great individual differences to about 1 μg per gland at term. Serum prolactin remained around 20 ng/ml between the 12th and 29th weeks of gestation, indicating that the pituitary is already under inhibitory control during this time. After about 200 days of gestation a striking increase in serum prolactin levels occurred, with individual values above 500 ng/ml. A small drop in prolactin secretion or an increased metabolic clearance may occur in late gestation, since prolactin levels of umbilical cord blood of full-term newborns were lower (about 170 ng/ml) than in late gestation (about 250 ng/ml). During the 1st day of life prolactin levels remained high, then declined to about 50–80 ng/ml within 7 days. After about 6 weeks serum prolactin levels were reduced to the normal prepubertal level (Aubert et al. 1977).

Until puperty no sex difference in prolactin serum levels was observed. In a longitudinal study with boys, mean basal prolactin levels rose only from 4.0 ng/ml at puberty stage P_1 to 5.2 ng/ml at P_5, while girls developed their serum prolactin levels from 4.5 to 8.3 ng/ml, probably as a consequence of increasing circulating estrogen (Aubert et al. 1977).

In young sexually mature women prolactin levels are higher than in men, but with older women these values were found to decrease to those of young men (Vekemans and Robyn 1975a; del Pozo et al. 1977b). In old men somewhat elevated prolactin levels have been found in one study (Vekemans and Robyn 1975a) but not in the other (del Pozo et al. 1977b). In the adult human, basal prolactin serum concentrations show episodic or pulsatile variations of low amplitude (Sassin et al. 1972; Parker et al. 1973), probably reflecting oscillations of tone in the central control of prolactin secretion. Prolactin secretion is increased during sleep (Fig. 2.1), and this occurs independently of daytime. Surges of prolactin release were observed during nonREM sleep phases (Parker et al. 1974) although del Pozo and Brownell (1979) were unable to find a correlation between the incidence of REM episodes and the sleep hPRL profile by studying individuals undergoing suppressive treatment with bromocriptine. Waking baseline levels are mostly only reached 1–2 h after waking (Frantz 1979). The sleep-related increase of serum prolactin levels (nycthemeral rhythm) is maintained in women continuously stimulated by estrogen administration (Vekemans and Robyn 1975b).

Considering the stimulatory effect of estrogens on prolactin secretion, one would expect basal prolactin serum levels to change with the ovarian cycle. Robyn et al.

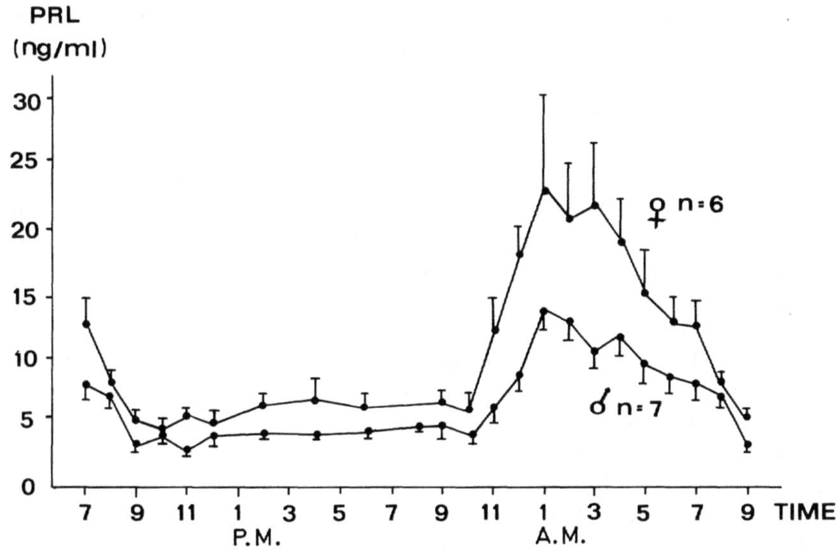

Fig. 2.1. Avereged prolactin profile of young and healthy volunteers. The subjects were monitored by EEG during night hours. Onset of sleep (EEG stage 2) was for the majority between 10.30 and 11.30 P.M., and the majority were awake at 7 A.M. (del Pozo and Clarenbach, unpublished)

(1973) have been able to demonstrate both a slight augmentation of basal prolactin levels during the follicular phase, and a small midcycle peak. During the luteal phase serum prolactin and estrogen levels followed a similar pattern. Also Franchimont et al. (1976) observed an increase in prolactin levels during the ovulatory and the luteal phase as compared to the follicular phase.

If women become pregnant, serum prolactin levels slowly increase, reaching about 150–200 ng/ml at term. A good parallelism was observed between the prolactin and estradiol concentrations measured in the same serum samples up to the 30th week. In late pregnancy, a dissociation in the increments occurred with estradiol levelling off (del Pozo et al. 1977b). After delivery the further development of prolactinemia depends on the nursing behavior of the mother, tactile stimulation of breast and nipple being the most powerful physiologic stimulus for prolactin release (Frantz et al. 1972; Konner and Worthman 1980). Development of prolactinemia in the postpartum and its consequences will be discussed in detail in Chap. 4 by del Pozo.

Physiologically, prolactin secretion may also be enhanced by physical exercise, stress, hypoglycemia, stimulation of the breast and by coitus (see Sect. 4.2).

2.1.2 Regulatory Factors

Physiologic prolactin secretion obviously varies with different environmental conditions, the actual rate of secretion reflecting the tone of an integrating control system. As with other pituitary hormones, the integrity of the connection between the median eminence and the anterior pituitary is critical for the physiologic control of prolactin secretion. Everett (1954) reported that after transplanting the pituitary gland of rats from the sella turcica to the kidney, signs of prolactin hypersecretion could be observed. Later, electrolytic lesions in the median eminence or transsection of the hypophyseal stalk were found also to increase prolactin secretion and synthesis (Bishop et al. 1971; Chen et al. 1970; MacLeod and Lehmeyer 1972). Thus, prolactin secretion by the mammalian pituitary, in contrast to the other anterior pituitary hormones, is under the predominantly inhibitory influence of the CNS. Apart from the mammals, the predominantly inhibitory influence of the hypothalamus on prolactin control was demonstrated in amphibians (Holmes and Ball 1974) and in teleost fishes whose prolactin cells receive direct catecholaminergic innervation from the hypothalamus (Batten and Ball 1976). In reptiles and birds, the inhibitory influence of the hypothalamus does not seem to be of such importance (see Ensor 1978).

Pasteels (1961) and Talwalker et al. (1963) were the first to demonstrate in the hypothalamus of mammals the presence of an extractable fraction which, when incubated in vitro with pituitaries, attenuated the release of prolactin. Thus, the CNS may exert its inhibitory influence on prolactin secretion by the release of one or several prolactin inhibitory factors (PIFs) into the hypothalamopituitary portal system. The hypothalamus, in addition, contains extractable fractions which enhance the release of prolactin (Nicoll et al. 1970; Valverde et al. 1972; Boyd et al. 1976; Clemens et al. 1978). The CNS may thus exert a stimulatory influence by releasing one or several prolactin-releasing factors (PRF's), TRH of hypothalamic origin being the first candidate for such a function. Besides the hypothalamus, PRL-inhibiting and PRL-releasing activity has also been demonstrated in extracts of the pineal gland. Although some fractions were shown to have in vitro PIF activity (Chang et al. 1979), nothing is known of the physiologic relevance of this.

Prolactin attenuates its own secretion (Voogt and Meites 1973) and this involves dopaminergic mechanisms (Olson et al. 1972; Gudelsky et al. 1976; Advis et al. 1977; Perkins and Westfall 1978; Voogt et al. 1979). The role of prolactin-sensitive neurons involved in the brain (Clemens et al. 1971; Yamada 1975; Poulain and Carette 1976; Perkins and Westfall 1978; Wood et al. 1980) in this short-loop feedback system is still unclear. It has been claimed that prolactin may also inhibit prolactin secretion by a direct action on the prolactin cell (Herbert et al. 1979), but the physiologic relevance of this is not clear.

As the isolated mammalian pituitary secretes prolactin autonomously, the inhibitory aspect of prolactin control in general is thought to be more important than the stimulatory aspect. Therefore, our present-day understanding of prolactin control in mammals is heavily biased towards PIF and we mostly explain changes in prolactin secretion as a consequence of modulation of the inhibitory tone.

The neuronal control of prolactin secretion involves not only classic neurotransmitters (cholinergic, noradrenergic, serotoninergic and dopaminergic) (Clemens 1976) but also GABA, histamine, enkephalins and several other polypeptides (McCann et al. 1978) and possibly prostaglandins (Ojeda et al. 1978). Their relative contribution to the physiologic regulation of prolactin secretion is in most cases unknown. Table 1.2 gives a survey of such peptide actions on prolactin secretion.

Of the various transmitter systems involved with prolactin control at the CNS level, the serotoninergic system merits special mentioning, because of the demonstration of its importance in suckling-induced prolactin release (Kordon et al. 1973/74, 1976) and because of its involvement with gonadotropin secretion (Pilotte

Table 2.1. Actions of natural peptides and derivatives on prolactin secretion

Inhibitory action		
AVP (in vitro)		Vaughan et al. (1975)
Histidyl-proline-diketopiperazine (a major metabolite of TRH)		Bauer et al. (1978)
Somatostatin		Yen et al. (1974), but Hall et al. (1978): no effect
Pyroglutamic acid		Lam et al. (1978)
Stimulatory action		
Calcitonins (salmon, porc, eel and human) (in vitro)		Iwasaki et al. (1979)
AVT (in vitro)		Vaughan et al. (1975)
CCK (in vitro)		Malarkey et al. (1980)
Substance P (in vivo)		Rivier et al. (1977 a)
TRH (in vitro)		Tashjian et al. (1971)
(in vivo)		Refetoff et al. (1974)
VIP (in vitro)		Gourdji et al. (1979)
LHRH (in vivo)		Giampetro et al. (1979)
Opioids:		
enkephalins (in vivo)		Bruni et al. (1977), Meltzer et al. (1978)
β-endorphin (in vivo)		Rivier et al. (1977 b)
met-enkephalin (in vitro)		Enjalbert et al. (1979 c)
β-endorphin (in vitro)		Enjalbert et al. (1979 c)
Stimulatory and inhibitory actions		
Gastrin	(i. v.: stimulatory) (i. c. v.: inhibitory) (in vitro: no effect)	Vijayan et al. (1978)
Bombesin	stimulatory in estrogen-primed rats inhibitory in stressed rats	Rivier et al. (1978) Taché et al. (1979)
Neurotensin	stimulatory in nonstressed, inhibitory in stressed rats	Rivier et al. (1977 a) Taché et al. (1979)

and Porter 1979; Marko and Flückiger 1980) at least in rats. Serotoninergic neurons also seem to be involved in the expression of the prolactin stimulatory effect of morphine (Koenig et al. 1979). MacIndoe and Turkington (1973) showed that in humans infusion of L-tryptophan induced a pronounced elevation of serum PRL and a slight reduction of LH levels. In rats, L-tryptophan or 5-hydroxytryptophan, as well as drugs which inhibit serotonin reuptake or stimulate serotonin receptors in the CNS, increased circulating prolactin (Quattrone et al. 1978). The question of how changes in central serotoninergic tone are translated into changes in prolactin secretion rate, is still unanswered. Certain findings indicate that serotonin alters the release of a hypothalamic releasing factor (Clemens et al. 1978; Garthwaite and Hagen 1979), but other observations point to the involvement of dopamine (PIF) in the portal system (Pilotte and Porter 1979) to translate changes in central serotoninergic tone.

Disregarding now the control of putative PRF's, the main final (inhibitory) common pathway of prolactin control consists of tuberoinfundibular dopaminergic neurons which terminate in the median eminence adjacent to the primary capillary plexus of the portal vessels, connecting the median eminence and the anterior pituitary. The question of whether additional neurons with different transmitters also contribute as effectors of the inhibitory control of prolactin secretion is not settled. Thus, GABA was considered by Schally's group as an additional inhibitory hypothalamic effector (Schally et al. 1977) (see below). Recently, a metabolite of TRH, histidyl-proline-diketo-piperazine was described as inhibiting prolactin release (Bauer et al. 1978) but the efficacy of this peptide, assessed in vitro, does not seem to be high (Enjalbert et al. 1979b).

Van Maanen and Smelik (1968) first postulated that the tuberoinfundibular neurons could exert their inhibitory control on prolactin secretion "via the release of the inhibitory neurotransmitter into the portal vessel system". Thus, prolactin cells were postulated to carry monoamine receptors, the stimulation of which leads to inhibition of prolactin secretion. Today it seems accepted that this view correctly predicted the quantitatively most important aspect of prolactin control. The evidence may by summarized as follows:

The pituitary has been shown to possess monoamine oxidase activity (MacLeod et al. 1970), which could anneal catecholamines reaching this organ. Prolactin cells have been shown to carry catecholamine receptors, the stimulation of which attenuates hormone release. Mainly through MacLeod's pharmacologic work, these receptors have been characterized as DA receptors (MacLeod 1976). Physicochemical studies in various laboratories with radio ligands have confirmed the existence of binding sites with high affinity to DA receptor agonists and antagonists (Brown et al. 1976a; Creese et al. 1977; Cronin and Weiner 1979). DA binding sites on prolactin cell membranes have been visualized by immunocytochemical staining, and with the same technique Weiner et al. (1979) demonstrated the internalization of such surface areas. This same process is probably also responsible for the fact that DA found in the anterior pituitary of the rat was associated with subcellular particles undistinguishable from prolactin secretory granules (Nansel et al. 1979; Gudelsky et al. 1980). The amount of DA associated with prolactin granules could be augmented by pretreatment with L-DOPA in vivo or by co-incubation with DA in vitro. Pretreatment with DA agonists or with a DA antagonist reduced the incorporation of DA, indicating that this incorporation involves DA-binding sites (Gudelsky et al. 1980). Prolactin cells show the phenomenon of supersensitivity to stimulation by dopaminomimetic agents after prolonged desinhibition of prolactin secretion by surgical destruction of the medial basal hypothalamus (Cheung and Weiner 1976, 1978) or by chronic treatment with α-methyl-p-tyrosine (Annunziato and Moore 1977), haloperidol or morphine (Lal et al. 1977). The phenomenon of supersensitivity suggests that, physiologically, attenuation of hormone release by a continuous flow of dopamine represents the basal state of the prolactin cell in situ. Dopamine has been demonstrated in effective concentrations in the blood coming from the rat median eminence (Ben-Jonathan et al. 1977; Gibbs and Neill 1978), and 17β-estradiol, which stimulates prolactin secretion, has been found to reduce DA release into the hypophyseal portal blood (Cramer et al. 1979). Reduction of DA in portal blood was also observed after injection of an analogue of met-enkephalin which stimulates prolactin secretion and both effects were antagonized by naloxone, an opioid antagonist (Gudelsky and Porter 1979).

It is evident that a strong set of observations exists which agrees with the postulate of Van Maanen and Smelik (1968) that prolactin secretion is controlled by an inhibitory neurotransmitter which reaches the prolactin cell via the portal system. This does not exclude the possible existence of further inhibitory and of stimulatory control systems. The clear picture obtained on the role of the inhibitory dopaminergic control in mammals is based on its being so obviously the dominant or most relevant prolactin controlling system. This role also includes a high sensitivity to drugs of various therapeutic groups which act by interfering with dopamine function (Del Pozo et al. 1976; Rolandi et al. 1979) as will be shown below (Sect. 4.1). The intracellular mechanisms through which dopamine acts on the prolactin cell are poorly understood. An up to date discussion of the various aspects may be found in MacLeod and Scapagnini (1980) (see also Sect. 2.2.1 below).

An important modulating factor in the control of prolactin secretion is the estrogens, especially 17β-estradiol. This effect is thought to take place at two levels, the hypothalamus and the pituitary (McCann et al. 1968; Dufy et al. 1979b). Recent in vitro studies with hypothalamic tissue of the rat have demonstrated an increased release of noradrenaline and dopamine in the presence of estradiol-17β or diethylstilbestrol (Paul et al. 1979). Prolonged treatment of rats with estrogens increased both prolactin secretion and dopamine turnover of the TIDA neurons but not of striatal neurons (Fuxe et al. 1969; Eikenburg et al. 1977). Recently, antidopaminergic effects of estrogens have also been described at the striatal level of rats (Euvrard et al. 1979). Hypophysectomy prevented the stimulation of transmitter turnover in the TIDA neurons, thus indicating that the change in the activity of TIDA neurons was probably secondary to the change in prolactin secretion. A delay of several hours between the change of prolactin level and the response of the TIDA neurons was usually found (Fuxe et al. 1969; Gudelsky et al. 1976; Annunziato and Moore 1977). The increased turnover of DA in the TIDA neurons indicates increased release of DA which should lead to attenuation of prolactin secretion. Under the acute influence of estrogens this had not been observed: although estrogen levels are highest in cycling rats during proestrus, DA was found to be lower in hypophyseal stalk blood (Cramer et al. 1979).

The direct action of estrogens on prolactin cells involves an initial binding to specific cytoplasmic receptors and a temperature-sensitive transport of the estrogen receptor complex to the nucleus (Hang et al. 1978). It is probably after this second step that prolactin synthesis becomes augmented by specifically increasing the incorporation of precursors into prolactin and an accumulation of preprolactin mRNA (Stone et al. 1977; Lieberman et al 1980). In contrast to stimulation of prolactin release by psychotropic agents, estrogen stimulation also implies an increase in cell metabolism (Nagy et al. 1979a; Nagy and MacLeod 1980).

Stimulation of prolactin secretion by estrogens also involves a desensitization of the prolactin cells toward the inhibitory action of DA and dopaminomimetics (Gala and Boss 1975; Labrie et al. 1978; Beaulieu et al. 1979; Ferland et al. 1979). Interestingly, estrogens were found to immediately reduce the response to dopamine stimulation in such cells in vitro (Dufy et al. 1979b). We have here the example of a peripheral hormone altering the receptors for hypothalamic control of a pituitary function. The desensitization of DA receptors on prolactin cells may involve a structural change, because it may persist after prolonged estrogen treatment has stopped: the estrogen-induced prolactin-secreting rat pituitary tumor 7315a was found not to respond to dopaminomimetic agents, except after preincubation with an antiestrogen, tamoxifen (Timmermans and Lamberts 1979).

As already indicated in Sect. 2.1, exposure to estrogens increases serum prolactin levels in the human also. Estrogens are responsible for the slightly higher prolactin levels in women during the reproductive phase of life and during pregnancy a close parallelism between plasma estradiol and prolactin concentrations has been observed (del Pozo et al. 1976) which fits with the increased pituitary weight (Erdheim and Stumme 1909) and prolactin cell hyperplasia in pregnant women (Goluboff and Ezrin 1969). Antiestrogens may reduce prolactin serum levels (Willis et al. 1977) (see Sect. 2.2.5, p. 40). Besides estrogens testosterone also, but not its 5α-reduced metabolite, may stimulate prolactin secretion in castrated rats, probably after conversion to estrogen (Martini et al. 1978). Recently, the nonuterotrophic metabolite of estradiol, 2-hydroxy-estrone, a catechol-estrogen, was claimed to decrease prolactin levels of postmenopausal, estrogen-primed women to less than 25% of the initial value (Schinfeld et al. 1980). This was an unexpected effect which was not observed in nonestrogenized postmenopausal women, and which had not been observed in an earlier, but similarly conducted study (Adashi et al. 1979). Another catecholestrogen, 4-hydroxy-estradiol has been described to be equipotent with estradiol in stimulating prolactin secretion in the rat, while 2-hydroxy-estradiol was found to be inactive (Franks and Naftolin 1980).

Progesterone, on the other hand, is known to inhibit prolactin secretion in cattle and goats (Schams et al. 1974; Hart 1975). Progesterone may antagonize the stimulatory effect of estrogen in rat pituitary tumor cells (Haug 1979) and in cattle, goats and sheep (Hart and Morant 1980). In rat pituitary tumor cells (GH_3) progesterone was in addition found to reduce the number of estrogen receptors (Haug 1979).

Apart from estrogens, there exists another group of endogenous compounds which was found to attenuate dopaminomimetic inhibition of prolactin secretion, i.e., various peptides with high affinity to morphine (opiate) -binding sites, so-called opioids. Compounds of this pharmacologic class, e.g., enkephalins, β-endorphin and morphine stimulate (see Table 2.1) while opiate antagonists like naloxone reduce prolactin secretion in vivo, but not in vitro when tested alone on anterior pituitaries or dispersed pituitary cells. It was the general opinion (McCann et al. 1978) that opiates act at the CNS level of prolactin control. Recently, Enjalbert et al. (1979c) observed in in vitro experiments that if prolactin secretion was inhibited by dopamine, met-enkephalin as well as β-endorphin and morphine attenuated this inhibitory effect in a concentration-dependent fashion. This opiate effect could be antagonized by naloxone to restore the inhibitory action of dopamine. Thus, it is possible that endogenous opiate peptides may play a modulatory role on prolactin secretion by also acting directly at the pituitary level to interfere with the dopaminergic control. Studies in healthy male volunteers with the met-enkephalin analogue FK 33–824 (SANDOZ) confirmed the prolactin stimulatory action of enkephalins also for man (von Graffenried et al. 1978; Brownell et al. 1980).

It was mentioned earlier that Schally's group had come to the conclusion that GABA might act as an additional PIF besides DA (Schally et al. 1977). Independently, others had also observed that the hypothalamus contained more PIF activity than could be accounted for by DA (Enjalbert et al. 1977a,b). The presence in the rat anterior pituitary lobe of GABA and of its main catabolizing enzyme, α-ketoglutarate-GABA transaminase, but not of its principal synthesizing enzyme, glutamic acid decarboxylase, has been demonstrated (Racagni et al. 1979), as well as an increase in anterior pituitary GABA content and a decrease in serum prolactin levels after inhibition in vivo of GABA degradation by ethanolamine-O-sulphate (EOS). These findings suggest that GABA, like dopamine, is brought to the anterior pituitary by the portal blood stream, probably from the hypothalamus. Recent studies with GABA mimetic and antagonistic drugs clearly demonstrated the existence of inhibitory GABA receptors at the pituitary level. These receptors show similar characteristics to those situated in the brain (Grandison and Guidotti 1979). The maximal prolactin inhibition obtained with GABA in vitro, however, was found to be less than half that obtained with DA, and the affinity of GABA was about two orders of magnitude lower (Enjalbert 1979a). Besides the inhibitory GABA-sensitive receptor system of prolactin cells, there exists a second, stimulatory system at the CNS level of rats (Locatelli et al. 1979). Interestingly, muscimol, a GABA receptor agonist, failed to lower plasma prolactin levels in hypophysectomized rats carrying an ectopic anterior pituitary under a kidney capsule (Locatelli et al. 1978, 1979), an observation which is unexplained. It is thus still quite unclear what the physiologic importance of the GABA-ergic control of prolactin secretion might be.

Finally, one other hypothalamic factor should be mentioned which acts directly at the prolactin cells to stimulate prolactin release: TRH. This tripeptide was first observed by Tashjian et al. (1971) to directly stimulate prolactin release using pituitary cell cultures. In man, this compound was also found to stimulate prolactin release besides TSH release (Bowers et al. 1971; Jacobs et al. 1971). Estrogens were recently observed in vitro to increase the number of TRH-binding sites on prolactin-producing cells of the GH_3 and GC lines of rat pituitary tumor cells (Gershengorn et al. 1979). This effect could be antagonized by the addition of tamoxifen, an estrogen receptor blocker. As prolactin and TSH serum levels in the healthy organism vary independently, TRH is not considered a serious candidate to act as PRF physiologically, but its high potency to release prolactin suggests that it may be structurally closely related to PRF (Meites 1973).

2.2 Pharmacologic Aspects

The review of the physiologic factors involved in the regulation of prolactin secretion has made it clear that this hormone is under multifactorial control. Theoretically, this situation also offers a number of different approaches to the inhibition of prolactin secretion by pharmacologic means. It will be seen that up to now most of these different approaches have not yet yielded useful drugs for the suppression of prolactin secretion.

The study of prolactin control with the aid of pharmacologic agents foreign to the organism was started by two independent observations: Shelesnyak (1954a,b), interested in the autonomic regulation of ovum implantation in the rat, found the ergot alkaloid ergotoxine to block this and to interrupt pseudopregnancy by interfering with the luteotrophic function of the pituitary. This then initiated the study of ergot compounds as prolactin-secretion inhibitors (Flückiger and Wagner 1968) but the involvement of dopaminergic mechanisms was only first mentioned in 1970 (Fuxe and Hökfelt 1970). On the other hand, Audibert et al. (1956), taking up the clinical observation of Winnik and Tennenbaum (1955) that galactorrhea may develop in patients treated by chlorpromazine, were able to induce lactation in rabbits with this drug. Soon afterwards, the first review on the pharmacologic

Table 2.2. Drugs which enhance prolactin secretion in man

Tranquillizers	Phenothiazines (type: chlorpromazine), Thioxanthenes, Azaphenothiazines, Dibenzooxazepines, Dibenothiazepines, Butyrophenones (type: haloperidol), Benzamides (type: sulpiride)
Antihistaminics (H_1, H_2)	Meclizine, Tripelenamine, Cimetidine
Antihypertensives	α-Methyl-DOPA, Reserpine
Hormones	TRH, Estrogens
Investigational tools	5-Hydroxy-tryptophan, Insulin, Opioids
Antidepressants	Amitriptyline, Imipramide
Antiemetics	Benzamides (type: metoclopramide), Phenothiazines (type: thiethylperazine)

control of prolactin secretion and lactation appeared (Meites 1962), but it concentrated on the stimulatory aspects only. In accordance with what has been said about the physiologic control mechanisms, prolactin secretion stimulation may be expected to occur under the influence of drugs interfering with catecholamine, especially dopamine neuron function. Thus, drugs which either reduce DA synthesis, such as α-methyl-p-tyrosine and α-methyl-DOPA, or storage, such as reserpine or tetrabenazine, or which block the action of DA on postsynaptic receptors, like antipsychotic drugs and related compounds, will enhance prolactin secretion and may induce lactation in experimental animals (Sulman 1970) or galactorrhea in patients (Apostolakis et al. 1972). In Table 2.2 a summary of clinically relevant drugs which enhance prolactin release is given. In the following section we shall concentrate on the inhibition of prolactin secretion by drugs. Organization of the material will be by basic pharmacologic aspects.

2.2.1 Inhibition by Dopaminergic Mechanisms

2.2.1.1 Directly Acting drugs

This mechanism of action has in the past 10 years received most interest, partly because of the physiologic importance of the dopaminergic control of prolactin secretion (MacLeod 1976), and partly because of the therapeutic relevance which one of the drugs showing dopamine-like actions, bromocriptine[1] (Flückiger and Wagner 1968), attained quite early in this period (Lutterbeck et al. 1971; Editorial 1971; Besser et al. 1972; Varga et al. 1972).

Therefore, compounds showing this type of action became very attractive to the analytical and inventive mind of many academic as well as industrial researchers, chemists and biologists alike. Drugs suppressing prolactin secretion by a dopaminergic mechanism were found in several chemical classes, which are more or less related to the structure of dopamine (see Fig. 2.2).

[1] Bromocriptine as the mesylate salt is the active principle in Parlodel®, Pravidel®

Dopamine

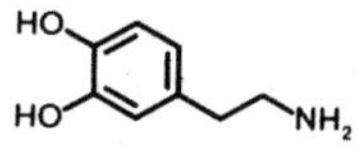

Phenyl-ethylamines

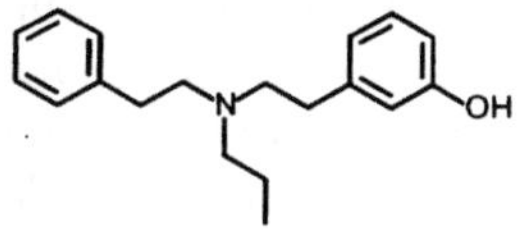

Raymond et al. (1978)
RU 24213

2-Amino-tetralines

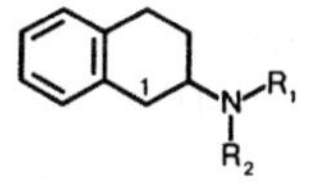

R_1 = Me, R_2 = Pr
R_1 = R_2 = Pr

Rusterholz et al. (1979)

2-Amino-indane

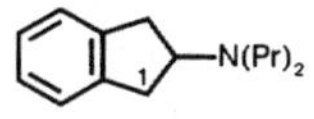

Rusterholz et al. (1979)

Amino-dihydroxy-tetralines (ADTN)

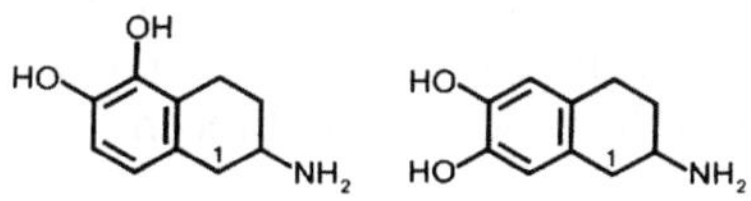

Rick et al. (1979)

Apomorphine

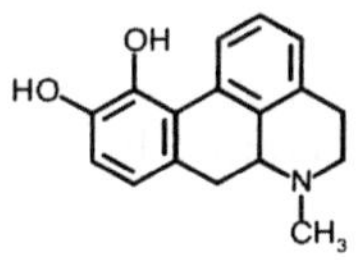

Smalstig et al. (1974)
Martin et al. (1974)

Ergolines

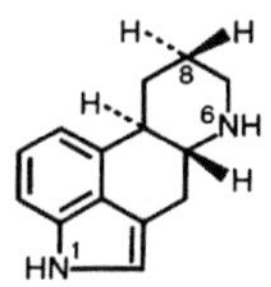

see separate Table for details

Benz [c,d] indoles

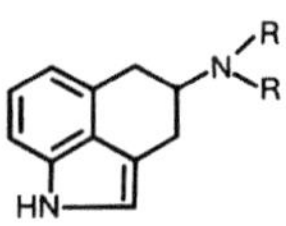

R = Me, Pr

Clemens and Smalstig (1979)

Pyrrolo-ethylamines

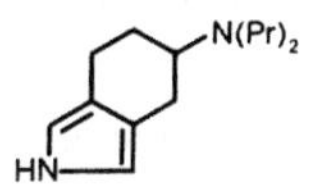

Bach et al. (1980)

Piperazines

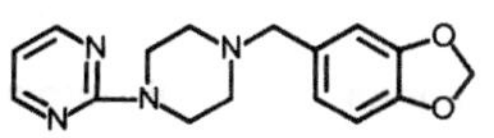

Piribedil
ET 495

MacLeod et al. (1976)

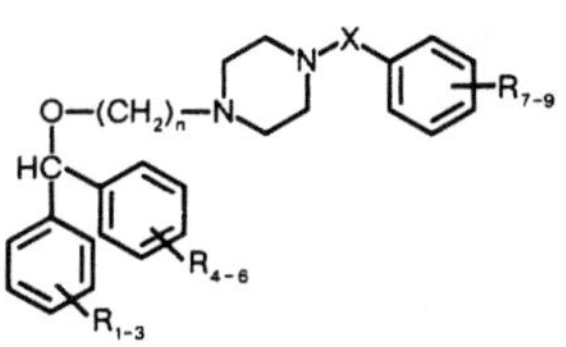

Gootjes (1978)

Fig. 2.2. Representative chemical structures which inhibit prolactin secretion by dopaminomimetic action. Abbreviations used: *Me*, *Pr* = methyl, n-propyl

The ergolines (ergot alkaloids and related compounds) have been the most successful class, with several compounds which are of clinical interest. Figure 2.3 shows the structural differences of the four important families of ergolines. Very recently a new ergoline family, showing potentially important prolactin secretion inhibitory potency was described, the 8-acetyl-ergolines (Beacco et al. 1980).

Structure activity relationships (SAR) have been discussed elsewhere (Flückiger and del Pozo 1978; Flückiger et al. 1978; Flückiger and Vigouret 1979; Hauth 1979; Bach et al. 1980) and need not be repeated here. Recently, one important additional piece of information on the relationship between structure and activity became known: it was found that the naturally occurring 6-methyl-substitution of the ergoline nucleus is less favorable for maximal dopaminergic potency in the clavine family than 6-propyl substitution (Crider et al. 1977; Krepelka et al. 1977; Krepelka and Semonski 1978; Clemens and Smalstig 1979). Pergolide is the first representative of the 6-propyl series of ergoline derivatives which is in clinical evaluation. It shows high activity and a long duration of action (Lemberger and Crabtree 1979; Kleinberg et al. 1980). The pharmacologic aspects of ergot alkaloids and related compounds have been comprehensively treated recently (Berde and Schild 1978).

Site of Action. Compounds of Fig. 2.3 all suppress prolactin secretion by a direct action at the pituitary level. Zeilmaker and Carlsen (1962), using rats with an ectopically implanted pituitary, were the first to present evidence that the ergopeptide ergocornine inhibits prolactin secretion by acting on the pituitary. In

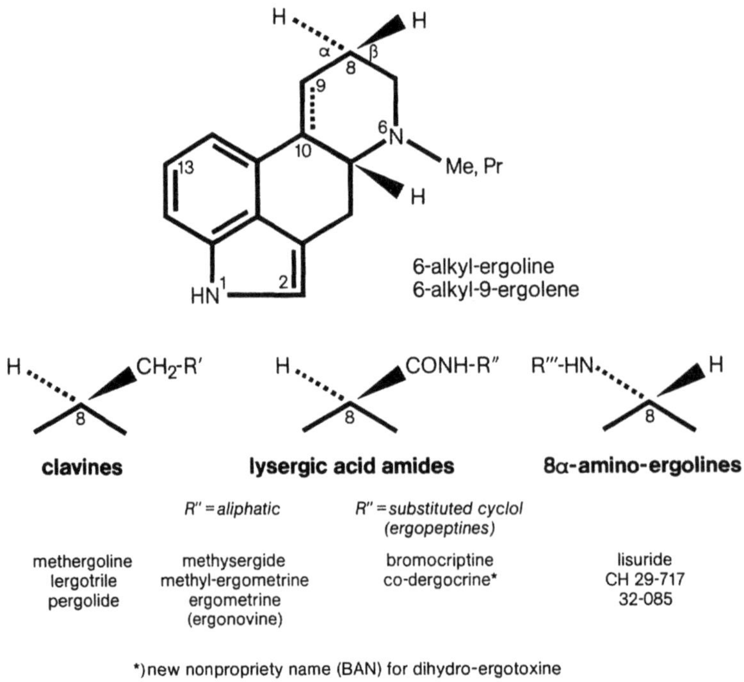

Fig. 2.3. Ergolines. The four pharmacologically and clinically important families of ergot compounds, the clavine derivatives, the aliphatic lysergic acid amides, the ergopeptines and the amino-ergolines

vivo, evidence of a direct inhibitory action is also given by the fact that such drugs antagonize TRF-induced prolactin release (Schams 1972; Delitala et al. 1979a) or that they suppress the hyperprolactinemia in rats produced by pretreatment with reserpine and α-methyl-p-tyrosine (Horowski and Gräf 1975; Flückiger 1978b). In vitro experiments by Pasteels' group (Pasteels et al. 1971; Ectors et al. 1972) and by that of Meites (Lu et al. 1971) produced direct evidence of the inhibitory action using normal prolactin cells. Other laboratories obtained similar results with further representatives of the ergoline families and related compounds (Clemens 1976; MacLeod 1976; Caron et al. 1978; Lamberts and MacLeod 1978a; Delitala et al. 1979b; Peillon et al. 1979). Such a direct action could also be shown using human pituitary adenoma cells (Mashiter et al. 1977; Adams et al. 1979a; de Camilli et al. 1979; Peillon et al. 1979). Thus, compounds of this group of agents inhibit prolactin secretion independently of the functional state of the hypothalamopituitary axis, because they act directly on the prolactin cell to suppress hormone release.

Mechanism of Action. Prolactin secretion inhibition is obtained by stimulation of DA receptors situated on the prolactin cell membrane. The action of ergot compounds could be antagonized like that of DA by DA receptor blockers (Clemens 1976; MacLeod 1976; Labrie et al. 1978; West and Dannies 1979). Several studies showed that these compounds also have a high affinity for the same binding sites on membranes isolated from pituitaries as do other DA-like drugs. In one study, comparing four ergot derivatives and apomorphine in vitro, the rank order of potency for displacing tritiated dihydroergocryptine from bovine pituitary membranes was the same as that for inhibiting prolactin release from dissociated rat pituitary cells (Caron et al. 1978). Bromocriptine, apomorphine and dopamine showed similar affinity for DA receptors of bovine caudate and bovine anterior pituitary membranes labeled with tritiated spiroperidol (Creese et al. 1977), the rank order being bromocriptine > apomorphine > dopamine. This sequence was confirmed in another study on bovine pituitary membranes, using tritiated DA (Calabro and MacLeod 1978), and in two other studies using tritiated spiroperidol and dihydroergokryptine with sheep anterior pituitary membranes (Cronin et al. 1978, Cronin and Weiner 1979).

If one considers the anatomic position of the DA receptors of prolactin cells relative to the hypothalamus, these receptors are postsynaptic. Denervation supersensitivity should therefore develop after prolonged dopaminergic deprivation of the prolactin cells. In rats, evidence for supersensitivity after destruction of the mediobasal hypothalamus was indeed observed in vivo for apomorphine (Cheung and Weiner 1976) and in vitro for DA and apomorphine (Cheung and Weiner 1978). A similar increase in sensitivity to apomorphine has also been described in rats after prolonged haloperidol or morphine treatment (Lal et al. 1977). No information exists on desensibilization of prolactin cells after prolonged treatment with dopaminomimetic agents.

Recently, a classification of DA receptors has been proposed (Kebabian and Calne 1979) on the basis of whether or not they are linked to adenylyl cyclase. DA receptors of prolactin cells were considered not to be linked to this enzyme (D_2-receptors) [target organ with c-AMP linked D_1-receptors: the parathyroid (Brown et al. 1977; Blum et al. 1980)]. On the basis of binding studies though, Sibley and Creese (1979) have pointed out that the bovine anterior pituitary contains two distinct DA receptors. Unfortunately, the available information on the relationship in prolactin cells between cAMP, dopaminergic stimulation and hormone release is not unanimous: prolactin release in vitro is stimulated by augmenting available cAMP (Nagasawa and Yanai 1972; Pelletier et al. 1972), which can be antagonized by dopamine (Kimura et al. 1976) or bromocriptine (Nagasawa et al. 1973). Dopamine and bromocriptine were observed to attenuate basal adenylate cyclase activity in homogenates of rat pituitaries (Markstein et al. 1978). Dopamine and the amino-ergoline CH 29–717 (SANDOZ) inhibited cAMP formation in homogenates of human pituitary adenoma (de Camilli et al. 1979). These findings agree with the definition of these DA receptors as D_2 receptors. Others observed (Ahn et al. 1979) an enhanced activity of adenylate cyclase by such drugs, using homogenates of rat, rabbit or monkey anterior pituitaries, while still others could not find a change (Schmidt and Hill 1977). It is possible that the different changes in adenylate cyclase activity after dopamine-like drugs just described, are secondary effects. In fact, bromocriptine inhibited prolactin release stimulated by a depolarizing concentration of potassium, which does not increase cAMP (Gautvik et al. 1973). In addition, the cation ionophore A 23197-stimulated prolactin release was inhibited by DA, bromocriptine and dihydroergokryptine, which effect could be antagonized by the DA receptor antagonists butaclamol and flupenthixol (Tam and Dannies 1980). These findings suggest that DA receptor stimulators inhibit the process of prolactin extrusion at a step beyond Ca^{2+} uptake.

Not all prolactin cells respond to the presence of dopamine-like agents with attenuated hormone secretion, such an exception being certain transplantable pituitary tumors of the rat. Lamberts and MacLeod (1979a) showed that tumor growth and prolactin secretion of the tumors MtTW15 and 7315a in host rats was not suppressed by bromocriptine. In contrast, both ergotamine and α-ergocryptine were active. Interestingly, their actions were not antagonized by concomitant haloperidol treatment. The authors suggested that both α-ergocryptine and ergotamine, which are ca. 50 and >100 times more potent vasoconstrictor agents (α-agonists) than bromocriptine (Flückiger 1976), reduced capillary blood supply to the tumors and thereby decreased tumor growth and prolactin release. The reason why these tumors are not sensitive to the action of bromocriptine is unexplained. A connection seems to exist between the phenomenon of desensitization and estrogens of DA receptors of normal prolactin cells (Labrie et al. 1978; Beaulieu et al. 1979). Timmermans and Lamberts (1979) found that the DA-refractive tumor 7315a, when treated with an antiestrogen, tamoxifen, became DA sensitive (see also de Quiada et al. 1980a). This suggests that the antiestrogen activate latent DA receptors, while estrogens either induce a reversible inactive state or deplete prolactin cells of DA receptors. For further discussion of these aspects see Cronin et al. (1980). De Quiada et al. (1980b) further observed that tamoxifen given alone could reduce the growth of the pituitary tumor 7315a in the rat and attenuate prolactin secretion of both the tumor and the pituitary. Working with subcutaneous tumors of the GH_3 cell line Prysor-Jones and Jenkins (1980) also found that bromocriptine did not affect growth or hormone secretion. Ergokryptine, ergocornine, lergotrile and CH 29–717 were similarly inactive, but ergotamine reduced hormone secretion and growth. In vitro, ergotamine inhibited DNA synthesis of tumor cells.

Prolactin-secreting cells do not only react to dopaminomimetic stimulation by attenuation of hormone release. It has been demonstrated that such drugs may also affect hormone synthesis and hormone degradation. Thus, up to 12 h of inhibition of prolactin release in vivo results in accumulation of the hormone in the pituitary (Flückiger and Kovacs 1974; Häusler et al. 1978), while prolonged treatment of mice with ergocornine or bromocriptine reduced the pituitary prolactin content and concentration (Yanai and Nagasawa 1970a,b). In lactating rats, after several days' treatment with bromocriptine, the secretory granules of prolactin cells were found in densely packed groups away from the cell membranes, and the number of lysosomes was increased (Häusler and Hodel 1975). Lysosomes fuse with the surplus secretory vesicles (crinophagy) to degrade the secretory granules (Farquhar 1977). Activation of lysosomal enzymes has been demonstrated in pituitary homogenates after treating the rats with L-DOPA or after treating the pituitary in vitro with DA (Nansel et al. 1980). This activation could be prevented by preincubation with chloroquine. Unexpectedly, chloroquine also prevented the release inhibitory action of DA, so that Nansel et al. (1980) suggested that the increase in lysosomal activity (crinophagy) may be involved in the inhibitory action of DA on prolactin release. In in vitro studies Dannies and Rudnick (1980) using bromocriptine as the dopaminomimetic, first observed an accumulation of prolactin, but after 4 days of release inhibition the total accumulation of hormone was greatly reduced while synthesis of new hormone was only partly attenuated, which indicated that an increased rate of hormone degradation must take place. When treating male rats continuously with estrogen to stimulate prolactin synthesis and prolactin cell proliferation, concomitant treatment with bromocriptine produced lower serum prolactine levels, diminished DNA synthesis and a reduced mitotic activity in the pituitary (Davies et al. 1974; Lloyd et al. 1975, 1978) suggesting that continued suppression of prolactin release may lead to attenuation of cell metabolism. This latter effect may bear clinical relevance (see Sect. 4.1.3).

Because of the great interest in the practical and theoretical relevance of ergot compounds, a number of symposia devoted to them have recently been published (Bayliss et al. 1976; Hökfelt and Nillius 1978; Colloque 1979; Fuxe and Calne 1979; Goldstein et al. 1980; MacLeod and Scapagnini 1980; Müller 1980) as well as some monographs on the most important of these derivatives, bromocriptine (Parkes 1978; Metha and Tolis 1979; Thorner et al. 1980a). These publications contain important additional information on pharmacologic and clinical aspects not covered in the present book.

2.2.1.2 Indirectly Acting Drugs

D-amphetamine, amantadine, nomifensine, methylphenidate and cocaine are agents which do not stimulate DA receptors themselves but facilitate neuronal DA release and/or inhibit DA reuptake. Such drugs therefore enhance dopaminergic neurotransmission and could be expected to lower prolactin secretion if the hypothalamopituitary axis is intact. Interestingly, one study reported a seven-fold increase of prolactin serum levels and a depletion of pituitary prolactin stores in female rats after d-amphetamine (Lu and Meites 1971). In another study, d-amphetamine lowered prolactin serum levels in intact rats as expected and was ineffective in reserpine and α-methyl-p-tyrosine-pretreated rats (Horowski and Gräf 1976). In a third study, d-amphetamine, methylphenidate and cocaine showed only weak activity in intact male rats after high doses but pretreatment with α-methyl-p-tyrosine increased the sensitivity of the animals to d-amphetamine and methylphenidate (Ravitz and Moore 1977). Nomifensine showed properties similar to amphetamine in rats (Cocchi et al. 1979). Its effect in humans and the possible use of this drug to differentiate between tumorous and non-tumorous hyperprolactinemia are discussed in Chaps. 4 and 5.

Monoamine-oxidase inhibitors increase available catecholamines by blocking their inactivation, and may thereby reduce prolactin secretion in vivo. This was found to be the case with pargyline, Lilly-15641 and iproniazide (Lu and Meites 1971) in female rats. In contrast, depressive patients showed an increase in prolactin levels after pargyline and clorgyline (Slater et al. 1977) and L-deprenyl (Mendlewicz and Youdim 1977). The discrepancy between the rat findings and the human reaction to these drugs remains to be explained.

L-DOPA was used quite early on to increase catecholamine activity in the hypothalamus and was found to increase PIF activity and reduce prolactin secretion in rats from in situ and ectopic pituitaries (Lu and Meites 1971, 1972). Also in man L-DOPA was found to suppress serum prolactin (Friesen et al. 1972). The action of L-DOPA is rather short lived and due to peripheral metabolic clearance massive doses are needed to increase effectively catecholamine (dopamine) activity in the CNS. Therefore, inhibitors of the aromatic aminoacid decarboxylase which do not readily enter the brain offer themselves for the protection of L-DOPA from peripheral metabolic conversion (Bartholini and Pletscher 1975; Pinder et al. 1976), e. g., benserazide (Ro 4–4602) or carbidopa (MK 486). Used by themselves they both increase serum prolactin levels in man (Brown et al. 1976b; Pontiroli et al. 1977). This effect is explained by the assumption that they decrease the available dopamine in the nerve terminals of the outer median eminence which is outside the blood-brain barrier. Of the two compounds benserazide is the more potent stimulant of prolactin release (Polleri et al. 1980). If given together with L-DOPA, the prolactin secretion inhibitory effect of L-DOPA is augmented (Frantz et al. 1973; Pontiroli et al. 1977). L-DOPA has been found to also inhibit prolactin secretion in monkeys after sectioning of the pituitary stalk (Diefenbach et al. 1976), thus suggesting that it may act, after decarboxylation to dopamine, directly at the prolactin cell. Pyridoxine (Vitamin B_6 200 mg t. i. d. orally was reported by Foukas (1973) to inhibit postpartum lactation. He had tried this compound with the view of facilitating DA synthesis, as pyridoxal phosphate is involved in DA formation, acting as co-enzyme of DOPA decarboxylase. Pyridoxine, given parenterally, was found to reduce the prolactin surge in the afternoon, of proestrus in rats, but it had a short duration of action (Harris et al. 1978). This vitamin , in the same study, was also able to suppress the stimulatory effects of TRH on prolactin, and was even active in rats in which DA synthesis had been inhibited by pretreatment with α-methyl-p-tyrosine (AMPT). This could be understood as indicating a direct effect of pyridoxine at the prolactin cell (Harris et al. 1978), but it is difficult to explain by a facilitation of DA synthesis in AMPT treated rats. Brambilla et al. (1979) did not find a lowering of prolactin levels after pyridoxine (300 mg i.v. to 1200 mg orally) in neuroleptic treated psychotic patients. In contrast Delitala et al. (1976, 1977) giving pyridoxine (300 mg) intravenously observed a lowering of serum prolactin levels in healthy subjects and in patients with primary hypothyroidism. In a study using male rats, pyridoxine, but not pyridoxal, was found to lower chlorpromazine-induced hyperprolactinemia. This effect was clearly less than that after bromocriptine (Rosenberg et al. 1977).

2.2.2 Inhibition by Serotoninergic Mechanisms

Although there is good evidence from experiments in rats and also in humans that serotonin is involved in prolactin release mechanisms, the use of serotonin (5HT) antagonists has produced seemingly contradictory results. Chemically, there are two classes of 5HT receptor blockers, ergoline derivatives and the so-called tricyclic compounds. Methysergide and methergoline are two widely used 5HT antagonists which are ergot derivatives. Besides blocking 5HT receptors, both compounds stimulate DA receptors in vivo as has recently been demonstrated (Lamberts and MacLeod 1978a, 1979b). Because of this DA-like activity, such drugs cannot be used as investigational tools to assess the contribution of serotoninergic mechanisms to prolactin regulation. Of the class of tricyclic compounds, cyproheptadine has been studied in vitro and in vivo; cyproheptadine was found to inhibit prolactin release by normal rat pituitary cells and by dispersed cells from a transplantable pituitary tumor (Lamberts and MacLeod 1978b). Since serotonin was shown to have no effect on prolactin release in vitro, and since the effect of cyproheptadine was not counteracted by DA antagonists, its inhibitory action was considered not to be due to interactions with 5HT or DA receptors. In man, cyproheptadine was found to augment TRH-stimulated prolactin secretion (Egge et al. 1977) but to antagonize insulin-induced hyperprolactinemia (Corenblum and Whitaker 1977). While the former effect is unexplained, the latter indicates the involvement of serotoninergic transmission in the system transducing acute hypoglycemia into prolactin release stimulation. Using pizotifen, another tricyclic 5HT antagonist, other studies could not demonstrate an effect on basal prolactin secretion during sleep (del Pozo and Lancranjan 1978; Clarenbach et al. 1980) nor on prolactin release due to breast stimulation in postpartal women (Lancranjan et al. 1979). From this it follows that serotonin antagonism is not a useful mechanism to obtain prolactin secretion inhibition in man.

2.2.3 Inhibition by Cholinergic Mechanisms

The cholinergic system of the CNS was suggested to have an inhibitory role in prolactin control (Grandison et al. 1974; Lichtensteiger and Keller 1974; Chen and Meites 1975; Subramanian and Gala 1976a, b, 1977). As pilocarpine (i. p.), which significantly reduced the elevated prolacting levels in estrogen-treated rats, did not reduce serum prolactin levels of reserpine, haloperidol or pimozide pretreated rats, it was concluded (Grandison and Meites 1976) that the cholinergic inhibitory effect was mediated via catecholaminergic neurons. In agreement with this, atropine was found to prolong the stimulatory effect of perphenazine on prolactin secretion in monkeys (Gala et al. 1976). Such indirect inhibitory action does not offer a practical approach to prolactin secretion control. On the other hand, Vale et al. (1975) have observed in long-term in vitro cultures of dissociated pituitary cells that cholinergic agonists, such as acetylcholine, carbachol and oxotremorine, inhibit prolactin secretion and that this effect is antagonized by low concentrations of atropine. This interesting observation does not seem to have been taken up by others. In a recent discussion contribution Vale remarked (Gotto et al. 1979) that the maximal inhibitory response of such dissociated pituitary cells to acetylcholine is less than that obtained with dopamine, but greater than that seen with GABA. These observations make it probable that prolactin cells carry cholinergic receptors of the muscarinic type besides dopamine receptors, but the existence of cholinergic binding sites has not yet been ascertained. It is thus clear that the potential of the cholinergic approach to prolactin suppression has not yet been fully assessed. In man, benztropine had not effect on basal prolactin secretion but it enhanced the stimulatory action of haloperidol (Lal et al. 1979). Thus in man, under dopamine receptor blockade, cholinergic mechanisms may exert an inhibitory tone on prolactin secretion, indicating a mechanism of prolactin control independent of dopamine.

2.2.4 Inhibition via Gaba-ergic Mechanisms

Although evidence exists to suggest that GABA should be considered besides dopamine, a hypothalamic prolactin inhibitory factor (PIF), (Fuxe et al. 1978a; Müller et al. 1978b), its relevance for a practical approach to prolactin suppression is far from clear. As we have seen in Sect. 2.1.2 (p. 31) GABA-mimetic drugs, by acting at the CNS level, may stimulate prolactin release, only compounds which do not pass through the blood-brain barrier being useful when clarifying the practical relevance of a direct inhibitory action on the pituitary (see also del Pozo and Brownell 1979).

2.2.5 Miscellaneous Approaches to Prolactin Suppression

2.2.5.1 Inhibition of Prostaglandin Synthesis

Various prostaglandins (E and F series) have been shown to stimulate prolactin release in rats, and it has been suggested that PGE_1 and PGE_2 of the hypothalamus may be involved in the increased secretion of prolactin after estrogens (Ojeda et al. 1978):pretreatment of rats with indomethacin reduced the effect of estradiol on prolactin serum levels dose dependently and sensibilized the animals for PGE_1 and PGE_2 injected intravenously. Hedge (1977) reported that prolonged treatment of rats with indomethacin lowered basal prolactin levels. In man, Dussault et al. (1976) observed that after prolonged treatment with aspirin basal prolactin level and its response to TRH was unaltered. A more recent study using indomethacin was also unable to find a change in basal or phenothiazine-stimulated prolactin levels in man (Buckman et al. 1979).

2.2.5.2 Steroids and Antagonists

It has been mentioned (Sect. 2.1.2, p. 29) that estrogens stimulate prolactin cells to increase hormone synthesis and release, both by a hypothalamic and by a direct action. This stimulatory effect on prolactin cell activity can be counteracted by the antiestrogens tamoxifen and nitromifen citrate (CI-628) (Nagy et al. 1979b). The estrogen-induced prolactin-secreting rat pituitary tumor 7315a was reported to respond to dopaminomimetic drugs only after pretreatment with the antiestrogen tamoxifen (Timmermans and Lamberts 1979). Tamoxifen was also found to reduce prolactin serum levels in hyperprolactinemic mammary cancer patients and to attenuate the prolactin response to TRH (Willis et al. 1977).

Testosterone and 5α-dihydrotestosterone were found not to influence prolactin secretion in rats, but the synthetic androgen R 1881 reduced serum prolactin levels

(Fuxe et al. 1978a). In contrast, prolonged treatment of healthy humans with cyproterone acetate, the antiandrogenic and progestative steroid derivative, induced increased basal levels of prolactin in both sexes (Gräf et al. 1978). In female rats, prolonged treatment with progesterone did not produce signs of increased prolactin secretion (increased mammary gland DNA levels) but norethisterone enanthate was active (Gräf et al. 1973). In ovariectomized women primed with ethinyl estradiol, progesterone (10 mg given intramuscularly) induced, after a few hours delay, an increase in serum prolactin levels (Rakoff and Yen 1978). Such observations are in contrast with the prolactin secretion inhibitory action of progesterone in domestic animals and the reported antiestrogen effects on prolactin cells (see p. 30 above). In contrast, no elevation of serum prolactin was observed in women after long-term use of a progesterone-releasing intrauterine device (P-IUD) (Spellacy and Buhi 1979). Medroxyprogesterone acetate, used as an injectable long-acting contraceptive did not elevate basal prolactin levels or increase sensitivity to a chlorpromazine stimulus significantly (Spellacy et al. 1975). A more recent study using norethisterone enanthate and medroxyprogesterone acetate intramuscularly also found no change in basal or stimulated prolactin levels in healthy women (Horowski et al. 1978).

2.3 **References** (also for Chap. 1)

Adams EF, Brajkovich IE, Mashiter K (1979a) Hormone secretion by dispersed cell cultures of human pituitary adenomas: Effects of theophylline, thyrotropin-releasing hormone, somatostatin, and 2-Br-α-ergokryptine. J Clin Endocrinol Metab 49:120–126

Adams ND, Garthwaite TL, Gray RW, Hagen TC, Lemann J (1979b) The interrelationships among prolactin, 1,25-dihydroxyvitamin D, and parathyroid hormone in humans. J Clin Endocrinol Metab 49:628–630

Adashi EY, Rakoff J, Divers W, Fishman J, Yen SSC (1979) The effect of acutely administered 2-hydroxyestrone on the release of gonadotropins and prolactin before and after estrogen priming in hypogonadal women. Life Sci 25:2051–2055

Advis JP, Hall TR, Hodson CA, Mueller GP, Meites J (1977) Temporal relationship and role of dopamine in "short-loop" feedback of prolactin. Proc Soc Exp Biol Med 155:567–570

Ahn HS, Gardner E, Makman MH (1979) Anterior pituitary adenylate cyclase stimulation by dopamine and other monoamines. Eur J Pharmacol 53:313–317

Ambrosi B, Travaglini P, Beck-Peccoz P, Bara R, Elli R, Paracchi A, Faglia G (1976) Effect of sulpiride-induced hyperprolactinemia on serum testosterone response to HCG in normal men. J Clin Endocrinol Metab 43:700–703

Annunziato L, Moore KE (1977) Increased ability of apomorphine to reduce serum concentrations of prolactin in rats treated chronically with α-methyl-tyrosine. Life Sci 21:1845–1850

Aono T, Shioji T, Shoda T, Kurachi K (1977) The initiation of human lactation and prolactin response to suckling. J Clin Endocrinol Metab 44:1101–1106

Apostolakis M, Kapetanakis S, Lazos G, Madena-Pyrgaki A (1972) Plasma prolactin activity in patients with galactorrhea after treatment with psychoactive drugs. In: Wolstenhome GEW, Knight J (eds) Lactogenic hormones. Livingstone, Edinburgh, pp 349–354

Aragona C, Priesen HG (1975) Specific prolactin binding sites in the prostate and testis of rats. Endocrinology 97:677–684

Armstrong DT, Knudsen KA, Miller LS (1970) Effects of prolactin upon cholesterol metabolism and progesterone biosynthesis in corpora lutea of rats hypophysectomised during pseudopregnancy. Endocrinology 86:634–641

Asawaroengchai H, Russell SM, Nicoll CS (1978) Electrophoretically separable forms of rat prolactin with different bioassay and radioimmunoassay activities. Endocrinology 102:407–414

Assies J, Schellekens APM, Touber JL (1978) Prolactin in human cerebrospinal fluid. J Clin Endocrinol Metab 46:576–586

Astwood EB (1941) The regulation of the corpus luteum function by hypophysial luteotrophin. Endocrinology 29:309–319

Aubert ML, Sizonenko PC, Kaplan SL, Grumbach MM (1977) The ontogenesis of human prolactin from fetal life to puberty. In: Crosignani PG, Robyn C (eds) Prolactin and human reproduction. Academic Press, London, pp 9–20

Audibert A, Forgue G, Gage C (1956) Influence de la chlorpromazine sur l'activité hypophyso-ovarienne. C R Soc Biol (Paris) 150:173–175

Aumüller G (1979) Prostate gland and seminal vesicles. Springer, Berlin Heidelberg New York

Bach NJ, Kornfeld EC, Jones ND, Chaney MO, Dorman DE, Paschal JW, Clemens JA, Smalstig EB (1980) Bicyclic and tricyclic ergoline partial structures. Rigid 3-(2-Aminoethyl)pyrroles and 3- and 4-(2-Aminoethyl)pyrazoles as dopamine agonists. J Med Chem 23:481–491

Barkey RJ, Shani J, Amit T, Barzilai D (1977) Characterization of the specific binding of prolactin to binding sites in the seminal vesicle of the rat. J Endocrinol 80:181–189

Barkey RJ, Shani J, Barzilai D (1979) Regulation of prolactin binding sites in the seminal vesicle, prostate gland, testis and liver of intact and castrated adult rats: effect of administration of testosterone, 2-bromo-α-ergocryptine and fluphenazine. J Endocrinol 81:11–18

Barnett AH, Chapman C, Gailer K, Hayter CJ (1980) Effect of bromocriptine on maturity onset diabetes. Postgrad Med J 56:11–14

Bartholini G, Pletscher A (1975) Decarboxylase inhibitors. Pharmacol Ther 3:407–421

Bartke A (1964) Histology of the anterior hypophysis, thyroid and gonads of two types of dwarf mice. Anat Rec 149:225–236

Bartke A (1965) Influence of luteotrophin on fertility of dwarf mice. J Reprod Fertil 10:93–103

Bartke A (1966 a) Influence of prolactin on male fertility in dwarf mice. J Endocrinol 35:419–420

Bartke A (1966 b) Reproduction of female dwarf mice treated with prolactin. J Reprod Fertil 11: 203–206

Bartke A (1969) Prolactin changes cholesterol stores in the mouse testes. Nature 224:700–701

Bartke A (1971) Effects of prolactin on spermatogenesis in hypophysectomized mice. J Endocrinol 49:311–316

Bartke A (1973) Plasma testosterone levels in male mice and rats treated with inhibitors of prolactin release. Acta Endocrinol [Suppl 177] (Copenh) 73:22

Bartke A (1974) Effects of inhibitors of pituitary prolactin on testicular cholesterol stores, seminal vesicle weight, fertility and lactation in mice. Biol Reprod 11:319–325

Bartke A, Dalterio S (1976) Effects of prolactin on the sensitivity of the testis to LH. Biol Reprod 15:90–95

Bartke A, Lloyd CW (1970 a) The influence of pituitary homografts on the weight of the accessory reproductive organs in castrated male mice and rats and on mating behaviour in male mice. J Endocrinol 46:313–320

Bartke A, Lloyd CW (1970 b) Influence of prolactin and pituitary isografts on spermatogenesis in dwarf mice and hypophysectomised rats. J Endocrinol 46:321–329

Bartke A, Goldman BD, Bex F, Dalterio S (1977) Effects of prolactin (PRL) on pituitary and testicular function in mice with hereditary PRL deficiency. Endocrinology 101:1760–1766

Bartosik DB, Romanoff EB (1969) The luteotrophic process; effects of prolactin and LH on sterol and progesterone metabolism in bovine luteal ovaries perfused in vitro. In: McKerns KW (ed) The gonads. North-Holland, Amsterdam: pp 55–92

Bartosik DB, Romanoff EB, Watson DJ, Scricco E (1967) Luteotrophic effects of prolactin in the bovine ovary. Endocrinology 81:186–194

Batten TF, Ball JN (1976) Ultrastructural studies on the prolactin cells and their innervation in the teleost Poecilia latipinna. J Endocrinol 69:35P–36P

Bauer AGC, Wilson JHP, Lamberts SWJ (1980) The kidney is the main site of prolactin elimination in patients with liver disease. J Clin Endocrinol Metab 51:70–73

Bauer K, Gräf KJ, Faivre-Bauman A, Beier S, Tixier-Vidal A, Kleinkauf H (1978) Inhibition of prolactin secretion by histidyl-proline-diketo-piperazine. Nature 274:174–175

Baumann G, Marynick GF, Winters SJ, Loriaux DL (1977) The effect of osmotic stimuli on prolactin secretion and renal water excretion in normal man and in chronic hyperprolactinemia. J Clin Endocrinol Metab 44:199–202

Bayliss RIS, Turner P, Maclay WP (eds) (1976) Pharmacological and clinical aspects of bromocriptine (Parlodel). MCS Consultants, Tunbridge Wells

Beacco E, Bernardi L, di Salle E, Patelli B, Temperilli A (1980) Antiprolactinic ergolines. Experientia 36:238–239

Beaulieu M, di Paolo T, Ferland L, Raymond V, Labrie F (1979) Antagonism between estrogens and dopamine at the anterior pituitary level on prolactin secretion. In: Usdin E, Kopin IJ, Barchas J (eds) Catecholamines: Basic and clinical frontiers. Pergamon, New York, pp 1263–1265

Beck JC, Gonda A, Hamid MA, Morgen RO, Rubinstein D, McGarry EE (1964) Some metabolic changes induced by primate growth hormone and purified ovine prolactin. Metabolism 13:1108–1134

Belanger A, Auclair C, Seguin C, Kelly PA, Labrie F (1979) Down-regulation of testicular androgen biosynthesis and LH receptor levels by an LHRH agonist: role of prolactin. Mol Cell Endocrinol 13:47–53

Ben-Jonathan N, Oliver C, Weiner HJ, Mical RS, Porter JC (1977) Dopamine in hyposial portal plasma of the rat during the estrous cycle and throughout pregnancy. Endocrinology 100:452–458

Bennett HP, McMartin C (1979) Peptide hormones and their analogues. Distribution, clearance from the circulation and inactivation in vivo. Pharmacol Rev 30:247–292

Berde B, Schild HO (eds) (1978) Ergot alkaloids and related compounds. Springer, Berlin Heidelberg New York: (Handbook of experimental pharmacology, vol 49)

Bergeron JJM, Posner BI, Josefsberg Z. Sikstrom R (1978) Intracellular polypeptide hormone receptors. J Biol Chem 253:4058–4066

Besser GM, Thorner MO (1976) Bromocriptine in the treatment of the hyperprolactinaemia-hypogonadism syndromes. Postgrad Med J 52:64–70

Besser GM, Parke L, Edwards CRW, Forsyth IA, McNeilly AS (1972) Galactorrhoea: successful treatment with reduction of plasma prolactin levels by brom-ergokryptine. Br Med J 3:669–672

Bevilacqua M, Norbiato G, Raggi U, Micossi P, Baggio E, Prandelli M (1980) Dopaminergic control of serum potassium. Metabolism 29:306–310

Bewley TA (1979) Circular dichroism of pituitary hormones. Recent Prog Horm Res 35:155–210

Bex FJ, Bartke A (1977) Testicular LH binding in the hamster: modification by photoperiod and prolactin. Endocrinology 100:1223–1227

Bhattacharya A, Vonderhaar BK (1979) Thyroid hormone regulation of prolactin binding to mouse mammary glands. Biochem Biophys Res Commun 88:1405–1411

Biales B, Dichter MA, Tischler A (1977) Sodium and calcium action potential in pituitary cells. Nature 267:172–173

Bigazzi M, Pollicino G, Nardi E (1979) Is human decidua a specialized endocrine organ? J Clin Endocrinol 49:847–850

Billeter L, Flückiger E (1971) Evidence for a luteolytic function of prolactin in the intact cycling rat using 2-Br-α-ergokryptine (CB 154). Experientia 27:464–465

Birkinshaw M, Falconer IR (1972) The localization of prolactin labelled with radioactive iodine in rabbit mammary tissue. J Endocrinol 55:323–334

Bishop W, Krulich L, Fawcett CP, McCann SM (1971) The effects of median eminence (ME) lesions on plasma levels of FSM, LH, and prolactin in the rat. Proc Soc Exp Biol Med 136:925–927

Biswas DK, Tashjian AM (1974) Intracellular site of prolactin synthesis in rat pituitary cells in culture. Biochem Biophys Res Commun 60:241–248

Blossey HC, Bartsch HH, Koebberling J, Nagel G (1979) Immunendocrinology related results of treatment in advanced breast cancer (Abstr). Cancer Treat Rep 63:1221

Blüm V (1977) Prolaktin: Phylogenetische Aspekte. Gynaekologe 10:51–61

Blum JW, Kunz P, Fischer JA, Binswanger U, Lichtensteiger W, Da Prada M (1980) Parathyroid hormone response to dopamine in cattle. Am J Physiol 239:E 255–E 264

Boass A, Tovernd SU, McCain TA, Pike JW, Haussler MR (1977) Elevated serum levels of 1α,25-dihydroxycholecalciferol in lactating rats. Nature 267:630–633

Bohnet HG, Friesen HG (1976) Effect of prolactin and growth hormone on prolactin and LH receptors in the dwarf mouse. J Reprod Fertil 48:307–311

Bohnet HG, Shiu RPC, Grinwich D, Friesen HG (1978) In vivo effects of antisera to prolactin receptors in female rats. Endocrinology 102:1657–1661

Boot LM (1969) Induction by prolactin of mammary tumours in mice. Noord-Hollandse Uitgevers Maatschappij, Amsterdam

Bowers CY, Friesen HG, Hwang P, Guyda HJ, Folkers K (1971) Prolactin and thyrotropin release in man by synthetic pyroglutamyl-histidyl-prolinamide. Biochem Biophys Res Commun 45:1033–1041

Bowman LA, Dilley SR, Keverne EB (1978) Suppression of oestrogen-induced LH surges by social subordination in talapoin monkeys. Nature 275:56–58

Boyd AE, Spencer E, Jackson IMD, Reichlin S (1976) Prolactin releasing factor (PRF) in porcine hypothalamic extract distinct from TRH. Endocrinology 99:861–871

Boyns AR, Cole EN, Golder MP, Danutra V, Harper ME, Brownsey B, Cowley T, Jones GE, Griffiths K (1972) Prolactin studies with the prostate. In: Boyns AR, Griffiths K (eds) Prolactin and carcinogenesis. Alpha Omega Alpha, Cardiff, pp 207–216

Brambilla F, Penati G, Freni S, Bottinelli S, Maffei C, Maiella V (1979) Failure of pyridoxine to affect neuroleptic-induced hyperprolactinaemia in psychotic patients. J Endocrinol Invest 2:299–302

Brisson GR, Volle MA, DeCarufel D, Desharnais M, Tanaka M (1980) Exercise-induced dissociation of the blood prolactin response in young women according to their sports habits. Horm Metab Res 12:201–205

Brown EM, Carroll RJ, Aurbach GD (1977) Dopaminergic stimulation of cyclic AMP accumulation and parathyroid hormone release from dispersed bovine parathyroid cells. Proc Natl Acad Sci USA 74:4210–4213

Brown GM, Seeman P, Lee T (1976a) Dopamine/neuroleptic receptors in basal hypothalamus and pituitary. Endocrinology 99:1407–1410

Brown GM, Garfinkel PE, Warsh JJ, Stancer HC (1976b) Effect of carbidopa on prolactin, growth hormone and cortisol secretion in man. J Clin Endocrinol Metab 43:236–239

Brownell J, del Pozo E, Donatsch P (1980) Inhibition of vasopressin secretion by a met-enkephalin (FK 33–824) in humans. Acta Endocrinol (Copenh) 94:304–308

Brun del Re R, del Pozo E, de Grandi O, Friesen HG, Hinselmann M, Wyss H (1973) Prolactin inhibition and suppression of puerperal lactation by 2-Br-α-ergocryptine (CB 154). A Comparison with oestrogen. Obstet Gynecol 41:884–887

Bruni JF, van Vugt D, Marshall S, Meites J (1977) Effects of naloxon, morphine and methionin-enkephalin on serum prolactin, luteinizing hormone, follicle stimulating hormone, thyroid stimulating hormone and growth hormone. Life Sci 21:461–466

Buckman MT, Peake GT, Srivastava S (1979) Indomethacin fails to alter basal or phenothiazine-induced prolactin concentrations in man. Horm Metab Res 11:395–398

Calabro MA, MacLeod RM (1978) Binding of dopamine to bovine anterior pituitary gland membranes. Neuroendocrinology 25:32–46

Camilli P de, Macconi D, Spada A (1979) Dopamine inhibits adenylate cyclase in human prolactin-secreting pituitary adenomas. Nature 278:252–254

Campbell GT, Wagoner J, Gregerson KA, Joyner WL (1979) Immunohistochemical visualization of prolactin, growth hormone, and a substance resembling placental lactogen in the in situ and ectopic pituitary in the hamster. Endocrinology 105:905–910

Carey RM, Johanson AJ, Seif SM (1977) The effects of ovine prolactin on water and electrolyte excretion in man are attributable to vasopressin contamination. J Clin Endocrinol Metab 44:850–858

Carey RM, Thorner MO, Ortt EM (1979) Effects of metoclopramide and bromocriptine on the renin-angiotensin-aldosterone system in man. Dopaminergic control of aldosterone. J Clin Invest 63:727–737

Carlo R di, Muccioli G, Conti G, Camilla R, di Carlo F (1980) Estrogen and prolactin receptor concentrations in human breast tumors. In: Genazzani E (ed). Pharmacological modulation of steroid action. Raven, New York, pp 261–266

Caron MG, Beaulieu M, Raymond V, Gagné B, Drouin J, Lefkowitz RJ, Labrie F (1978) Dopaminergic receptors in the anterior pituitary gland. J Biol Chem 253:2244–2253

Carr D, Friesen HG (1976) Growth hormone and insulin binding to human liver. J Clin Endocrinol Metab 42:484–493

Castracane VD, Shaikh AA (1979) Synergism of oestrogen and bromoergocryptine in the induction of luteolysis in cynomolgus monkeys (M. fascicularis). 61st Annual Meeting of the Endocrine Society, Anaheim, June 13–15, Abstr no 706

Chang N, Ebels I, Benson B (1979) Preliminary characterization of bovine pineal prolactin releasing (PPRF) and release-inhibiting (PPIF) activity. J Neural Transm 46:139–151

Charreau EH, Attramadal A, Torjesen PA, Purvis K, Calandra R, Hansson V (1977) Prolactin binding in rat testis: specific receptors in interstitial cells. Mol Cell Endocrinol 6:303–307

Chen CL, Amenomori Y, Lu KH, Voogt JL, Meites J (1970) Serum prolactin levels in rats with pituitary transplants or hypothalamic lesions. Neuroendocrinology 6:220–227

Chen HJ, Meites J (1975) Effects of biogenic amines and TRH on release of prolactin and TSH in the rat. Endocrinology 96:10–14
Cheung CY (1980) Calmodulin plays a pivotal role in cellular regulation. Science 207:19–27
Cheung CY, Weiner RI (1976) Supersensitivity of anterior pituitary dopamine receptors involved in the inhibition of prolactin secretion following destruction of the medial basal hypothalamus. Endocrinology 99:914–917
Cheung CY, Weiner RI (1978) In vitro supersensitivity of the anterior pituitary to dopamine inhibition of prolactin secretion. Endocrinology 102:1614–1620
Chi HJ, Shin SH (1978) The effect of exposure to ether on prolactin secretion and the half-life of endogenous prolactin in normal and castrated male rats. Neuroendocrinology 26:193–201
Chirito E, Gonda A, Friesen H (1972) Prolactin in renal failure. Clin Res 20:423
Clarenbach P, del Pozo E, Brownell J, Heredia E, Spiegel R, Cramer H (1980) Characterization of ergot and non-ergot serotonin antagonists by prolactin and growth hormone profiles during wakefulness and sleep. Brain Res 202:357–363.
Clemens JA (1976) Neuropharmacological aspects of the neural control of prolactin secretion. In: Labrie F, Meites J, Pelletier G (eds) Hypothalamus and endocrine functions. Plenum, New York London, pp 283–302
Clemens JA, Sawyer BD (1974) Identification of prolactin in cerebrospinal fluid. Exp Brain Res 21:399–402
Clemens JA, Smalstig EB (1979) Effects of some new dopamine agonists on prolactin secretion and turning in 6-hydroxydopamine lesioned rats. In: Usdin E, Kopin IJ, Barchas J (eds). Catecholamines: Basic and clinical frontiers. Pergamon, New York, pp 1248–1250
Clemens JA, Gallo RV, Whitmoyer DI, Sawyer CH (1971) Prolactin responsive neurons in the rabbit hypothalamus. Brain Res 25:371–379
Clemens JA, Roush ME, Fuller RW (1978) Evidence that serotonin releases prolactin by stimulating the release of a prolactin releasing factor. 60th Annual Meeting of the Endocrine Society, Miami, June 14–16, Abstr no 526
Clements JA, Reyes FI, Winter JSD, Faiman C (1977) Studies on human sexual development IV. Fetal pituitary and serum, and amniotic fluid concentrations of prolactin. J Clin Endocrinol Metab 44:408–413
Cocchi D, Locatelli V, Picciolini E, Genazzani AR, Müller EE (1979) Prolactin-lowering effect of nomifensine in the rat. Proc Soc Exp Biol Med 162:38–43
Colby HD (1979) Mechanism of action of prolactin on adrenocortical steroid secretion in hypophysectomized female rats. Endocrinology 104:1299–1303
Colby HD, Gaskin JA, Kitay JI (1973) Requirement of the pituitary gland for gonadal hormone effects on hepatic corticosteroid metabolism in rats and hamsters. Endocrinology 92:769–774
Colloque (1979) sur la pharmacologie des dérivés de l'ergot de seigle. J Pharmacol (Paris) 10:377–587
Corenblum B, Whitaker M (1977) Inhibition of stress-induced hypoprolactinaemia. Br Med J 2:1328
Costlow ME, McGuire WL (1977) Autoradiographic localization of the binding of ^{125}I-labelled prolactin to rat tissues in vitro. J Endocrinol 75:221–226
Costlow ME, Buschow RA, McGuire WL (1975) Prolactin stimulation of prolactin receptors in rat liver. Life Sci 17:1457–1466
Cowden EA, Ratcliffe WA, Ratcliffe JG, Dobbie JW, Kennedy AC (1978) Hyperprolactinaemia in renal disease. Clin Endocrinol (Oxf) 9:241–248
Cowden EA, Dobbie JW, Mountjoy K, Ratcliffe JG, Richardson B, Kennedy AC (1981) Prolactin status in chronic nephropathy in the rat. Nephron 27:84–88
Cowie AT, Tindal JS (1971) The physiology of lactation. Arnold, London
Cowie AT, Forsyth IA, Hart IC (1980) Hormonal control of lactation. Springer, Berlin Heidelberg New York
Cramer OM, Parker CR, Porter JC (1979) Estrogen inhibition of dopamine release into hypophysial portal blood. Endocrinology 104:419–422
Crane WA, Loomes RS (1967) Effect of age, sex and hormonal state on tritiated thymidine uptake by rat pituitary. Br J Cancer 21:787–792
Creese I, Schneider R, Snyder SH (1977) ^{3}H-spiroperidol labels dopamine receptors in pituitary and brain. Eur J Pharmacol 46:377–381
Crider AM, Robinson JM, Floss HG, Cassady JM, Clemens JA (1977) Ergot alkaloids. Synthesis of 6-alkyl-8-ergolenes and 6-methyl-8-aminoergolines as potential prolactin inhibitors. J Med Chem 20:1473–1477

Cronin MJ, Weiner RI (1979) [^{3}H] Spiroperidol (spiperone) binding to a putative dopamine receptor in sheep and steer pituitary and stalk median eminence. Endocrinology 104:307–312

Cronin MJ, Roberts JM, Weiner RI (1978) Dopamine and dihydroergocryptine binding to the anterior pituitary and other brain areas of the rat and sheep. Endocrinology 103:302–309

Cronin MJ, Cheung CY, Beach JE, Faure N, Goldsmith PC, Weiner RI (1980) Dopamine receptors on prolactin secreting cells. In: MacLeod RM, Scapagnini U (eds). Central and peripheral regulation of prolactin function. Raven, New York, pp 43–58

Czernichow P, Dauzet MC, Broyer M, Rappaport R (1976) Abnormal TSH, Prl and GH response to TSH releasing factor in chronic renal failure. J Clin Endocrinol Metab 43:630–637

Dannies PS, Rudnick MS (1980) 2-Bromo-α-ergokryptine causes degradation of prolactin in primary cultures of rat pituitary cells after chronic treatment. J Biol Chem 255:2776–2781

Dattatreyamurty B, Sheth AR (1977) Size heterogeneity and specific binding property of immunoreactive prolactin in human seminal plasma. Mol Cell Endocrinol 7:253–259

Davies C, Jacobi J, Lloyd HM, Meares JD (1974) DNA synthesis and the secretion of prolactin and growth hormone by the pituitary gland of the male rat: Effects of diethylstilboestrol and 2-Br-α-ergokryptine methanesulfonate. J Endocrinol 61:411–417

Davies TF, Katekineni M, Chan V, Harwood JP, Dufan ML, Catt KJ (1980) Lactogenic receptor regulation in hormone-stimulated steroidogenic cells. Nature 283:863–865

Davis SL, Borger ML (1973) Metabolic clearance rates of prolactin in sheep. Endocrinology 92:1414–1418

Delitala G, Masala A, Alagna S, Devilla L (1976) Effect of pyridoxine on human hypophyseal trophic hormone release: A possible stimulation of hypothalamic dopaminergic pathway. J Clin Endocrinol Metab 42:603–606

Delitala G, Rovasio P, Lotti G (1977) Suppression of thyrotropin (TSH) and prolactin (PRL) release by pyridoxine in chronic primary hypothyroidism. J Clin Endocrinol Metab 45:1019–1022

Delitala G, Wass JAH, Stubbs WA, Jones A, Williams S, Besser GM (1979 a) The effect of lisuride hydrogen maleate, an ergot derivative on anterior pituitary hormone secretion in man. Clin Endocrinol (Oxf) 11:1–9

Delitala G, Stubbs WA, Yeo T, Jones A, Besser GM (1979 b) Inhibition of prolactin secretion by lisuride hydrogen maleate in vitro. Evidence for a dopaminergic mechanism. Acta Endocrinol [Suppl 184] (Copenh) 225:397

Delouis C, Denamur R (1972) Induction of lactose synthesis by prolactin in rabbit mammary gland explants. J Endocrinol 52:311–319

Delvoye P, Taubert H-D, Jürgensen O, L'Hermite M, Delogne J, Robyn C (1974) Influence of circulating prolactin increased by a psychotic drug on gonadotrophin and progesterone secretion. Acta Endocrinol [Suppl 184], (Copenh) 75:110

Delvoye P, Badawi M, Demaegd M, Robyn C (1978) Long lasting lactation is associated with hyperprolactinemia and amenorrhea. In: Robyn C, Harter M (eds). Progress in prolactin physiology and pathology. Elsevier/North-Holland Biomedical Press, Amsterdam, pp 213–232

Denamur R, Martinet J, Short RV (1973) Pituitary control of the ovine corpus luteum. J Reprod Fertil 32:207–220

Deutsch S, Sherman L (1979) Hypoprolactinemia in men with secondary sexual impotence and men with premature ejaculation. 61st Annual Meeting of the Endocrine Society, Anaheim, June 13–15, Abstr no 350

Diamond EJ, Khan S, Paul MA, Koprak S, Hollander VP (1980) Chronic administration of pharmacological doses of ovine prolactin decreases hormone disappearance rate. Endocrinology 106:952–958

Diefenbach WP, Carmel PW, Frantz A, Ferin M (1976) Suppression of prolactin secretion by L-DOPA in the stalk-sectioned rhesus monkey. J Clin Endocrinol Metab 43:638–642

Djiane J, Durand P (1977) Prolactin-progesterone antagonism in self regulation of prolactin receptors in the mammary gland. Nature 266:641–643

Djiane J, Clauser H, Kelly PA (1979 a) Rapid down-regulation of prolactin receptors in mammary gland and liver. Biochem Biophys Res Commun 90:1371–1378

Djiane J, Delouis C, Kelly PA (1979 b) Prolactin receptors in organ culture of rabbit mammary gland: effect of cycloheximide and prolactin. Proc Soc Exp Biol Med 162:342–345

Dogliotti L, Mussa A, DiCarlo F (1979) Treatment of advanced breast cancer with high doses of MPA and bromocriptine: correlation with steroid receptor content (Abstr). Cancer Treat Rep 63: 1218

Donatsch P, Richardson B (1975) Localization of prolactin in rat kidney tissue using a double antibody technique. J Endocrinol 66:101–106

Douglas WW, Taraskevich PS (1977) Action potentials (probably calcium spikes) in normal and adenomatous cells of the anterior pituitary and the stimulant effect of thyrotropin releasing hormone. J Physiol (Lond) 41P–43P

Duello TM, Halmi NS (1979) Ultrastructural-immunocytochemical localization of growth-hormone and prolactin in human pituitaries. J Clin Endocrinol Metab 49:189–196

Dufy B, Vincent JD, Fleury H, DuPasquier P, Gourdji D, Tixier-Vidal A (1979 a) Membrane effects of thyrotropin-releasing hormone and estrogen shown by intracellular recording from pituitary cells. Science 204:509–511

Dufy B, Vincent JD, Fleury H, DuPasquier P, Gourdji D, Tixier-Vidal A (1979 b) Dopamine inhibition of action potentials in a prolactin secreting cell line is modulated by oestrogen. Nature 282: 855–857

Dular R, LaBella F (1978) Actions of releasing factors on isolated secretory granules mediated by calcium. Life Sci 21:1527–1534

Dussault JH, Turcotte R, Guyda H (1976) The effect of acetylsalicylicacid on TSH and Prl secretion after TRH stimulation in the human. J Clin Endocrinol Metab 43:232–235

Ectors F, Danguy A, Pasteels JL (1972) Ultrastructure of organ cultures of rat hypophyses exposed to ergocornine. J Endocrinol 52:211–212

Editorial (1971) Human prolactin. Br Med J 3:201

Edwards CRW, Jeffcoate WJ (1976) Bromocriptine and the adrenal cortex. In: Bayliss RIS, Turner P, Maclay WP (eds). Pharmacological and clinical aspects of bromocriptine (Parlodel). MCS Consultants, Tunbridge Wells, pp 43–51

Edwards CRW, Thorner MO, Miall PA, Al-Dujaili EAS, Hanker JP, Besser GM (1975) Inhibition of the plasma-aldosterone response to frusemide by bromocriptine. Lancet 2:903–904

Egge AC, Rogol AD, Varna MM, Blizzard RM (1977) Effect of cyproheptadine on TRH-stimulated prolactin and TSH release in man. J Clin Endocrinol Metab 44:210–213

Eikenburg DC, Ravitz AJ, Gudelsky GA, Moore KE (1977) Effects of estrogen on prolactin and tuberoinfundibular dopaminergic neurons. J Neural Transm 40:235–244

El Etreby MF (1980) Influence of contraceptive steroids on tumor development in experimental animals and man: A short review. In: Genazzani E (ed). Pharmacological modulation of steroid action. Raven, New York, pp 239–260

El Etreby MF, Günzel P (1973) Prolaktinzell-Tumoren im Tierexperiment und beim Menschen. Arzneim Forsch 23:1768–1790

Eneroth P, Gustafsson J-A, Skett P, Stenberg A (1977) The effects on hepatic steroid metabolism of an ectopic pituitary graft: A time study. Mol Cell Endocrinol 7:167–175

Enjalbert A, Priam M, Kordon C (1977 a) Evidence in favour of the existence of a dopamine-free prolactin inhibiting factor (PIF) in rat hypothalamic extracts. Eur J Pharmacol 41:243–244

Enjalbert A, Moos F, Carbonell L, Priam M, Kordon C (1977 b) Prolactin inhibiting activity of dopamine-free subcellular fractions from rat mediobasal hypothalamus. Neuroendocrinology 24:147–161

Enjalbert A, Ruberg M, Arancibia S, Fiore L, Priam M, Kordon C (1979 a) Independent inhibition of prolactin secretion by dopamine and γ-aminobutyric acid in vitro. Endocrinology 105:823–826

Enjalbert A, Ruberg M, Arancibia, S, Priam M, Bauer K, Kordon C (1979 b) Inhibition of in vitro prolactin secretion by histidyl-proline-diketopiperazine, a degradation product of TRH. Eur J Pharmacol 58:97–98

Enjalbert A, Ruberg M, Arancibia S, Priam M, Kordon C (1979 c) Endogenous opiates block dopamine inhibition of prolactin secretion in vitro. Nature 280:595–597

Ensor DM (1978) Comparative endocrinology of prolactin. Chapman & Hall, London, pp 91, 126–128

Erdheim J, Stumme E (1909) Über Schwangerschaftsveränderungen bei der Hypophyse. Beitr Pathol 46:1–132

Eskildsen PC, Lund B, Sørensen OH, Lund B, Bishop JE, Norman AW (1979) Acromegaly and vitamin D metabolism: effect of bromocriptine treatment. J Clin Endocrinol 49:484–486

European Breast Cancer Group (1972) Clinical trial of 2-Br-α-ergocryptine (CB-154) in advanced breast cancer. Eur J Cancer 8:155–156

Euvrard C, Labrie F, Boissier JR (1979) Antidopaminergic effect of estrogens at the striatal level. In: Usdin E, Kopin IJ, Barchas J (eds) Catecholamines. Basic and clinical frontiers. Pergamon, New York, pp 1269–1271

Evans HM, Simpson ME, Lyons WR, Turpeinen K (1941) Anterior pituitary hormones which favour the production of traumatic uterine placentomata. Endocrinology 28:933–945
Everett JW (1954) Luteotrophic function of autografts of the rat hypophyses. Endocrinology 54:685–690
Fang VS, Refertoff S, Rosenfield RL (1974) Hypogonadism induced by a transplantable, prolactin-producing tumor in male rats: hormonal and morphological studies. Endocrinology 95:991–998
Fang VS, Shian L-R, Worsley IG (1979) Generation of the big prolactin by sephadex column chromatography. In: MacIntyre I, Szelke M (eds) Molecular endocrinology. Elsevier/North-Holland Biomedical Press, Amsterdam, pp 133–140
Farkouh NH, Parker MG, Frantz AG (1979) Large molecular size prolactin with reduced receptor activity in human serum: high proportion in basal state and reduction after thyrotropin-releasing hormone. J Clin Endocrinol Metab 48:1026–1030
Farquhar MG (1977) Secretion and crinophagy in prolactin cells. Adv Exp Med Biol 80:37–86
Felgenhauser P (1974) Protein size and cerebrospinal fluid composition. Klin Wochenschr 52:1158–1164
Ferland L, Labrie E, Euvrard C, Raynaud J-P (1979) Antidopaminergic activity of estrogens on prolactin release at the pituitary level in vivo. Mol Cell Endocrinol 14:199–204
Flückiger E (1976) The pharmacology of bromocriptine. In: Bayliss RIS, Turner P, Maclay WP (eds) Pharmacological and clinical aspects of bromocriptine (Parlodel). MCS Consultants, Tunbridge Wells, pp 12–26
Flückiger E (1978 a) Lactation inhibition by ergot drugs. In: Yokoyama A, Mizuno H, Nagasawa H (eds) Physiology of mammary glands. Japan Scientific Societies Press, Tokyo, pp 71–82
Flückiger E (1978 b) Ergot alkaloids and the modulation of hypothalamic function. In: Cox B, Morris ID, Weston AH (eds). Pharmacology of the hypothalamus. Macmillan, London, pp 137–160
Flückiger E, del Pozo E (1978) Influence on the endocrine system. In: Berde B, Schild HO (eds). Ergot alkaloids and related compounds. Springer Berlin Heidelberg New York: (Handbook of experimental pharmacology, vol 49, pp 615–690)
Flückiger E, Kovacs E (1974) Inhibition by 2-Br-α-ergokryptine-mesylate (CB 154) of suckling-induced pituitary prolactin depletion in lactating rats. Experientia 30:1173
Flückiger E, Vigouret J-M (1979) Drugs, dopamine and pituitary secretion. In: Simkins MA (ed). Medicinal chemistry VI. Proceedings of the 6th international symposium on medicinal chemistry, Brighton, U.K., Sept 4–7, 1978. Cotswold, Oxford, pp 469–477
Flückiger E, Wagner HR (1968) 2-Br-α-ergokryptin: Beeinflussung von Fertilität und Laktation bei der Ratte. Experientia 24:1130–1131
Flückiger E, Lutterbeck PM, Wagner HR, Billeter E (1972) Antagonism of 2-Br-α-ergokryptine-methanesulfonate (CB 154) to certain endocrine actions of centrally active drugs. Experientia 28:924–925
Flückiger E, Markó M, Doepfner W, Niederer W (1976) Effects of ergot alkaloids on the hypothalamo-pituitary axis. Postgrad Med J [Suppl 1] 52:57–61
Flückiger E, Vigouret JM, Wagner H-R (1978) Ergot compounds and prolactin secretion. In: Robyn C, Harter M (eds). Progress in prolactin physiology and pathology. Elsevier/North-Holland Biomedical Press, Amsterdam, pp 383–396
Forbes AP, Henneman PH, Griswold GC, Albright F (1954) Syndrome characterized by galactorrhea, amenorrhea and low urinary FSH: comparison with acromegaly and normal lactation. J Clin Endocrinol Metab 14:265–271
Forsyth AI, Hayden TJ (1977) Comparative endocrinology of mammary growth and lactation. In: Peaker M (ed). Comparative aspects of lactation. Academic Press, London, pp 135–163
Fossati P, L'Hermite M, Buvat J, Cappouen JP, Grenier JL, Linquette M (1976 a) Etude de la fonction prolactinique dans 14 cas de galactorrhée avec hypogonadisme et micro-déformation de la selle turcique. Ann Endocrinol (Paris) 36:279–280
Fossati P, Strauch G, Tourniaire J (1976 b) Etude de l'activité de la bromocriptine dans les états d'hyperprolactinémie. Nouv Presse Med 5:1687–1691
Foukas MD (1973) An antilactogenic effect of pyridoxyne. J Obstet Gynaecol Br Commonwealth 80:718–720
Franchimont P, Dourcy C, Legros JJ, Reuter A, Vrindts-Gevaert Y, van Cauwenberge JR, Gaspard U (1976) Prolactin levels during the menstrual cycle. Clin Endocrinol (Oxf) 5:643–650
Franks S, Naftolin F (1980) Stimulation of prolactin secretion by 4-hydroxyestradiol. 62nd Annual Meeting of the Endocrine Society, Washington, June 18–20, Abstr no 767
Frantz AG (1979) Rhythms in prolactin secretion. In: Krieger DT (ed). Endocrine rhythms. Raven, New York, pp 175–186

Frantz AG, Kleinberg DL (1970) Prolactin: Evidence that it is separate from growth hormone in human blood. Science 170: 745–747

Frantz AG, Kleinberg DL, Noel GL (1972) Studies on prolactin in man. Recent Prog Horm Res 28:527–573

Frantz AG, Habif DV, Hyman GA, Suh HK, Sassin F, Zimmerman EA, Noel GL, Kleinberg DL (1973) Physiological and pharmacological factors affecting prolactin secretion, including its suppression by L-DOPA in the treatment of breast cancer. In: Pasteels JL, Robyn C (eds) Human prolactin. Excerpta Medica, Amsterdam, pp 273–290

Friedgood HB, Dawson AB (1940) Physiological significance and morphology of the carmine cell in the cat's anterior pituitary. Endocrinology 26:1022–1031

Friesen HG (1973) Placental protein and polypeptide hormones. In: Greep RO, Astwood EB (eds). Female reproductive system. (Handbook of physiology, sect 7, vol II/2, American Physiological Society, Washington, pp 295–309)

Friesen HG, Hwang P (1973) Human prolactin. Annu Rev Med 24:251–270

Friesen HG, Hwang P, Guyda H, Tolis G, Tyson J, Myers R (1972) A radioimmunoassay for human prolactin. In: Boyns AR, Griffiths K (eds). Prolactin and carcinogenesis. Fourth Tenovus Workshop. Alpha Omega Alpha, Cardiff, pp 64–80

Furuhashi N, Fang VS (1979) Sex difference in the induction of lactogenic receptors in rat livers. Acta Endocrinol (Copenh) 92:532–541

Fuxe K, Calne DB (eds) (1979) Dopaminergic ergot derivatives and motor function. Pergamon, Oxford New York

Fuxe K, Hökfelt T (1970) Central monoaminergic systems and hypothalamic function. In: Martini L, Motta M, Fraschini F (eds). The hypothalamus. Academic Press, New York, pp 123–138

Fuxe K, Hökfelt T, Nilsson O (1969) Castration, sex hormones and tubero-infundibular dopamine neurons. Neuroendocrinology 5:107–120

Fuxe K, Hökfelt T, Eneroth P, Gustafsson J-A, Skett P (1977) Prolactin-like immunoreactivity: localization in nerve terminals of rat hypothalamus. Science 196:899–900

Fuxe K, Andersson K, Ögren S-O, Perez de la Mora M, Schwarcz R, Hökfelt T, Eneroth P, Gustafsson J-A, Skett P (1978 a) GABA neurons and their interaction with monoamine neurons. An anatomical, pharmacological and functional analysis. In: Krogsgaard-Larsen P, Scheel-Kruger J, Kofold H (eds). GABA-neurotransmitters. Alfred Benzon Symposium 12. Munksgaard, Copenhagen, pp 74–94

Fuxe K, Ferland L, Andersson K, Eneroth P, Gustafsson J-A, Skett P (1978 b) On the functional role of hypothalamic catecholamine neurons in control of the secretion of hormones from the anterior pituitary, particularly in the control of LH and prolactin secretion. In: Scott DE, Kozlowski GR, Weindl A (eds). Brain-endocrine interaction II. Neural hormones and reproduction. Karger, Basel, pp 172–182

Gala RR, Boss RS (1975) Serum prolactin levels of rats under continuous estrogen stimulation and 2-Br-α-ergocryptine (CB 154) injection. Proc Soc Exp Biol Med 149:330–332

Gala RR, Singhakowinta A, Brennan MJ (1975) Studies on prolactin in human serum, urine and milk. Horm Res 6:310–320

Gala RR, Subramanian MG, Peters JA, Sandford J (1976) Influence of cholinergic receptor blockade on drug-induced prolactin release in the monkey. Horm Res 7:118–128

Garnier PE, Aubert ML, Kaplan S, Grumbach MM (1978) Heterogeneity of pituitary and plasma prolactin in men: Decreased affinity of "big" prolactin in a radioreceptor assay and evidence for its secretion. J Clin Endocrinol Metab 47:1273–1281

Garthwaite TL, Hagen TC (1979) Evidence that serotonin stimulates a prolactin-releasing factor in the rat. Neuroendocrinology 29:215–220

Gautvik KM, Hoyt RF, Tashjian AH (1973) Effects of colchicine and 2-Br-α-ergokryptine-methanesulfonate (CB 154) on the release of prolactin and growth hormone by functional tumour cells in culture. J Cell Physiol 82:401–410

Gershengorn MC, Marcus-Samuels BE, Geras E (1979) Estrogens increase the number of thyrotropin-releasing hormone receptors on mammotropic cells in culture. Endocrinology 105:171–176

Giampetro O, Moggi G, Chisci R, Coluccia A, Dalle Luche A, Simonini N, Brunori I (1979) Unusual prolactin response to luteinizing hormone-releasing hormone in some anovulatory women. J Clin Endocrinol Metab 49:141–143

Giannopoulos G (1980) Glucocorticoids and fetal lung development. In: Kaye AM, Kaye M (eds). The development of responsiveness to steroid hormones. Pergamon, Oxford, (Advances in the biosciences, vol 25, pp 241–261

Gibbs DM, Neill JD (1978) Dopamine levels in hypophysial stalk blood in the rat are sufficient to inhibit prolactin secretion in vivo. Endocrinology 102:1895–1900

Gibori G, Richards JS (1978) Dissociation of two distinct luteotropic effects of prolactin: regulation of luteinizing hormone receptor secretion during pregnancy. Endocrinology 102:767–774

Gibori G, Richards JS, Keyes PL (1979) Synergistic effects of prolactin and estradiol in the luteotropic process in the pregnant rat: regulation of estradiol receptor by prolactin. Biol Reprod 21:419–423

Gláz E, Lada G, Kiss R, Rácz K, Vida S, Wolf I (1979) Effect of metoclopramide on plasma aldosterone level in subjects receiving bromoergocryptine and cyproheptadine. Acta Endocrinol [Suppl 225] (Copenh) 91:344

Gluckman PD, Ballard PL, Kaplan SL, Liggins GC, Grumbach MM (1978) Prolactin in umbilical cord blood and the respiratory distress syndrome. J. Pediatr. 93:1011–1014

Gnudi A, Lugari R, Cavazzini MG (1977) Variation of blood glucose and serum growth hormone, prolactin and insulin in subjects with insulin-dependent diabetes, after OGTT and pretreatment with 2-Br-α-ergokryptine. Acta Diabetol Lat 14:119–128

Golander A, Barret J, Hurley T, Barry S, Handwerger S (1979) Failure of bromocriptine, dopamine and thyrotropin-releasing hormone to affect prolactin secretion by human decidual tissue in vitro. J Clin Endocrinol Metab 49:787–789

Goldstein M, Calne DB, Lieberman AN, Thorner M (eds) (1980) Ergot alkaloids in neurologic, neuropsychiatric and endocrine disorders. Raven, New York

Goluboff LG, Ezrin C (1969) Effect of pregnancy on the somatotroph and the prolactin cell of the human adenohypophysis. J Clin Endocrinol 29:1533–1538

Gootjes J (1978) Piperazinderivate, Verfahren zu ihrer Herstellung und ihre Verwendung. Deutsches Patentamt, *Offenlegungsschrift* 2755752 BRD: Deutsches Patentamt

Gotto AM, Peck EJ, Boyd AE (eds) (1979) Brain peptides: A new endocrinology. Elsevier/North-Holland Biomedical Press, Amsterdam, p 302

Gourdji D, Bataille D, Vauchin N, Grouselle D, Rosselin G, Tixier-Vidal A (1979) Vasoactive intestinal peptide (VIP) stimulates prolactin (Prl) release and cAMP production in a rat pituitary cell line (GH_3/B_6). Additive effects of VIP and TRH on Prl release. FEBS Lett 104:165–169

Gräf K-J, Neumann F, Nishino Y, Mehring M, Hasan SH (1973) The effect of different progestogens on prolactin secretion in rats. In: Pasteels JL, Robyn C (eds). Human prolactin. Excerpta Medica, Amsterdam, pp 160–162

Gräf K-J, Schmidt-Gollwitzer M, Koch UJ, Lorenz F, Hammerstein J (1978) Hyperprolactinaemia induced by cyproterone acetate in human subjects. Acta Endocrinol [Suppl 215] (Copenh) 87:96

Graffenried B von, del Pozo E, Roubicek J, Krebs E, Poldinger W, Burmeister P, Kerp L (1978) Effects of the synthetic enkephaline analogue FK 33-824 in man. Nature 272:729–730

Grandison L, Guidotti A (1979) γ-aminobutyric acid receptor function in rat anterior pituitary: Evidence for control of prolactin release. Endocrinology 105:754–759

Grandison L, Meites J (1972) Luteolytic action of prolactin during oestrous cycle of the mouse. Proc Soc Exp Biol Med 140:323–325

Grandison L, Meites J (1976) Evidence for adrenergic mediation of cholinergic inhibition of prolactin release. Endocrinology 99:775–779

Grandison L, Gelato M, Meites J (1974) Inhibition of prolactin secretion by cholinergic drugs. Proc Soc Exp Biol Med 145:1236–1239

Grayhack JT (1963) Pituitary factors influencing the growth of the prostate. Natl Cancer Inst Monogr 12:189–199

Grinwich DL, Hichens M, Behrman HR (1976) Control of the LH receptor by prolactin and prostaglandin $F_{2\alpha}$ in rat corpora lutea. Biol Reprod 14:212–218

Grosvenor CE (1967) Disappearance rate of exogenous prolactin from serum of female rats. Endocrinology 80:195–200

Grosvenor CE, Turner CW (1958) Pituitary lactogenic hormone concentration and milk secretion in lactating rats. Endocrinology 63:535–539

Grosvenor CE, Whitworth NS (1976) Incorporation of rat prolactin into rat milk in vivo and in vitro. J Endocrinol 70:1–9

Grosvenor CE, Whitworth NS (1979) Secretion rate and metabolic clearance rate of prolactin in the rat during mid- and late-lactation. J Endocrinol 82:409–415

Grosvenor CE, Mena F, Whitworth NS (1979 a) The secretion rate of prolactin in the rat during suckling

and its metabolic clearance rate after increasing intervals of nonsuckling. Endocrinology 104:372–376
Grosvenor CE, Mena F, Whitworth NS (1979 b) Ether releases large amounts of prolactin from rat pituitaries previously "depleted" by short-term suckling. Endocrinology 105:884–887
Gudelsky GA, Porter JC (1979) Opioid-peptide induced inhibition of the release of dopamine into pituitary stalk blood (Abstr). American Physiological Society Specialty Meeting, East Lansing, August 22–24, p 22
Gudelsky GAJ, Simkins GP, Mueller GP, Meites J, Moore KE (1976) Selective actions of prolactin on catecholamine turnover in the hypothalamus and on serum LH and FSH. Neuroendocrinology 22:206–215
Gudelsky GA, Nansel DD, Porter JC (1979) Effects of L-DOPA and bromocriptine on the apparent association of dopamine with the prolactin secretory granules in the anterior pituitary gland of the rat. 61st Annual Meeting of the Endocrine Society, Anaheim, June 13–15, Abstr no 515
Gudelsky GA, Nansel DD, Porter JC (1980) Uptake and processing of dopamine by cells of the anterior pituitary gland. Endocrinology 107:30–34
Guyda HJ (1975) Heterogeneity of human growth hormone and prolactin secreted in vitro: Immunoassay and radioreceptor assay correlations. J Clin Endocrinol Metab 41:953–967
Häusler A, Hodel C (1975) Morphological observations on pituitary mammotropes of lactating rats treated with 2-Bromo-α-ergokryptine (CB 154). Acta Endocrinol [Suppl 193] (Copenh) 78:66
Häusler A, Rohr HP, Marbach P, Flückiger E (1978) Changes in prolactin secretion in lactating rats assessed by correlative morphometric and biochemical methods. J Ultrastruct Res 64:74–84
Hafiez AA, Philpott JE, Bartke A (1971) The role of prolactin in the regulation of testicular function; the effect of prolactin and luteinising hormone on 3 hydroxy- steroid dehydrogenase activity in the testes of mice and rats. J Endocrinol 50:619–623
Hafiez AA, Lloyd CW, Bartke A (1972) The role of prolactin in the regulation of testis function, synergistic effects of prolactin and luteinising hormone on the plasma levels of testosterone and andosterone in hypophysectomised rats. J Endocrinol 52:327–332
Hall R, Snow M, Scanlon M, Mora B, Gomez-Pan A (1978) Pituitary effects of somatostatin. Metabolism [Suppl 1] 27:1257–1262
Hamosh M, Hamosh P (1977) The effect of prolactin on the lecithin content of fetal rabbit lung. J Clin Invest 59: 1002–1005
Hang E, Naess O, Gautvik KM (1978) Receptors for 17β-estradiol in prolactin secreting rat pituitary cells. Mol Cell Endocrinol 12: 81–95
Harper ME, Danutra V, Chandler JA, Griffiths K (1976) The effect of 2-bromo-α-ergocryptine (CB 154) administration on the hormone levels, organ weights, prostatic morphology and zinc concentrations in the male rat. Acta Endocrinol (Copenh) 83:211–224
Harris ARC, Smith MS, Alex S, Salhanick HA, Vagenakis AG, Braverman LE (1978) Pyridoxine (B_6)-induced inhibition of prolactin release in the female rat. Endocrinology 102:362–366
Hart IC (1973) Effect of 2-Br-α-ergokryptine on milk yield and the level of prolactin and growth hormone in the blood of the goat at milking. J Endocrinol 57:179–180
Hart IC (1975) Seasonal factors affecting the release of prolactin in goats in response to suckling. J Endocrinol 64:313–322
Hart IC, Morant SV (1980) Roles of prolactin, growth hormone, insulin and thyroxine in steroid-induced lactation in goats. J Endocrinol 84:343–351
Hartmeyer H, Kramer G, Schulz K-D (1975) The effect of prolactin inhibition and/or ovariectomy on mammary ribosomes of the rat. Acta Endocrinol [Suppl 193] (Copenh) 78:61
Haug E (1979) Progesterone suppression of estrogen-stimulated prolactin secretion and estrogen receptor levels in rat pituitary cells. Endocrinology 104:429–437
Hauth H (1979) Chemical aspects of ergot derivatives with central dopaminergic activity. In: Fuxe K, Calne DB (eds). Dopaminergic ergot derivatives and motor function. Pergamon, Oxford, pp 23–32
Hauth JC, Parker CR, MacDonald PC, Porter JC, Johnston JM (1978) A role of fetal prolactin in lung maturation. Obst Gynec 51:81–88
Hayden TJ, Bonney RC, Forsyth IA (1979) Ontogeny and control of prolactin receptors in the mammary gland and liver of virgin, pregnant and lactating rats. J Endocrinol 80:259–269
Healy DL, Burger HG, Muller HK (1978) Hypothesis: placental membranes produce prolactin. Mol Cell Endocrinol 11:1–6
Healy DL, Kimpton WG, Muller HK, Burger HG (1979) The synthesis of immunoreactive prolactin by decidua-chorion. Br J Obstet Gynaecol 86:307–313

Healy DL, Rattigan S, Hartmann PE, Herrington AC, Burger HG (1980) Prolactin in human milk: correlation with lactose, total protein, and α-lact-albumin levels. Am J Physiol 238:E83–E86
Hedge GA (1977) Roles for the prostaglandins in the regulation of anterior pituitary secretion. Life Sci 20:17–34
Herbert DC, Ishikawa H, Rennels EG (1979) Evidence for the autoregulation of hormone secretion by prolactin. Endocrinology 104:97–100
Herington AC, Veith NM (1977) The presence of lactogen but not growth hormone binding sites in the isolated rat hepatocyte. J Endocrinol 74:323–334
Herlant M (1960) Etude critique de deux techniques nouvelles destinées à mettre en évidence les différentes catégories cellulaires présentes dans la glande pituitaire. Bull Microsc Appl 10: 37–44
Heuson JC, Waelbroeck-vanGaver C, Legros N (1970) Growth inhibition of rat mammary carcinoma and endocrine changes produced by 2-Br-alpha-ergocryptine, a suppressor of lactation and nidation. Eur J Cancer 6:353–356
Hiba J, del Pozo E, Genazzani A, Pusterla E, Lancranjan I, Sidiropoulos D, Gunti J (1977) Hormonal mechanism of milk secretion in the newborn. J Cell Endocrinol Metab 44:973–976
Hilliard J, Spies HG, Sawyer CH (1969) Hormonal factors regulating ovarian cholesterol mobilisation and progestin secretion in intact and hypophysectomised rabbits. In: McKerns KW (ed). The gonads. North-Holland, Amsterdam, pp 55–92
Hobbs JR, Salih H, Flax H, Brander W (1973) Prolactin dependence among human breast cancers. In: Pasteels JL, Robyn C (eds). Human prolactin. Excerpta Medica, Amsterdam, pp 249–258
Hökfelt B, Nillius SJ (eds) (1978) The dopamine agonist bromocriptine. Theoretical and clinical aspects. Acta Endocrinol [Suppl 216] (Copenh) 88:1–230
Holder AT, Wallis M (1977) Actions of growth hormone, prolactin and thyroxine on serum somatomedin-like activity and growth in hypopituitary dwarf mice. J Endocrinol 74:223–229
Holmes RL, Ball JN (1974) The pituitary gland: A comparative account. Cambridge University Press, London, pp 26–53
Holt JA, Richards JS, Midgley AR (1976) Effect of prolactin on LH receptor in rat luteal cells. Endocrinology 98:1005–1013
Horowski R, Gräf K-J (1975) Prolactin secretion in rats under the influence of different agents acting on the dopaminergic system. Acta Endocrinol [Suppl 199] (Copenh) 80:203
Horowski R, Gräf K-J (1976) Influence of dopaminergic agonists and antagonists on serum prolactin concentrations in the rat. Neuroendocrinology 22:273–286
Horowski R, Gräf K-J, Nieuweboer B, Wendt H, Chiodini PG, Liuzzi A (1978) Influence of depot progestagens on anterior pituitary and adrenocortical hormones. Acta Endocrinol [Suppl 215] (Copenh) 87:98
Horrobin DF (1974) Prolactin 1974. Medical and Technical Publishing, Lancaster, p 96
Houdebine L-M, Djiane J (1980) Effects of lysomotropic agents, and of microfilament- and microtubule-disrupting drugs on the activation of casein-gene expression by prolactin in the mammary gland. Mol Cell Endocrinol 17:1–15
Hwang P, Guyda H, Friesen HG (1971) A radioimmunoassay for human prolactin. Proc Natl Acad Sci USA 68:1902–1906
Hwang P, Guyda H, Friesen HG (1972) Purification of human prolactin. J Biol Chem 247:1955–1958
Iwasaki Y, Chihara K, Iwasaki J, Abe H, Fujita T (1979) Effect of calcitonin on prolactin release in rats. Life Sci 25:1243–1248
Jacobs LS, Snyder PJ, Wilber JF, Utiger RD, Daughaday WH (1971) Increased serum prolactin after administration of synthetic TRH in man. J Clin Endocrinol Metab 33:996–998
Johke T (1969) Radioimmunoassay for bovine prolactin in plasma. Endocrinol Jpn 16:581–589
Jordan RM, Kendall JW (1978) Dissociation of plasma and CSF prolactin heterogeneity. Acta Endocrinol (Copenh) 89:38–47
Josimovich JB (1974) Human placental lactogen: further evidence of placental mimicry of pituitary function. Am J Obstet Gynecol 120:550–552
Josimovich JB, Merisko K, Boccella L, Tobon H (1977) Binding of prolactin by fetal rhesus cell membrane fractions. Endocrinology 100:557–563
Kalbermann LE, Szijan I, Jahn CA, Kraviec L, Burdman JA (1979 a) DNA synthesis in the pituitary gland of the rat. Neuroendocrinology 29:42–48
Kalbermann LE, Szijan I, Burdman JA (1979 b) Cell proliferation in the rat pituitary gland: a mechanism of control in prolactin cells. Experientia 35:689–690

Kann G, Carpentier M-C, Fevre J, Martinet J, Maubon M, Neusnier C, Paly J, Vermeire N (1978) Lactation and prolactin in sheep, role of prolactin in initiation of milk secretion. In: Robyn C, Harter M (eds). Progess in prolactin physiology and pathology. Elsevier/North-Holland Biomedical Press, Amsterdam, pp 201–212

Kaplan SL, Grumbach MM (1965) Serum chorionic 'growth hormone-prolactin' and serum pituitary growth hormone in mother and fetus at term. J Clin Endocrinol Metab 25:1370–1374

Karg H, Schams D (1974) Prolactin release in cattle. J Reprod Fertil 39:463–472

Kebabian JW, Calne DB (1979) Multiple receptors for dopamine. Nature 277:93–96

Kelly PA, Bradley C, Shiu RPC, Meites J, Friesen HG (1974 a) Prolactin binding to rat mammary tumor tissue. Proc Soc Exp Biol Med 146:816–819

Kelly PA, Posner BI, Tsushina T, Friesen HG (1974 b) Studies of insulin, growth hormone and prolactin binding: Ontogenesis, effects of sex and pregnancy. Endocrinology 95:532–539

Kelly PA, Posner BI, Friesen HG (1975) Effects of hypophysectomy, ovariectomy, and cycloheximide on specific binding sites for lactogenic hormones in rat liver. Endocrinology 97:1408–1415

Kendall J, Orwoll E (1980) Anterior pituitary hormones in the brain and other extrapituitary sites. In: Martini L, Ganong WF (eds) Frontiers in neuroendocrinology, vol 6. Raven, New York, pp 33–65

Kidokoro Y (1975) Spontaneous Ca action potentials in a clonal pituitary cell line and their relation to prolactin secretion. Nature 258:741–742

Kimura H, Calabro MA, MacLeod RM (1976) Suppression of prolactin secretion and cAMP accumulation by dopamine in the pituitary. Fed Proc 35:305

Kleinberg DL, Todd J (1979) Comparison of lactogenic activity of human growth hormone and ovine prolactin in primates. 61st Annual Meeting of the Endocrine Society, Anaheim, June 13–15, Abstr no 586

Kleinberg DL, Lieberman A, Todd J, Greising J, Neophytides A, Kuppersmith M (1980) Pergolide mesylate: a potent day-long inhibitor of prolactin in Rhesus monkeys and patients with Parkinson's disease. J Clin Endocrinol Metab 51:152–154

Koenig JI, Mayfield MA, McCann SM, Krulich L (1979) Stimulation of prolactin secretion by morphine: role of the central serotonergic system. Life Sci 25:853–864

Komisaruk BR, Adler NT, Hutchison J (1972) Genital sensory field: enlargement by estrogen treatment in female rats. Science 178:1295–1298

Komisaruk BR, Adler NT, Hutchison J (1973) Genital sensory field. Science 181:366

Konner M, Worthman C (1980) Nursing frequency, gonadal function, and birth spacing among Kung hunter-gatherers. Science 207:788–791

Kordon C, Blake CA, Terkel J, Sawyer CH (1973/74) Participation of serotonin containing neurons in the suckling-induced rise in plasma prolactin levels in lactating rats. Neuroendocrinology 13:213–223

Kordon C, Hery M, Enjalbert A (1976) Neurotransmitters and control of pituitary function. In: Labrie F, Meites J, Pelletier G (eds). Hypothalamus and endocrine function. Plenum, New York, pp 51–61

Křepelka J, Semonský M (1978) Some N-nitrosubstituted derivatives of ergoline I. Coll Czech Chem Commun 43:1723–1726

Křepelka J, Cerný A, Kotva R, Semonský M (1977) Some 6-alkyl derivatives of D-8-cyanomethyl and D-8-methyl ergoline-I. Coll Czech Chem Commun 42:1209–1215

Kruse V (1973) Binding of bovine prolactin to serum proteins and its influence on the catabolism of prolactin. In: Pasteels JL, Robyn C (eds). Human prolactin. Excerpta Medica, Amsterdam, pp 51–55

Labrie F, Beaulieu M, Caron MG, Raymond V (1978) The adenohypophyseal dopamine receptor: specificity and modulation of its activity by estradiol. In: Robyn C, Harter M (eds). Progress in prolactin physiology and pathology. Elsevier/North-Holland Biomedical Press, Amsterdam, pp 121–135

Lal H, Brown W, Drawbaugh R, Hynes M, Brown G (1977) Enhanced prolactin inhibition following chronic treatment with haloperidol and morphine. Life Sci 20:101–106

Lal S, Mendis T, Cervantes P, Guyda H, deRivera, JL (1979) Effect of benztropine on haloperidol-induced prolactin secretion. Neuropsychobiology 5:327–331

Lam Y-K, Knudsen R, Folkers K (1978) Isolation of pyroglutamic acid from hypothalamic tissue and significance of its inhibition of prolactin release. Biochem Biophys Res Commun 81:680–683

Lamberts SWJ, MacLeod RM (1978 a) The interaction of serotoninergic and dopaminergic systems in prolactin secretion in the rat. Endocrinology 103:287–295

Lamberts SWJ, MacLeod RM (1978 b) Effects of cyproheptadine on prolactin synthesis and release by normal and suppressed pituitary glands and by dispersed pituitary tumor cells. Endocrinology 103:1710–1717

Lamberts SWJ, MacLeod RM (1979 a) The inability of bromocriptine to inhibit prolactin secretion by transplantable rat pituitary tumors: Observations on the mechanism of dynamics of the autofeedback regulation of prolactin secretion. Endocrinology 104:65–70

Lamberts SWJ, MacLeod RM (1979 b) Metergoline and other peripheral serotonin antagonists inhibit prolactin secretion through mechanisms unrelated to serotonin. Proc Soc Exp Biol Med 162: 75–79

Lamprecht SA, Lindner HR, Strauss JF (1969) Induction of 20 α-hydroxysteroid dehydrogenase in rat corpora lutea by pharmacological blockade of pituitary prolactin secretion. Biochim Biophys Acta 187:133–143

Lancranjan I, del Pozo E, Picciolini E, d'Antona N, Genazzani AR (1979) Effect of two serotonin antagonists on prolactin release induced by breast stimulation in postpartum women. Horm Res 10:14–19

Landgraf R, Weissmann A, Landgraf-Leurs M, von Werder K (1975) Glucose tolerance and glucose-induced insuline release in patients with hyperprolactinaemia. Acta Endocrinol [Suppl 193] (Copenh) 78:65

Landgraf R, Landgraf-Leurs MMC, Weissmann A, Hörl R, von Werder K, Scriba PC (1977) Prolactin: A diabetogenic hormone. Diabetologia 13:99–104

Lemberger L, Crabtree RE (1979) Pharmacological effects in man of a potent, long-acting dopamine receptor agonist. Science 205:1151–1153

Lesniak MA, Gorden P, Roth J (1977) Reactivity of non-primate growth hormones and prolactins with human growth hormone receptors on cultured human lymphocytes. J Clin Endocrinol Metab 44: 838–849

Lewis UJ, Singh RNP, Sinha YN, Vanderlaan WP (1971) Electrophoretic evidence for human prolactin. J Clin Endocrinol 33:153–156

Lewis UJ, Peterson SM, Bonewald LF, Seavey BK, Vanderlaan WP (1977) An interchain disulfide dimer of human growth hormone. J Biol Chem 252:3697–3702

Lichtensteiger W, Keller PJ (1974) Tubero-infundibular dopamine neurons and the secretion of luteinizing hormone and prolactin: extrahypothalamic influences, interaction with cholinergic systems and the effect of urethane anesthesia. Brain Res 74:279–303

Lieberman ME, Maurer RA, Stone RT, Shupnik MA, Vician L, Baxter LA, Ryan R, Gorski J (1980) Regulation of prolactin synthesis by estrogen. In: Kaye AM, Kaye M (eds). The development of responsiveness to steroid hormones. Pergamon, Oxford, pp 21–36

Lindner HR, Shelesnyak MC (1967) Effect of ergocornine on ovarian synthesis of progesterone and 20-α-hydroxy-pregn-4-en-3-one in the pseudopregnant rat. Acta Endocrinol (Copenh) 56:27–34

Lloyd HM, Meares JD, Jacobi J (1975) Effects of oestrogen and bromocriptine on in vivo secretion and mitosis in prolactin cells. Nature 255:497–498

Lloyd HM, Jacobi JM, Meares JD (1978) DNA synthesis and depletion of prolactin in the pituitary gland of the male rat. J Endocrinol 77:129–136

Locatelli V, Cocchi D, Racagni G, Cattabeni F, Maggi A, Krogsgaard-Larsen P, Müller EE (1978) Prolactin-inhibiting activity of γ-aminobutyric acid-mimetic drugs in the male rat. Brain Res 145:173–179

Locatelli V, Cocchi D, Frigerio C, Betti R, Krogsgaard-Larsen P, Racagni G, Müller EE (1979) Dual γ-aminobutyric acid control of prolactin secretion in the rat. Endocrinology 105:778–785

Login IS, MacLeod RM (1977) Prolactin in human and rat serum and cerebrospinal fluid. Brain Res 132:477–483

Lu KH, Meites J (1971) Inhibition by L-DOPA and monoamine oxidase inhibitors of pituitary prolactin release, stimulation by methyldopa and d-amphetamine. Proc Soc Exp Biol Med 137:480–483

Lu KH, Meites J (1972) Effects of L-DOPA on serum prolactin and PIF in intact and hypophysectomized, pituitary-grafted rats. Endocrinology 91:868–872

Lu KH, Koch Y, Meites J (1971) Direct inhibition by ergocornine of pituitary prolactin release. Endocrinology 89:229–233

Lund B, Selnes A (1979) Plasma 1,25-dihydroxyvitamin D levels in pregnancy and lactation. Acta Endocrinol (Copenh) 92:330–335

Lutterbeck PM, Pryor S, Varga L, Wenner R (1971) Treatment of non-puerperal galactorrhoea with an ergot alkaloid. Br Med J 3:228–229

MacIndoe JH, Turkington RW (1973) Stimulation of human prolactin secretion by intravenous infusion of L-tryptophan. J Clin Invest 52:1972–1978

MacIntyre I, Colston KW, Robinson CJ, Spanos E (1977) Vitamin D: Introduction and recent studies on regulation. In: MacIntyre I, Szelke N (eds). Molecular endocrinology. Elsevier/North-Holland Biomedical Press, Amsterdam, pp 73–85

MacLeod RM (1976) Regulation of prolactin secretion. In: Martini L, Ganong WF (eds). Frontiers in neuroendocrinology, vol IV. Raven, New York, pp 169–194

MacLeod RM, Lehmeyer JE (1972) Regulation of the synthesis and release of prolactin. In: Wolstenholme GEW, Knight J (eds). Lactogenic hormones. Livingstone, London, pp 53–82

MacLeod RM, Scapagnini U (eds) (1980) Central and peripheral regulation of prolactin function. Raven, New York

MacLeod RM, Fontham EH, Lehmeyer JE (1970) Prolactin and growth hormone production as influenced by catecholamines and agents that affect brain catecholamines. Neuroendocrinology 6:283–294

MacLeod RM, Kumura H, Login I (1976) Inhibition of prolactin secretion by dopamine and piribedil (ET-495) In: Pecile A, Müller EE (eds). Growth hormones and related peptides. Excerpta Medica, Amsterdam, pp 443–453

Magrini G, Ebiner JR, Burckhardt P, Felber JP (1976) Study on the relationship between plasma prolactin levels and androgen metabolism in man. J Clin Endocrinol Metab 43:944–947

Mahajan KK, Robinson CJ, Horrobin DF (1974) Actions of prolactin on plasma calcium and calcium excretion in normal, parathyroid-ectomised and thyroparathyroid-ectomised rats. Lancet 1:1237–1238

Mainoya JR (1975) Analysis of the role of endogenous prolactin on fluid and sodium chloride absorption by the rat jejunum. J Endocrinol 67:343–349

Malarkey WB, O'Dorisio TM, Cataland S (1980) Effects of VIP and CCK on Prl release in rat and human pituitary monolayer cultures. 62nd Annual Meeting of the Endocrine Society, Washington, June 18–20, Abstr no 701

Malven PV, Cousar GJ, Row EH (1969) Structural luteolysis in hypophysectomised rats. Am J Physiol 216:421–424

Manns JD, Boda JM (1965) Effects of ovine growth hormone and prolactin on blood glucose, serum insulin, plasma non-esterified fatty acid, and amino-nitrogen in sheep. Endocrinology 76:1109–1114

Mantero F, Opocher G, Armanini D, Boscaro M, Edwards CW (1979) Effect of serotonin (5-HT) on plasma aldosterone in man. Acta Endocrinol [Suppl 225] (Copenh) 91:345

Marek J, Horky K (1976) Bromocriptine and plasma-aldosterone. Lancet 2:1409

Markó M, Flückiger E (1980) Role of serotonin in the regulation of cyclic LH-secretion: evidence from pharmacological studies. Neuroendocrinology 30:228–231

Markstein R, Herrling P, Wagner H: Bromocriptine mimics dopamine effects on the cyclic AMP system of rat pituitary gland. In: Proc. of the 7th Int. Congress on Pharmacology, Paris, 16–21 July 1978, pp. 848, Abstract No. 2623, J.R. Boissier, P. Lechat. Soc. frc. Pharmacol. (Eds.), Pergamon Press, Oxford, Paris, 1978

Marshall S, Gelato M, Meites J (1975) Serum prolactin levels and prolactin binding activity in adrenals and kidneys of male rats after dehydration, salt loading and unilateral nephrectomy. Proc Soc Exp Biol Med 149:185–188

Marshall S, Huang HH, Kledzik GS, Campbell GA, Meites J (1978) Glucocorticoid regulation of prolactin receptors in kidneys and adrenals of male rats. Endocrinology 102:869–875

Marshall S, Bruni JF, Meites J (1979) Effects of hypophysectomy, thyroidectomy, and thyroxine on specific prolactin receptor sites in kidneys and adrenals of male rats. Endocrinology 104:390–395

Martin JB, Lal S, Tolis G, Friesen HG (1974) Inhibition by apomorphine of prolactin secretion in patients with elevated serum prolactin. J Clin Endocrinol Metab 39:180–182

Martini L, Celotti F, Juneja H, Motta M, Zanisi M (1978) Feedback effects on central mechanisms controlling neuroendocrine functions. In: Fuxe K, Hökfelt T, Luft R (eds). Central regulation of the endocrine system. Plenum, New York, London, pp 273–295

Mashiter K, Adams E, Beard M, Holley A (1977) Bromocriptine inhibits prolactin and growth hormone release by human pituitary tumours in culture. Lancet 2:197

Maslar IA, Kaplan BM, Luciano AA, Riddick DH (1980) Prolactin production by the endometrium of early human pregnancy. J Clin Endocrinol Metab 51:78–83

Maurer RA, Stone R, Gorski J (1976) Cell-free synthesis of a large translation product of prolactin messenger RNA. J Biol Chem 251:2801–2807

Maurer RA, Gorski J, McKean DJ (1977) Partial aminoacid sequence of rat pre-prolactin. Biochem J 161:189–192

McCann SM, Dhariwal APS, Porter JC (1968) Regulation of the adenohypophysis. Annu Rev Physiol 30:589–640

McCann SM, Krulich L, Ojeda SR, Negro-Vilar A, Vijayan E (1978) Neurotransmitters in the control of anterior pituitary function. In: Fuxe K, Hökfelt T, Luft R (eds). Central regulation of the endocrine system. Plenum, New York London, pp 329–347

McFarland RJ (1979) Gastrointestinal hormones and disease of the gastrointestinal tract. Clin Endocrinol Metab 8:331–347

McGarry EE, Beck JC (1972) Biological effects of nonprimate prolactin and human placental lactogen. In: Wolstenholme GEW, Knight J (eds). Lactogenic hormones. Ciba Foundation Symposium. Livingstone, London, pp 361–383

McGrath P (1968) Prolactin activity and human growth hormone in pharyngeal hypophyses from embalmed cadavers. J Endocrinol 42:205–212

McNatty KP, Sawyers RS, McNeilly AS (1974) A possible role for prolactin in control of steroid secretion by the human graafian follicle. Nature 250:653–655

McNatty KP, Neal P, Baker TG (1976) Effect of prolactin on the production of progesterone by mouse ovaries in vitro. J Reprod Fertil 47:155–156

Meites J (1962) Pharmacological control of prolactin secretion and lactation. In: Proceedings, First Internat. Pharmacol. Meeting, vol I, part 2, Guillemin R (ed): Pharmacol Control of Release of Hormones. Pergamon, Oxford, pp 151–180

Meites J (1973) Control of prolactin secretion in animals. In: Pasteels JL, Robyn C (eds) Human prolactin. Excerpta Medica, Amsterdam, pp 105–118

Meltzer HY, Miller RJ, Fessler RG, Simonovic M, Fang VS (1978) Effects of enkephalin analogues on prolactin release in the rat. Life Sci 22:1931–1938

Mena F, Grosvenor CE (1971) Release of prolactin in rats by exteroceptive stimulation: sensory stimuli involved. Horm Behav 2:107–116

Mendlewicz J, Youdim MBH (1977) Monoamine-oxidase inhibitors and prolactin secretion. Lancet 2:507

Metha AE, Tolis G (1979) Pharmacology of bromocriptine in health and disease. Drugs 17: 313–325

Midgley AR (1973) Autoradiographic analysis of gonadotropin binding to rat ovarian tissue section. In: O'Malley BW, Means AR (eds). Receptors for reproductive hormones. Plenum, New York, pp 365–371

Morishige WK, Rothchild I (1974) Temporal aspects of the regulation of corpus luteum function by luteinizing hormone, prolactin and placental luteotrophin during the first half of pregnancy in the rat. Endocrinology 95:260–274

Mühlbock O, Boot LM (1967) The mode of action of ovarian hormones in the induction of mammary cancer in mice. Biochem Pharmacol 16:627–630

Müller EE (ed) (1980) Neuroactive drugs in endocrinology. Elsevier/North-Holland, Amsterdam

Müller EE, Genazzani AR, Murru S (1978 a) Nomifensine: Diagnostic test in hyperprolactinaemic states. J Clin Endocrinol Metab 47:1352–1357

Müller EE, Cocchi D, Locatelli V, Krogsgaard-Larsen P, Bruno F, Racagni G (1978 b) GABA-ergic neurotransmission and the secretion of prolactin in the rat. In: Krogsgaard-Larsen P, Scheel - Kruger J, Kofold H (eds). GABA-neurotransmitters. Alfred Benzon symposium 12. Munksgaard, Copenhagen, pp 518–532

Mulloy AL, Malven PV (1979) Prolactin and colibacillus-induced fluid transport in ligated intestinal segments. Am J Vet Res 40:971–973

Nagasawa H (1977) Mammary gland DNA synthesis as a limiting factor for mammary tumorigenesis. IRCS J Med Sci 5:405–408

Nagasawa H, Yanai R (1972) Promotion of pituitary prolactin release in rats by dibutyryl-adenosine-3′,5′-monophosphate. J Endocrinol 55:215–217

Nagasawa H, Yanai R (1978) Normal and abnormal growth of the mammary gland. In: Yokoyama A, Mizuno H, Nagasawa H (eds). Physiology of mammary glands. Japan Scientific Societies Press, Tokyo, pp 121–159

Nagasawa H, Yanai R, Flückiger E (1973) Counteraction by 2-Br-α-ergocryptine of pituitary prolactin release promoted by dibutyryl-adenosine-3′, 5′-monophosphate. In: Pasteels JL, Robyn C (eds). Human prolactin. Excerpta Medica, Amsterdam, pp 313–315

Nagasawa H, Yanai R, Taniguchi H (1976) Importance of mammary gland DNA synthesis on carcinogen-induced mammary tumorigenesis in rats. Cancer Res 36:2223–2226

Nagasawa H, Sakai S, Banerjee MR (1979) Prolactin receptors. Life Sci 24:193–208

Nagel TC, Freinkel N, Bell RH, Friesen H, Wilber JF, Metzger BE (1973) Gynaecomastia, prolactin and other peptide hormones in patients undergoing chronic haemodialysis. J Clin Endocrinol Metab 36:428–432

Nagy I, MacLeod RM (1980) Metabolic changes in the rat anterior pituitary and prolactin production following estradiol treatment. Metabolism 29:429–434

Nagy I, Valdenegro CA, Login IS, MacLeod RM (1979a) Anterior pituitary enzyme activity and prolactin secretion in male rats treated with psychotropic drugs. Proc Soc Exp Biol Med 162:191–193

Nagy I, Valdenegro CA, Login IS, MacLeod RM (1979b) Antioestrogens decrease prolactin synthesis and enzyme activities in the anterior pituitary of rats. 61st Annual Meeting of the Endocrine Society, Anaheim, June 13–15, Abstr no 175

Nansel DD, Gudelsky GA, Porter JC (1979) Subcellular localization of dopamine in the anterior pituitary gland of the rat: apparent association of dopamine with prolactin secretory granules. Endocrinology 105:1073–1077

Nansel DD, Gudelsky GA, Porter JC (1980) Dopamine-induced stimulation of lysosomal enzyme activity in the anterior pituitary. 62nd Annual Meeting of the Endocrine Society, Washington, June 18–20, Abstr no 460

Nicoll CS (1972) Some observations and speculations on the mechanism of "depletion", "repletion", and release of adenohypophyseal hormones. Gen Comp Endocrinol [Suppl] 3:86–96

Nicoll CS (1973) 'Spinoff' from comparative studies on prolactin physiology of significance to the clinical endocrinologist. In: Pasteels JL, Robyn C (eds). Human prolactin. Excerpta Medica, Amsterdam, pp 119–136

Nicoll CS (1974) Physiological actions of prolactin. In: Knobil E, Sawyer WH (eds). The pituitary gland and its neuroendocrine control IV. American Physiological Society, Washington, pp 253–292

Nicoll CS (1975) Radioimmunoassay and radioreceptor assays for prolactin and growth hormone: a critical appraisal. Am Zool 15:881–903

Nicoll CS, Fiorindo RP, McKennee CT, Parsons JA (1970) Assay of hypothalamic factors which regulate prolactin secretion. In: Meites J (ed). Hypophysiotropic hormones of the hypothalamus: Assay and chemistry. Williams & Wilkins, Baltimore, pp 115–150

Nicoll CS, Mena F, Nichols CW, Green SH, Tai M, Russell SM (1976) Analysis of suckling-induced changes in adenohypophyseal prolactin concentration in the lactating rat by three assay methods. Acta Endocrinol (Copenh) 83:512–521

Niederer W, Richardson BP, Donatsch P (1975) Hormonal control of aqueous humour production. Exp Eye Res 20:329–340

Nolin JM (1978) Intracellular prolactin in rat corpus luteum and adrenal cortex. Endocrinology 102:402–406

Nolin JM, Witorsch RJ (1976) Detection of endogenous immunoreactive prolactin in rat mammary epithelia cells during lactation. Endocrinology 99:949–958

Noth RH, McCallum RW, Coutino C, Havelick J (1980) Tonic dopaminergic suppression of plasma aldosterone. J Clin Endocrinol Metab 51:64–69

Ojeda SR, Jameson HE, McCann SM (1978) Effects of indomethacin and prostaglandin injections on plasma prolactin and growth hormone levels in rats. Endocrinology 102:531–539

Ølgaard K, Hagen C, Madsen S, Hummer L (1976) Aldosterone and prolactin. Lancet 2:959

Olson L, Fuxe K, Hökfelt T (1972) The effect of pituitary transplants on the tubero-infundibular dopamine neurons in various endocrine states. Acta Endocrinol (Copenh) 71:233–244

Ontjes DA (1980) The pharmacologic control of adrenal steroidogenesis. Life Sci 26:2032–2035

Orgebin-Crist M-C, Djiane J (1979) Properties of a prolactin receptor from the rabbit epididymis. Biol Reprod 21: 135–139

Ozawa S, Sand O (1978) Electrical activity of rat anterior pituitary cells in vitro. Acta Physiol Scand 102:330–341

Parker DC, Rossman LG, van der Laan EF (1973) Sleep related, nyctohemeral and briefly episodic variation in human plasma prolactin concentration. J Clin Endocrinol Metab 36:1119–1124

Parker DC, Rossman LG, van der Laan EF (1974) Relation of sleep-entrained human prolactin release to REM-nonREM cycles. J Clin Endocrinol 38:646–651

Parkes D (1978) Bromocriptine. Adv Drug Res 12:247–344

Pasteels JL (1961) Sécrétion de prolactine par l'hypophyse en culture de tissu. CR Acad Sci [D] (Paris) 253:2140

Pasteels JL (1972) Tissue culture of human hypophyses: evidence of a specific prolactin in man. In: Wolstenholme GEW, Knight J (eds). Lactogenic hormones. Livingstone, London, pp 269–277

Pasteels JL, Danguy A, Frérotte M, Ectors F (1971) Inhibition de la sécrétion de prolactine par l'ergocornine et la 2-Br-α-ergokryptine: action directe sur l'hypophyse en culture. Ann Endocrinol (Paris) 32:188–192

Paul SM, Axelrod J, Saavedra JM, Skolnick P (1979) Estrogen-induced efflux of endogenous catecholamines from the hypothalamus in vitro. Brain Res 178:499–505

Pawlikowski M, Stepien H (1979) Effects of fibroblast growth factor and bromocriptine on the mitotic activity of the anterior pituitary gland in organ culture. Cell Tissue Res 202:165–169

Peillon F, Cesselin F, Bression D, Zygelman N, Brandi AM, Nousbaum A, Mauborgne A (1979) In vitro effect of dopamine and L-DOPA on prolactin and growth hormone release from human pituitary adenomas. J Clin Endocrinol Metab 49:737–741

Pelletier G, Lemay A, Béraud G, Labrie F (1972) Ultrastructural changes accompanying the stimulatory effect of N6-monobutyryl-adenosine-3', 5'-monophosphate on the release of growth hormone in rat anterior pituitary gland in vitro. Endocrinology 91:1355–1371

Perkins NA, Westfall TC (1978) The effect of prolactin on dopamine release from rat striatum and medial basal hypothalamus. Neuroscience 3:59–63

Perks AM, Vizsolyi E, Holt WF, Cassin S (1978) Hormonal influences on the movement and composition of amniotic fluid. In: Gaillard PJ, Boer HH (eds). Comparative endocrinology. Elsevier/North-Holland Biomedical Press, Amsterdam, pp 231–234

Pilotte NS, Porter JC (1979) Circulating luteinizing hormone and prolactin concentrations in intact or castrated male rats treated with 5-hydroxytryptamine. Endocrinology 105:875–878

Pinder RM, Brogden RN, Sawyer PR, Speight TM, Avery GS (1976) Levodopa and decarboxylase inhibitors: a review of their clinical pharmacology and use in the treatment of parkinsonism. Drugs 11:329–377

Piva F, Gagliano P, Motta M, Martini L (1973) Adrenal progesterone: factors controlling its secretion. Endocrinology 93:1178–1184

Polleri A, Masturzo P, Murialdo G, Carolei A (1980) Dose and sex related effects of aromatic decarboxylase inhibitors on serum prolactin in humans. Acta Endocrinol (Copenh) 93:7–12

Pontiroli AE, Castegnaro E, Vettaro MP, Verberti GC, Pozza G (1977) Stimulatory effect of DOPA-decarboxylase inhibitor Ro 4-4602 on prolactin release: inhibition by L-DOPA, metergoline, Acta Endocrinol (Copenh) 84:36–44

Posner BI, Kelly PA, Shiu RPC, Friesen GH (1974) Studies on insulin, growth hormone and prolactin binding: tissue distribution, species variation and characterization. Endocrinology 95: 521–531

Posner BI, Kelly PA, Friesen HG (1975) Prolactin receptors in rat liver: possible induction by prolactin. Science 188:57–59

Poulain P, Carette B (1976) Actions of iontophoretically applied prolactin on septal and preoptic neurons in the guinea-pig. Brain Res 116:172–176

Pozo E del, Brownell J (1979) Prolactin. I. Mechanism of control, peripheral actions and modification by drugs. Horm Res 10:143–172

Pozo E del, Lancranjan I (1978) Clinical use of drugs modifying the release of anterior pituitary hormones. In: Ganong WF, Martini L (eds). Frontiers in neuroendocrinology, vol 5. Raven, New York, pp 207–247

Pozo E del, Ohnhaus EE (1976) Lack of effect of acute prolactin suppression on renal water, sodium and potassium excretion during sleep. Horm Res 7:11–15

Pozo E del, Flückiger E, Lancranjan I (1976) Endogenous control of prolactin release and its modification by drugs. In: Charro Salgado AL, Fernández Durango R, López del Campo JG (eds). Basic applications and clinical uses of hypothalamic hormones. Excerpta Medica, Amsterdam, pp 137–150

Pozo E del, Darragh A, Lancranjan I, Ebeling D, Burmeister P, Bühler F, Marbach P, Braun P (1977 a) Effect of bromocriptine on the endocrine system and fetal development. Clin Endocrinol [Suppl] (Oxf) 6:47 s–55 s

Pozo E del, Hiba J, Lancranjan I, Künzig HJ (1977 b) Prolactin measurements throughout the life cycle; endocrine correlations. In: Crosignani PG, Robyn C (eds). Prolactin and human reproduction. Academic Press, London, pp 61–69

Pozo E del, Schulz KD, Wyss H, Lancranjan I, Braun P (1978) Effect of prolactin on the mechanisms of ovulation and on pregnancy: some recent findings. In: Robyn C, Harter M (eds). Progress in prolactin physiology and pathology. Elsevier/North-Holland Biomedical Press, Amsterdam, pp 281–292

Pozo E del, Wyss H, Tolis G, Alcaniz J, Campana A, Naftolin F (1979) Prolactin and deficient luteal function. Obstet Gynecol 53:282–286

Pozo E del, Bigazzi M, Calaf J (1980) Induced human gestational hypoprolactinaemia. Lack of action on fetal adrenal androgen synthesis. J Clin Endocrinol Metab 51:936

Prysor-Jones RA, Jenkins JS (1980) Effect of bromocriptine, ergotamine and other ergot alkaloids on the hormone secretion and growth of a rat pituitary tumour. J Endocrinol 86:147–153

Purvis K, Clausen OPF, Olsen A, Haug E, Hansson V (1979) Prolactin and Leydig cell responsiveness of LH/hCG in the rat. Arch Androl 3:219–230

Quattrone A, diRenzo G, Schettini G, Tedeschi G, Scopacasa F (1978) Increased plasma prolactin levels induced in rats by d-fenfluramine: relation to central serotoninergic stimulation. Eur J Pharmacol 49:163–167

Quiada M de, Timmermans HAT, Lamberts SWJ, MacLeod RM (1980a) Tamoxifen enhances the sensitivity of dispersed prolactin secreting tumour cells to dopamine and bromocriptine. Endocrinology 106:702–706

Quiada M de, Timmermans HAT, Lamberts SWJ (1980b) Tamoxifen suppressed both the growth of prolactin secreting tumours and normal prolactin synthesis in the rat. J Endocrinol 86: 109–116

Racagni G, Apud JA, Locatelli V, Cocchi D, Nistico G, di Giorgio RM, Müller EE (1979) GABA of CNS origin in the rat anterior pituitary inhibits prolactin secretion. Nature 281:575–578

Rakoff JS, Yen SSC (1978) Progesterone induced acute release of prolactin in estrogen primed ovariectomized women. J Clin Endocrinol Metab 47:918–921

Ramsay DH, Bern HA (1972) Stimulation by ovine prolactin of fluid transfer in everted sacs of rat small intestine. J Endocrinol 53:453–459

Ranke MB, Stanley CA, Rodbard D, Baker L, Bongiovanni AM, Parks JS (1976) Sex differences in binding of human growth hormone to isolated rat hepatocytes. Proc Natl Acad Sci USA 73: 874–851

Rathgeb IB, Winkler B, Steele R, Altszuler N (1971) Effect of ovine prolactin administration on glucose metabolism and plasma insulin levels in the dog. Endocrinology 88:718–722

Ravault JP, Courot M, Garnier D, Pelletier J, Terqui M (1977) Effect of 2-Bromo-α-ergocryptine on plasma prolactin, LH and testosterone levels, accessory reproductive glands and spermatogenesis in lambs during puberty. Biol Reprod 17:192–197

Ravitz AJ, Moore KE (1977) Effects of amphetamine, methylphenidate and cocaine on serum prolactin concentrations in the male rat. Life Sci 21:267–272

Raymond V, Beaulieu M, Labrie F, Boissier J (1978) Potent antidopaminergic activity of estradiol at the pituitary level on prolactin release. Science 200:1173–1175

Rees LH, Bloomfield GA, Rees GM, Corrin B, Franks LM, Ratcliffe JG (1974) Multiple hormones in a bronchial tumor. J Clin Endocrinol Metab 38:1090–1097

Refetoff S, Fang VS, Ropoport B, Friesen HG (1974) Interrelationship in the regulation of TSH + prolactin secretion in man: Effects of L-DOPA, TRH + thyroid hormone in various combinations. J Clin Endocrinol Metab 38:450–457

Richardson B, Luginbühl H (1976) The role of prolactin in the development of chronic progressive nephropathy in the rat. Virchows Arch [Pathol Anat] 370:13–19

Rick J, Szabo M, Payne P, Kovathana N, Cannon JG, Frohman LA (1979) Prolactin-suppressive effects of two aminotetralin analogs of dopamine: Their use in the characterization of the pituitary dopamine receptor. Endocrinology 104:1235–1242

Riddick DH, Luciano AA, Kusmik WF, Maslar IA (1978) De novo synthesis of prolactin by human decidua. Life Sci 23:1913–1922

Riddle O (1963) Prolactin in vertebrate function and organisation. J Natl Cancer Inst 31:1039–1110

Rivier C, Brown M, Vale W (1977a) Effect of neurotensin, substance P and morphine sulfate on the secretion of prolactin and growth hormone in the rat. Endocrinology 100:751–754

Rivier C, Vale W, Ling N, Brown M, Guillemin R (1977b) Stimulation in vivo of the secretion of prolactin and growth hormone by β-endorphin. Endocrinology 100:238–241

Rivier C, Rivier J, Vale W (1978) The effect of bombesin and related peptides on prolactin and growth hormone secretion in the rat. Endocrinology 102:519–522

Robyn C, Delvoye P, Nokin J, Vekemans M, Badawi M, Perez-Lopez FR, L'Hermite M (1973) Prolactin and human reproduction. In: Pasteels JL, Robyn C (eds). Human prolactin. Excerpta Medica, Amsterdam, pp 167–188

Robyn C, Vekemans M, Delvoye P, Joostens-Defleur V, Caufriez A, L'Hermite M (1976) Prolactin and fertility control in women. In: Pecile A, Müller EE (eds). Growth hormone and related peptides. Excerpta Medica Amsterdam, pp 396–406

Rodway RG, Kuhn NJ (1975) Luteal 20-α-hydroxy steroid dehydrogenase and the formation of Δ^4-3-oxo steroids in the rat after weaning or treatment with 2-Br-α-ergocryptine during lactation. Biochem J 152:445–448

Röpcke G (1975) Interaction of hypophyseal isografts and ovarian hormones in mammary tumour-development in mice. *Academisch Proefschrift.* Mondeel, Amsterdam

Rogol AD (1975) Three components of immunoreactive and receptor active plasma prolactin in man. Univ Milan Ric Sci Educaz Permamente [Suppl] 1:139–144

Rogol AD, Rosen SW (1974) Prolactin of apparent large molecular size: The major immunoactive prolactin component in plasma of a patient with a pituitary tumor. J Clin Endocrinol Metab 38:714–717

Rolandi E, Barreca T, Gallamini A, Gianrossi R, Masturzo P, Murialdo G, Nizzo MC (1979) Physiolocigal, pharmacological and pathological hyperprolactinemias. In: Polleri A, MacLeod RM (eds). Neuroendocrinology: Biological and clinical aspects. Academic Press, London New York, pp 257–286

Rolland R, Gunsalus GL, Hammond JM (1976) Demonstration of specific binding of prolactin by porcine corpora lutea. Endocrinology 98:1083–1091

Rosen JM, Guyette WA, Matusik RJ (1979) Hormonal regulation of casein gene expression in the mammary gland. In: Hamilton TH, Clark JH, Sidler WA (eds). Ontogenesis of receptors and reproductive hormone action. Raven, New York, pp 249–279

Rosenberg JM, Lau-Cam CA, McGuire H (1977) Effects of pyridoxine hydrochloride (vitamin B_5) on chlorpromazine-induced serum prolactin rise in male rats. J Pharmacol Sci 68:1179–1181

Rubin RT, Poland RE, Tower BB (1976) Prolactin-related testosterone secretion in normal adult men. J Clin Endocrinol Metab 42:112–116

Rusterholz DB, Long JP, Flym JR, Cannon JG, Lee T, Pease JP, Clemens JA, Wong DT, Bymaster FP (1979) Dopaminergic effects of non-hydroxylated rigid analogs of apomorphine. Eur J Pharmacol 55:73–82

Saito T, Saxena BB (1975) Specific receptors for prolactin in the ovary. Acta Endocrinol (Copenh) 80:126–137

Sakai S, Bowman PD, Yang J, McCormick K, Nandi S (1979) Glucocorticoid regulation of prolactin receptors on mammary cells in culture. Endocrinology 104:1447–1449

Sassin JF, Frantz AG, Weitzman ED, Kapen S (1972) Human prolactin: 24 hour pattern with increased release during sleep. Science 177:1205–1207

Schally AV, Redding TW, Arimura A, Dupont A, Linthicum GL (1977) Isolation of gamma-amino butyric acid from pig hypothalami and demonstration of its prolactin release-inhibiting (PIF) activity in vivo and in vitro. Endocrinology 100:681–691

Schams D (1972) Prolactin release effects of TRH in the bovine and their depression by a prolactin inhibitor. Horm Metab Res 4:405

Schams D (1976) Hormonal control of lactation. Ciba Found Symp 45: 27–48

Schams D, Reinhardt V, Karg H (1974) The immediate response of the plasma prolactin level to oestrogen and progesterone infusion in bulls. Acta Endocrinol (Copenh) 76:242–247

Schinfeld JS, Tulchinsky D, Schiff I, Fishman J (1980) Suppression of prolactin and gonadotropin secretion in post-menopausal women by 2-hydroxyoestrone. J Clin Endocrinol Metab 50:408–410

Schmidt MJ, Hill LE (1977) Effects of ergots on adenylate cyclase activity in the corpus striatum and pituitary. Life Sci 20:789–798

Schneider AB, Kowalski K, Sherwood LM (1975) Big human placental lactogen: disulfide peptide chains. Biochem Biophys Res Commun 64:717–724

Schulz K-D, Czygan P-J, del Pozo E, Friesen HG (1973) Varying response of human metastasizing breast cancer to the treatment with 2-Br-α-ergocryptine (CB-154). Case report. In: Pasteels JL, Robyn C (eds) Human prolactin. Excerpta Medica, Amsterdam, pp 268–271

Schulz K-D, Geiger W, del Pozo E, Lose KH, Künzig HJ, Lancranjan I (1976) The influence of the prolactin-inhibitor bromocriptine (CB 154) on human luteal function in vivo. Arch Gynaekol 221:93–96

Schulz K-D, Geiger W, del Pozo E, Künzig HJ (1978) Pattern of sexual steroids, prolactin and gonadotropic hormones during prolactin inhibition in normally cycling women. Am J Obstet Gynecol 132:561–566

Shelesnyak MC (1954 a) Ergotoxine inhibition of deciduoma formation, and its reversal by progesterone. Am J Physiol 179:301–304

Shelesnyak MC (1954 b) The action of selected drugs on deciduoma formation. Endocrinology 55:85–89

Sheth AR, Mugatwala PO, Shah GV, Rao SS (1975) Occurrence of prolactin in human semen. Fertil Steril 26:905–907

Shiu RPC, Friesen HG (1974 a) Properties of a prolactin receptor from the rabbit mammary gland. Biochem J 140:301–311

Shiu RPC, Friesen HG (1974 b) Solubilization and purification of a prolactin receptor from the rabbit mammary gland. J Biol Chem 249:7812–7911

Shiu RPC, Friesen HG (1976 a) Blockade of prolactin action by an antiserum to its receptor. Science 192:259–261

Shiu RPC, Friesen HG (1976 b) Prolactin receptors. In: Blecher M (ed). Methods in receptor research, part II. Dekker, New York Basel, pp 565–598

Shome B, Parlow AF (1977) Human pituitary prolactin (hPRL): the entire linear amino acid sequence. J Clin Endocrinol Metab 45:1112–1115

Short RV (1976) Lactation – the central control of reproduction. Ciba Found Symp 45:73–86

Short RV, Drife JO (1977) The aetiology of mammary cancer in man and animals. In: Peaker M (ed). Comparative aspects of lactation. Academic Press, London, pp 211–230

Shupnik MA, Gorski J (1980) Isolation and characterization of ovine prolactin and growth hormone messenger RNAs. Mol Cell Endocrinol 17:181–193

Sibley DR, Creese I (1979) Guanine nucleotides regulate anterior pituitary dopamine receptors. Eur J Pharmacol 55:341–343

Sinha YN, Baxter SR (1979) Metabolism of prolactin in mice with a high incidence of mammary tumours: evidence for greater conversion into a non-immunoassayable form. J Endocrinol 81:299–314

Sinha YN, Selby FW, Vanderlaan WP (1973) Radioimmunoassay of prolactin in the urine of mouse and man. J Clin Endocrinol Metab 36:1039–1042

Skett P, Eneroth P, Gustafsson J-A (1978) The effect of 2 Br-α-erokryptine on the hepatic steroid metabolism and serum pituitary hormone levels in normal rats and rats with an ectopic pituitary. Biochem Pharmacol 27:1713–1716

Slater SL, Lipper S, Shiling DJ, Murphy DL (1977) Elevation of plasma-prolactin by monoamine-oxidase inhibitors. Lancet 2:275–276

Smalstig EB, Sawyer BD, Clemens JA (1974) Inhibition of rat prolactin release by apomorphine in vivo and in vitro. Endocrinology 95:123–129

Sombre ER de, Kledzik G, Marshall S (1976) Estrogen and prolactin receptor concentrations in rat mammary tumours and response to endocrine ablation. Cancer Res 36:3830–3833

Spanos E, Colston KW, Evans IMS, Galante LS, Macauley SJ, MacIntyre I (1976 a) Effect of prolactin on vitamin D metabolism. Mol Cell Endocrinol 5:163–167

Spanos E, Pike JW, Haussler MR, Colston KW, Evans IMA, Goldner AM, McCain TA, MacIntyre I (1976 b) Circulating 1α,25-dihydroxy-vitamin D in the chicken. Life Sci 19:1751–1756

Spellacy WN, Buhi WC (1979) A prospective study of plasma prolactin levels in women using the progesterone releasing intrauterine device (P-IUD). Contraception 19:91–94

Spellacy WN, Buhi WC, Birk SA (1975) Stimulated plasma prolactin levels in women using medroxyprogesterone acetate or an intrauterine device for contraception. Fertil Steril 26:970–972

Spencer EM (1980) Lack of response of serum somatomedin to hyperprolactinaemia in humans. J Clin Endocrinol Metab 50:182–185

Spitz IM, Landau H, Almaliach U, Rosen E, Brautbar N, Russell A (1977) Diminished prolactin reserve: a case report. J Clin Endocrinol Metab 412–418

Stähelin H, Burckhardt-Vischer B, Flückiger E (1971) Rat mammary cancer inhibition by a prolactin suppressor, 2-Br-α-ergokryptine (CB 154). Experientia 27:915–916

Steichen JJ, Tsang RC, Gratton TL, Hamstra A, DeLuca HF (1980) Vitamin D homeostasis in the perinatal period. 1,25-dihydroxyvitamin D in maternal cord, and neonatal blood. N Engl J Med 302:315–319

Stepien H, Wolaniuk A, Pawlikowski M (1978) Effects of pimozide and bromocriptine on anterior pituitary cell proliferation. J Neural Transm 42:239–244.

Stevens E, Helfenstein JE (1966) Some effects of certain naturally occurring oestrogens on the anterior pituitary glands of rats. Nature 211:879–880

Stone RT, Maurer RA, Gorski J (1977) Effect of estradiol-17β on preprolactin messenger ribonucleic acid activity in the rat pituitary gland. Biochemistry 16:4915–4921

Stricker P, Grüter F (1928) Action du lobe antérieur de l'hypophyse sur la montée laiteuse. CR Soc Biol (Paris) 99:1978–1980

Subramanian MG, Gala RR (1976a) The influence of cholinergic, adrenergic, serotonergic drugs on the afternoon surge of prolactin in ovariectomized, estrogen-treated rats. Endocrinology 98: 842–848

Subramanian MG, Gala RR (1976b) Further studies on the effects of adrenergic, serotoninergic and cholinergic drugs on the afternoon surge of plasma prolactin in ovariectomized, estrogen-treated rats. Neuroendocrinology 22:240–249

Subramanian MG, Gala RR (1977) Specificity of arecoline and apomorphine and the site of action of arecoline in inhibiting the diurnal prolactin surge. Proc Soc Exp Biol Med 155:353–356

Suh HK, Frantz AG (1974) Size hetereogeneity of human prolactin in plasma and pituitary extracts. J Clin Endocrinol Metab 39:928–935

Sulman FG (1970) Hypothalamic control of lactation. Springer, Berlin Heidelberg New York Monographs in endocrinology, vol 3, p 235

Sundberg DK, Fawcett CP, McCann SM (1976) The involvement of cyclic-3'5'-AMP in the release of hormones from the anterior pituitary in vitro. Proc Soc Exp Biol Med 151:149–154

Taché Y, Brown M, Collu R (1979) Effects of neuropeptides on adenohypophyseal hormone response to acute stress in male rats. Endocrinology 105:220–224

Takahara J, Yunoki S, Yamane Y, Yamauchi J, Yakushiji W, Kageyama N, Fajino K, Ofuji T (1979) Effects of nifedipine on prolactin, growth hormone, and luteinizing hormone release by rat anterior pituitaries. Proc Soc Exp Biol Med 162:31–33

Talwalker PK, Ratner A, Meites J (1963) In vitro inhibition of pituitary prolactin synthesis and release by hypothalamic extracts. Am J Physiol 205:213–218

Tam WS, Dannies PS (1980) Dopaminergic inhibition of ionophore A23187-stimulated release of prolactin from rat pituitary cells. J Biol Chem 255:6595–6599

Tanaka T, Shiu RPC, Gout PW, Beer CT, Noble RL, Friesen HG (1980) Rapid, sensitive and specific bioassay for lactogenic hormones using a lymphoma cell line. 62nd Annul a Meeting of the Endocrine Society, Washington, June 18–20, Abstr no 769

Taraskevich PS, Douglas WW (1978) Catecholamines of supposed inhibitory hypophysiotrophic function suppress action potentials in prolactin cells. Nature 276:832–834

Tashjian AH, Barowsky NJ, Jensen DK (1971) Thyrotropin releasing hormone; direct evidence for stimulation of prolactin production by pituitary cells in culture. Biochem Biophys Res Commun 43:516–523

Tashjian AH, Lomedico ME, Maina D (1978) Role of calcium in the thyrotropin-releasing hormone-stimulated release of prolactin from pituitary cells in culture. Biochem Biophys Res Commun 81:798–806

Thomas JA, Manandhar M (1974) Effect of prolactin and/or testosterone on cyclic AMP in the rat prostate gland. Horm Metab Res 6:529–530

Thomas JA, Manandhar M (1975) Effects of prolactin and/or testosterone on nucleic acid levels in prostate glands of normal and castrated rats. J Endocrinol 65:149–150

Thorner MO (1977) Prolactin: Clinical physiology and the significance and management of hyperprolactinemia. In: Martini L, Besser GM (eds). Clinical neuroendocrinology. Academic Press, New York, pp 319–361

Thorner MO, Flückiger E, Calne DB (1980a) Bromocriptine. Raven, New York

Thorner MO, Rogol AD, Evans WS, Nunley WC, MacLeod RM (1980b) The effects of prolactin on gonadal function in man. In: MacLeod RM, Scapagnini U (eds). Central and peripheral regulation of prolactin function. Raven, New York, pp 271–285

Thorner MO, Hackett JT, Murad F, MacLeod RM (1980c) Calcium rather than cAMP as the physiological intracellular regulator of prolactin release. Neuroendocrinology 39:390–402

Timmermans HAT, Lamberts SWJ (1979) Anti-oestrogens increase for sensitivity to bromocriptine of prolactin secreting pituitary cells in vitro. Acta Endocrinol [Suppl 225] (Copenh) 91:185

Tindal JS (1978) Central pathways in oxytocin and prolactin release. In: Yokoyama A, Mizuno H, Nagasawa H (eds). Physiology of mammary glands. Japan Scientific Societies Press, Tokyo, pp 305–322

Tindal JS, Knaggs GS (1972) Pathways in the forebrain of the rabbit concerned with the release of prolactin. J Endocrinol 52:253–262

Toubeau G, Desclin J, Parmentier M, Pasteels JL (1979) Cellular localization of a prolactin-like antigen in the rat brain. J Endocrinol 83:261–266

Tourniaire J, Pallo D, Pousset G, Bizollon C, Bachelot I (1974) Diminution de la tolérance glucidique et hyperinsulinisme dans l'adénome à prolactine. Nouv Presse Med 3:1705–1707

Trifaró JM (1977) Common mechanisms of hormone secretion. Annu Rev Pharmacol Toxical 17: 27–47

Turkington RW (1971) Ectopic production of prolactin. N Engl J Med 285:1455–1458

Turkington RW (1972) Molecular biological aspects of prolactin. In: Wolstenholme GEW, Knight J (eds). Lactogenic hormones. Livingstone, London, pp 111–127

Turkington RW, Frantz WL, Majumder GC (1973) Effector-receptor relations in the action of prolactin. In: Pasteels JL, Robyn C (eds). Human prolactin. Excerpta Medica, Amsterdam, pp 24–34

Tyson JE, Hwang P, Guyda H, Friesen H (1972 a) Studies of prolactin secretion in human pregnancy. Am J Obstet Gynecol 113:14–20

Tyson JE, Friesen HG, Anderson MS (1972 b) Human lactational and ovarian response to endogenous prolactin. Science 177:897–899

Tyson JE, Andreasson FB, Huth J, Smith B, Zacur H (1975) Neuroendocrine dysfunction in galactorrhea-amenorrhea after oral contraceptive use. J Obstet Gynecol 46:1–11

Tyson JE, Freedman RS, Perez A, Zacur HA, Zanartu J (1976) Significance of the secretion of human prolactin and gonadotropin for puerperal lactational infertility. Ciba Found Symp 45:49–64

Uberti EC degli, Fabbri BL, Margutti AR, Fersini CM, Pansini R (1979) Effect of bromocriptine on the control of plasma aldosterone diurnal variation in normal supine man. Horm Res 10:64–78

Vale W, Rivier C, Brown M, Chan L, Ling N, Rivier J (1975) Application of adenohypophyseal cell cultures to neuroendocrine studies. In: Labrie F, Meites J, Pelletier G (eds). Hypothalamus and endocrine functions. Plenum, New York, pp 397–429

Valverde RC, Chieffo V, Reichlin S (1972) Prolactin-releasing factor in porcine and rat hypothalamic tissue. Endocrinology 91:982–993

Van der Gugten AA, Kwa HG (1970) Disappearance rate of bovine prolactin from plasma of female rats studied at intervals of up to 160 minutes by radioimmunoassay. Acta Endocrinol (Copenh) 64:718–725

Van Maanen JH, Smelik PG (1968) Induction of pseudopregnancy in rats following local depletion of monoamines in the median eminence of the hypothalamus. Neuroendocrinology 3:177–186

Varga L, Lutterbeck PM, Pryor JS, Wenner R, Erb H (1972) Suppression of puerperal lactation with an ergot alkaloid: a double-blind study. Br Med J 2:743–744

Vaughan MK, Blask DE, Johnson LY, Reiter RJ (1975) Prolactin releasing activity of arginine-vasotocin in vitro. Horm Res 6: 342–350

Vekemans M, Robyn C (1975 a) Influence of age on serum prolactin levels in women and men. Br Med J 3:738–739

Vekemans M, Robyn C (1975 b) The influence of exogenous estrogen on the circadian periodicity of circulating prolactin in women. J Clin Endocrinol Metab 40:886–889

Veldhuis JD, Klase P, Hammond JM (1980) Divergent effects of prolactin upon steroidogenesis by porcine granulosa cells in vitro: influence of cytodifferentiation. Endocrinology 107:42–46

Vijayan E, Samson WK, McCann SM (1978) Effects of intraventricular injection of gastrin on release of LH, prolactin, TSH and GH in conscious ovariectomized rats. Life Sci 23:2225–2232

Vonderhaar BK, Greco AE (1979) Lobuloalveolar development of mouse mammary glands is regulated by thyroid hormones. Endocrinology 104:409–418

Voogt JL, Meites J (1973) Suppression of prooestrus and suckling induced increase in serum prolactin by hypothalamic implant of prolactin. Proc Soc Exp Biol Med 142:1056–1058

Voogt JL, Carr L, Levin J (1979) Self regulation of prolactin via hypothalamic catecholamines. In: Usdin E, Kopin IJ, Barchas J (eds). Catecholamines: Basic and clinical frontiers. Pergamon, New York, pp 1257–1259

Vorherr H (1979) Renal and vascular activity of prolactin preparations. Klin Wochenschr 57:101–109

Walsh RJ, Posner BI, Kopriva BM, Brawer JR (1978) Prolactin binding sites in the rat brain. Science 201:1041–1042

Waters MJ, Friesen HG, Bohnet HG (1978) Regulation of prolactin receptors by steroid hormones and use of radioligand assays in endocrine research. In: Birnbaumer L, O'Malley BW (eds). Receptors and hormone actions, vol III. Academic Press, New York, pp 457–477

Weiner RI, Cronin MJ, Cheung CY, Annunziato L, Faure N, Goldsmith PC (1979) Dopamine: a prolactin inhibitory hormone. In: Müller EE, Agnoli A (eds). Neuroendocrine correlates in neurology and psychiatry. Elsevier/North-Holland Biomedical Press, Amsterdam, pp 41–55

Welsch CW (1978) Prolactin and the development and progression of early neoplastic mammary gland lesions. Cancer Res 38:4054–4058

Welsch CW, Nagasawa H, Meites J (1970) Increased incidence of spontaneous mammary tumors in female rats with induced hypothalamic lesions. Cancer Res 30:2310–2313

Werder K von, Clemm C (1974) Evidence for "big" and "little" components of circulating immunoreactive prolactin in humans. FEBS Lett 47:181–184

West B, Dannies PS (1979) Antipsychotic drugs inhibit prolactin release from rat anterior pituitary cells in culture by a mechanism not involving the dopamine receptor. Endocrinology 104:877–880

Whitworth NS, Grosvenor CE (1978) Transfer of milk prolactin to the plasma of neonatal rats by intestinal absorption. J Endocrinol 79:191–199

Willis KJ, London DR, Ward HWC, Butt WR, Lynch SS, Rudd BT (1977) Recurrent breast cancer treated with the antioestrogen tamoxifen: correlation between hormonal changes and clinical course. Br Méd J 1:425–528

Winnik MHZ, Tennenbaum L (1955) Apparition de galactorrhée au cours du traitement de largactil. Presse Med 63:1062

Winters AJ, Colston C, MacDonald PC, Porter JC (1975) Fetal plasma prolactin levels. J Clin Endocrinol Metab 41:626–629

Witorsch RJ, Edwards JT (1976) Comparison of effects of prolactin and growth hormone on adrenal 5-α-reductase in hypophysectomized rats. Proc Soc Exp Biol Med 151:684–693

Witorsch RJ, Kitay JI (1972) Pituitary hormones affecting adrenal 5-α-reductase activity: ACTH, growth hormone and prolactin. Endocrinology 91:764–769

Witorsch RJ, Smith JP (1977) Evidence for androgen-dependent intracellular binding of prolactin in rat ventral prostate gland. Endocrinology 101:929–938

Wood PL, Cheney DL, Costa E (1980) A prolactin action on acetylcholine metabolism in striatum, hippocampus and thalamus. J Neurochem 34:1053–1057

Wuttke W, Döhler KD (1973) Partial and total blockade of ergocornine on serum prolactin release in rats. Effects on pregnancy and serum progesterone. In: Pasteels JL, Robyn C (eds). Human prolactin. Excerpta Medica, pp 156–159

Wuttke W, Meites J (1971) Luteolytic role of prolactin during the estrous cycle of the rat. Proc Soc Exp Biol Med 137:988–991

Yamada Y (1975) Effects of iontophoretically-applied prolactin on unit activity of the rat brain. Neuroendocrinology 18:263–271

Yanai R, Nagasawa H (1970 a) Suppression of mammary hyperplastic nodule formation and pituitary prolactin secretion in mice induced by ergocornine and 2-Br-α-ergokryptine. J Natl Cancer Inst 45:1105–1112

Yanai R, Nagasawa H (1970 b) Effects of ergocornine and 2-Br-α-ergocryptine (CB 154) on the formation of mammary hyperplastic alveolar nodules and the pituitary prolactin levels in mice. Experientia 26:649–650

Yen S, Siler T, de Vane G (1974) Effect of somatostatin in patients with acromegaly: Suppression of growth hormone, prolactin, insulin and glucose levels. N Engl J Med 290:935–938

Zeilmaker GH, Carlsen RA (1962) Experimental studies on the effect of ergocornine methanesulfonate on the luteotrophic function of the rat pituitary gland. Acta Endocrinol (Copenh) 41:321

Zinder O, Hamosh M, Fleck TRC, Scow RO (1974) Effect of prolactin on lipoprotein lipase in mammary gland and adipose tissue of rats. Am J Physiol 226:744–748

3 Physiology and Pharmacology of Human Lactation

Conditioning of breast tissue for milk production has as a background a complicated hormonal interplay synchronized with the progress of pregnancy and gradual takeover by the placenta. The biologic effect of the endocrine changes is the development of functionally differentiated secretory cells able to synthesize the constituents of milk. Changes in hormonal secretion, mechanical factors, and neurogenic reflexes will then facilitate the excretion of the formed product in order to cover the caloric demands of the newborn during the early stage of extrauterine life. Throughout this chapter it will become obvious that, although considerable progress has been made in the understanding of lactational mechanisms, marked species differences will curtail the extrapolation to man of data recorded in other mammals. Moreover, socioeconomic background, traditional habits, and geographical factors should also be considered when dealing with human lactation as will be shown further down.

3.1 Preparation of Mammary Tissue for Lactation: Hormonal Mechanisms

In lower mammals, preparation of breast tissue for lactation demands the presence of estrogen, progesterone, thyroxine, corticosteroids and insulin (Cowie and Tindal 1971). More recently, however, the role of lactogenic proteohormones has come to light since they not only enhance the production of intermediary enzymes for lactoprotein synthesis but also stimulate mammary cell proliferation (Rivera et al. 1972; Vician et al. 1979). In humans, the contribution of isolated steroidal or peptidic hormones to mammary cell growth is uncertain. The fact that pregnant women exhibit increasing plasma concentrations of estrogens, placental lactogen (PL) and prolactin (PRL) renders interpretation of animal data difficult since elevations of estrogen and PRL during gestation are missing in most species whereas high concentrations of PL can be measured (Kelly et al. 1976).

Studies in the rat (Yoshinaga et al. 1969) have revealed low estrogen production during pregnancy. Only shortly before parturition is there a rise in circulating estrogen coincidental with an elevation in plasma PRL (Amenomori and Meites 1970). In the rhesus monkey, estradiol plasma levels do not increase during pregnancy: similar to findings in the rat, a moderate rise in plasma PRL is noticed before parturition (Seron-Ferre et al. 1979) concomitant with an increase in estrogen concentration (Resko et al. 1975). Also in the ewe, goat, and in bovine blood serum PRL is high only at the end of gestation, in good correlation with the elevation of circulating estrogens which occurs near term (Kann and Denamur 1974; Hart 1974; Karg and Schams 1974). Thus, it is possible that in these animals the placental lactogen replaces the complex estrogen-PRL-PL present in human pregnant plasma. In contrast, selective elevation of PRL secretion by dopamine blockade in nonpregnant female rats leads to profuse mammary acini proliferation in the presence of low estrogen and lack of PL (del Pozo et al., unpublished data).

More recently, Raymond et al. (1978) have shown that estrogens can reverse the inhibitory effect of DA agonists on PRL release by the cultured pituitary, thus establishing a direct link between sex steroids and lactotrope cell function. The estrogen dependence of pituitary lactotropes during gestation has been documented in humans. It is well known that the pituitary gland increases in size throughout pregnancy to almost double its weight at term. At the beginning of this century, Erdheim and Stumme (1909) demonstrated with tissue staining techniques that this increase in weight is almost exclusively due to lactotrope cell hyperplasia. These findings have been confirmed later (Goluboff and Ezrin 1969; Pasteels et al. 1972) and are supposed to be mediated by the stimulatory effect of placental estrogens, which reach extremely high concentrations in plasma of pregnant women (Tulchinsky et al. 1972; Abraham et al. 1972). In agreement with these findings, a number of authors have reported increasing PRL values during human gestation (Hwang et al. 1971; Jacobs et al. 1972; L'Hermite et al. 1972; Tyson et al. 1972; Jaffe et al. 1973; Schenker et al. 1975; Biswas and Rodeck 1976; del Pozo et al. 1976a, 1977; Kletzky et al. 1980). A high correlation has been found between basal estradiol and PRL concentrations in human pregnant serum (Barberia et al. 1975; del Pozo et al. 1977; Kletzky et al. 1980). Figure 3.1 depicts this parallelism in a group of gestational women from conception to term. The correlation is maintained during the first and second trimester but there is a slight divergence towards term through levelling of the PRL curve, suggesting a reduction in lactotrope cell sensitivity. Indeed, Kletzky et al. (1980) have reported a steady state situation for PRL reserve in the third trimester of pregnancy. Among primates, only the chimpanzee exhibits gestational estrogen and PRL plasma elevations similar to humans (Reyes et al. 1975).

As will be illustrated in the next section, the human fetal pituitary is also exposed to the stimulatory effect of placental estrogens. It seems that the enhancing effect of

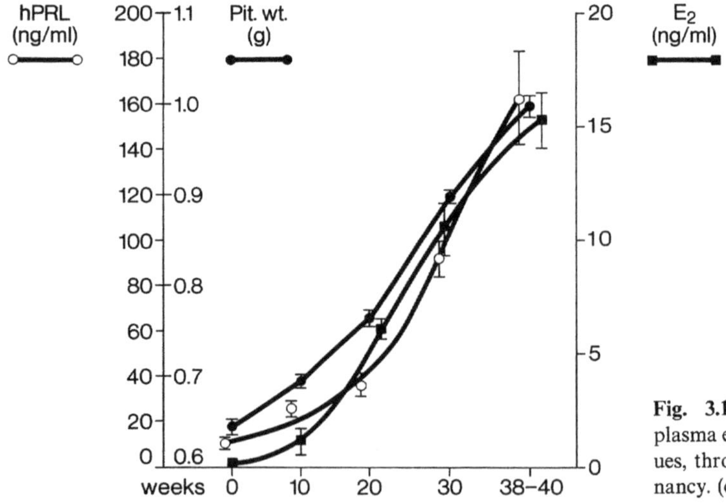

Fig. 3.1. Pituitary weights, plasma estradiol and PRL values, throughout normal pregnancy. (del Pozo et al. 1976a)

estrogen can take place without hypothalamic mediation since plasma PRL in anencephalic fetuses at term is found within the normal endgestational range (Aubert et al. 1975; Forsbach et al. 1976).

The role of progesterone in the development of the mammary gland during pregnancy is well established but an action at the level of the pituitary lactotropes has not been clearly defined. Bohnet et al. (1978) and Rakoff and Yen (1978) have demonstrated a weak but significant stimulatory effect of progestogenic compounds on PRL release in nonpregnant women but no reports are available as to the possible effect of gestational progesterone.

3.2 Maternofetal Correlations; Concentration of Lactogens in a Three-Compartment System Throughout Pregnancy

Using material free of contamination, and being able to measure accurately extremely low hormone concentrations, the maternofetal relationship can be contemplated from the point of view of lactogenic hormones as a model constituted by three compartments, mother, fetus and amniotic fluid, each with marked secretory activities and with practically no intercompartmental exchange.

Figure 3.2 presents parallel measurements of estradiol (E_2 17β), PRL, GH, and placental lactogen (PL) in a group of ten pregnant women and in dead fetuses of comparable gestational age. It can be clearly seen, in agreement with previous reports, that $E_2 17\beta$ and PRL increase gradually in maternal and fetal blood (Aubert et al. 1975; del Pozo et al. 1976a). Amniotic fluid PRL also increases to reach extremely high concentrations at term (Friesen et al. 1972; Tyson et al. 1972; Badawi et al. 1973; Schenker et al. 1975; Chochinov et al. 1976; Soria et al. 1977; Clements et al. 1977). It is assumed that the fetal pituitary lactotropes grow under the same steroid stimulation as in the mother. Lactogenic cells have been identified in the fetal pituitary by Pasteels (1963). Although the situation of real hyperplasia is difficult to recognize in a growing gland, Baker and Yu (1977) have localized lactotrope cells by immunocytochemical techniques at 14 weeks of gestational age, and Siler-Khodr et al. (1974) as early as the 5th week. These authors noticed a direct correlation between the secretory capacity of these cells and the weight of the anterior pituitary gland. It is interesting to note that basal maternal PRL shows a more pronounced increase in twin pregnancies than in single ones (Bellmann 1977). PRL values in maternal and cord blood are not very different at term (Tyson et al. 1972; Schenker et al. 1975; Fang and Kim 1975; Clements et al. 1977; Soria et al. 1977) although there have been discrepancies among some reports.

Some controversy existed about a possible transfer of PRL from the maternal to the amniotic compartment based on experiments conducted in monkeys (Josimovich 1974). This author was able to recover 1% of radioactive PRL injected into the amniotic space of the mother animal whereas no evidence of fetal transfer to the maternal circulation was disclosed. The authors concluded that the main source of amniotic PRL was located in the maternal compartment. Fang and Kim (1975), and Clements et al. (1977), however, suggested a fetal origin, whereas Soria et al. (1977) found, in humans, no correlation between the PRL concentrations in the three compartments and proposed independent sources for each of them. Indeed, intercompartmental transfer seems unlikely in the light of recent work by del Pozo

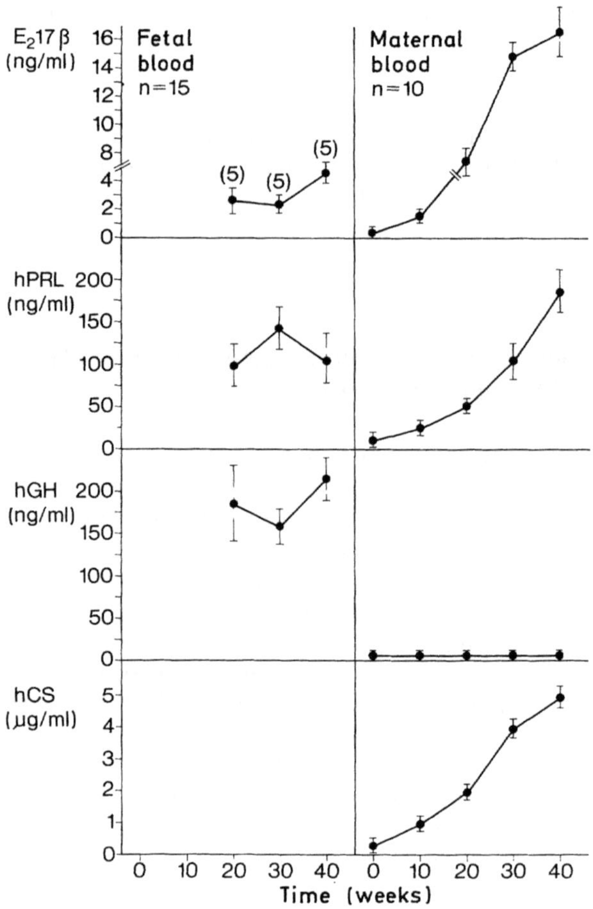

Fig. 3.2. Parallel measurements of $E_2 17\beta$, PRL, GH and PL (here as HCS) in ten pregnant women and in dead fetuses of comparable gestational age. (del Pozo et al. 1977)

et al. (1976b) and Bigazzi et al. (1979a) which showed that PRL inhibition with bromocriptine administered to the mother is accompanied by substantial reductions in fetal and maternal plasma PRL but not change in amniotic fluid levels (Table 3.1) despite measurable concentrations of the drug in the three compartments. It was assumed that the amniotic fluid space was supplied with PRL originating in the placental structures. There is, however, experimental evidence to support a lack of membrane transfer for PRL by using an in situ perfusion system in sheep (Alexander et al. 1973). Elevation of maternal PRL with reserpine did not result in changes in perfusate lactogen concentrations. In humans, Weiss et al. (1977) have also provided evidence for an independent origin of placental PRL by studying serum and amniotic fluid obtained from midtrimester abortions. Recently, amnion and decidua have been found to secrete PRL under experimental conditions (Healy et al. 1977; Riddick and Kusmik 1977; Frame et al. 1979; Golander et al. 1979), and Bigazzi et al. (1979b) have clearly shown that the decidual source of PRL is not controlled by dopaminergic mechanisms as reflected in the inability of bromocriptine, a dopamine agonist, to reduce the synthesis of this lactogen by cultured decidual tissue. This would explain the lack of effect of this compound in modifying

Table 3.1. Maternal (M), newborn (N) and amniotic fluid (AF) PRL concentrations (ng/ml) at delivery after bromocriptine therapy throughout pregnancy (mean $\pm$ SEM)

	n	M	N	AF
Controls	6	153 ± 13	136 ± 16	710 ± 82
Bromocriptine	8	19 ± 6	23 ± 7	670 ± 68

the PRL content of amniotic fluid in human pregnancy (del Pozo et al. 1976b; Bigazzi et al. 1979).

Being the first growth-promoting proteohormone to be identified, growth hormone (GH) has also been studied in pregnancy and in fetal life. GH cells have been localized in the fetal pituitary between the 7th and 9th gestational week using different techniques (Matsuzaki et al. 1971; Kaplan et al. 1972; Siler-Khodr et al. 1974), and cultured pituitary cells actively secrete GH from the 5th week onwards (Pasteels 1963; Gitlin and Biasucci 1969). As pregnancy progresses fetal blood GH increases to reach maximal values during midgestation declining slowly toward term (Kaplan et al. 1972). Onset of suprasellar control of GH secretion can be assumed to occur toward the end of the first trimester as a consequence of maturation of central neuroendocrine function (Kaplan et al. 1976). This control, however, seems to be rather stimulatory since circulating GH in anencephalic newborns has been found to be low in contrast to normal children (Kaplan et al. 1976).

Tyson et al. (1972) and Geiger (1974) have reported GH values in gestational serum within the range found in normal, nonpregnant subjects. This is in contrast with the high average values in fetal blood between midpregnancy and term, and it suggests lack of placental transfer of GH, supporting previous data by Laron et al. (1965, 1966), Gitlin and Biasucci (1969), and King et al. (1971) using labeled GH. Also differences in GH contents between the umbilical artery and the vein, in favor of the former, support an increased secretory activity of the fetal pituitary (Tyson and Friesen 1972). Growth hormone concentrations in amniotic fluid are slightly above levels measured in pregnant plasma but the origin of the extra growth hormone is uncertain.

Placental lactogen (PL) is present in very high concentrations in maternal blood. As with the other two proteohormones in the same molecular weight range, the transfer of PL to the fetal side is extremely low: concentrations in this compartment are 1/20 to 1/300 of that in maternal blood (Kaplan et al. 1976). In most instances fetal plasma exhibits PL levels below the sensitivity of radioimmunoassay procedures in clinical use. PL in amniotic fluid is less than that in the maternal circulation but higher than that in fetal plasma throughout gestation. It is completely cleared from maternal blood within a few hours following delivery (Spellacy et al. 1966; Geiger 1974).

Results of PRL, GH and PL measurements at term in maternal and fetal plasma and in amniotic fluid of ten normal control pregnancies are presented in Table 3.2. There is a fivefold elevation in amniotic fluid PRL in comparison with the other compartments. Here there is a decline in PRL concentrations toward the end of pregnancy probably due to a dilutional effect (Friesen et al. 1972). GH remains

Table 3.2. Basal PRL, GH, and PL (ng/ml) recorded in a group of ten normal pregnancies at delivery (mean ± SEM)

	PRL	GH	PL
Maternal	126 ± 8	3.1 ± 0.7	5.2 ± 0.4
Fetal	122 ± 9	47.6 ± 9	< 0.02
Amniotic fluid	630 ± 86	5.5 ± 2	–

suppressed in the maternal organism, presumably through the action of PRL and PL. Indeed, there is poor or absent GH reaction to stimulating agents such as insulin or arginine during gestation (Tyson et al. 1972). In contrast, GH is elevated in fetal plasma probably due to active hypersecretion and not to a delay in metabolic clearance (Spellacy et al. 1973). There is some uncertainity as to the biologic activity of the hormone at the different stages of pregnancy.

As discussed previously, PRL remains elevated in both maternal and fetal circulations under the action of estrogens. A multiphasic secretory pattern has been described for the maternal compartment around the time of parturition, reaching a nadir approximately 2 h before initiation of labor (Rigg and Yen 1977; Mancuso et al. 1979).

3.3 Biologic Effects of Lactogenic Hormones During Pregnancy

Available evidence suggests that neither maternal nor fetal GH is essential for normal fetal any growth since the birth length of the apituitary, anencephalic newborn, is usually within the normal range, nor do children born to hypophysectomized women exhibit growth deficiencies (Kaplan et al. 1976). Despite its potential lactogenic power, there is no evidence either that GH may play a primary role in physiologic lactation. Thus, normal lactogenesis has been observed in the absence of circulating GH (Rimoin et al. 1968).

Although both PRL and PL can initiate lactoprotein synthesis in the mammary alveolar epithelium that eventually results in full lactation there are considerable species differences. Thus, whereas PL can induce clear signs of endocellular activity at mammary tissue level in the rabbit (Tobon et al. 1972) it remains to be proven that this lactogen is equally active in man after large doses of PL enough to induce blood levels comparable to midgestation were unable to provoke lactogenesis in human beings (Stock et al. 1971). In addition, extremely low PL values or the absence of PL from pregnant plasma seems to be compatible with uneventful pregnancy and lactation (Gaede et al. 1978; Nielsen et al. 1979). Since mammals other than some primates fail to exhibit plasma PRL elevations during pregnancy, placental lactogen being the only potent galactogenic agent found, the question has been raised as to its importance in preparing the mammary gland for lactation in other species (Josimovich 1974). Nevertheless, the fact that a hormone of placental origin circulates almost exclusively in maternal blood indicates a main regulatory action in this compartment. Thus, PL may facilitate the utilization of carbohydrates in the fetus by enhancing fat breakdown in the maternal compartment (Kaplan and

Grumbach 1974), although this mechanism is not free of criticism (Josimovich and Archer 1977).

Prolactin not only prepares breast tissue for lactation but also induces milk secretion. Full lactation does not usually occur in human pregnancy before term due to the overriding inhibition by estrogens and progesterone at the level of the acini and ductal epithelium (Yokoyama et al. 1969; Bruce and Ramirez 1970). It seems that the primary action of estrogens is to stimulate peripheral PRL binding (Kelly et al. 1975, 1979), an effect that can be counteracted by simultaneous treatment with antiestrogens. On the other hand, Kledzik et al. (1976) have reported a 30% fall in peripheral PRL receptors in rats after administration of high doses of estrogens. Progesterone seems to oppose primarily the synthesis of PRL receptors since it can block the stimulatory effect of PRL on binding sites (Djiane and Durand 1977), suggesting that the absence of milk secretion during pregnancy is the result of the interplay between sex steroids and PRL at mammary receptor level. That PRL has a primary role in the initiation of human lactation seems evident after milk letdown has been prevented by selectively inhibiting its secretion in postpartum women (Brun del Re et al. 1973). Moreover, lactation failed to occur in a women undergoing intrapartum pituitary necrosis and exhibiting low basal plasma PRL (Tyson et al. 1975).

There is experimental and clinical evidence (see Chap. 1) that PRL is required for normal function of the corpus luteum and the indispensability of this lactogen for maintenance of pregnancy as suggested by older work (Pencharz and Long 1933) has been recently confirmed in rats (Morishige and Rotchild 1974). Moreover, results of clinical work in non pregant women indicate that hypoprolactinemia may cause luteal inadequacy (Schulz et al. 1978), but the assessment of such a state after human conception is difficult. The fact that ovulation and pregnancy can occur in the hypophysectomized woman treated with exogenous gonadotropins indicates that at least pituitary PRL is not essential to maintain gestation. However, no reports are available on cases of gonadotropin-induced ovulation and subsequent pregnancy in the complete absence of circulating PRL. Data collected during pregnancy in females with an intact pituitary undergoing continuous PRL inhibition for different reasons have shown that a reduction in this lactogen up to 10% of the expected plasma concentrations had no influence on the progress of pregnancy (del Pozo et al. 1976a; Bigazzi et al. 1979a; Ylikorkala et al. 1979, 1980a). In a retrospective comprising 448 pregnancies in women conceiving under bromocriptine, Griffith et al. (1978) reported 10.9% spontaneous abortions, well within the expected incidence in the nontreated pregnant population. This is in agreement with recent data reported by Ranta et al. (1980) and Ylikorkala et al. (1980b) providing further evidence for a lack of action of endogenous PRL in the maintenance of the corpus luteum graviditatis.

There are experimental data to suggest that PRL may be a factor in fetal lung maturation. Ovine PRL injected into fetal rabbits increased the lecithin content of their lungs (Hamosh and Hamosh, 1977). Hauth et al. (1978) found that when the PRL level in cord plasma was less than 200 ng/ml, the incidence of development of neonatal respiratory distress was high. Of infants born at 25–29 gestation weeks, all with cord PRL levels of less than 200 ng/ml, 89% developed respiratory distress syndrome. From 29.5–33 gestational weeks 80% of infants with PRL values below that level also exhibited signs of respiratory distress. It appeared that fetal lung maturation and augmented surfactant formation are physiologic functions of fetal PRL. Gluckman et al. (1978) measured PRL in the umbilical cord of 77 newborns. They found lower values in infants developing respiratory distress syndrome at birth. It is interesting to note that in another group in which mothers were treated with betamethasone, the incidence of respiratory distress was lower but plasma PRL was not different from the first group. More recently, del Pozo et al. (unpulished) have failed to record any signs of respiratory ailment in seven children born to women treated with bromocriptine throughout pregnancy despite cord blood PRL concentrations down to 10% of average values recorded in non treated pregnancies. Therefore, it is difficult to make a definitive statement on the role of this lactogen in fetal lung maturation. An important finding is the inverse relationship between PRL concentration in amniotic fluid and the actual fluid volume found in the experimental animal (Josimovich et al. 1977), suggesting a regulatory role of this lactogen in water transport to the maternal compartment. Indeed, recent work by Bigazzi et al. (personal communication) has clearly shown that changes in amniotic fluid osmolality induced by injection of mannitol into this space were followed by parallel changes in PRL concentrations. In this context it is interesting to mention that Spellacy et al. (1971) have found an inverse relationship between blood concentrations of glucose and placental lactogen.

Links have been established between hypertensive disease of pregnancy and PRL concentrations in peripheral blood and in amniotic fluid (Josimovich and Archer 1977). Along these lines, Redman et al. (1975) were unable to find a clear relationship between plasma PRL and arterial pressure in hypertensive pregnant women but there was a positive correlation between circulating PRL and both proteinuria and severe edema. However, further studies concerning basal plasma PRL levels in hyperemesis gravidarum, preeclamptic toxemia and hypertension have failed to demonstrate a PRL dependency (Ylikorkala et al. 1976; Sadovsky et al. 1977; Kivinen et al. 1979).

Based on similarity of the PRL secretory pattern during gestation and increasing fetal adrenal weight, a regulatory function of PRL on the growth of the adrenal cortex has been proposed (Winters et al. 1975). Indeed, Bassi et al. (1977) and Giusti et al. (1978) (see Chap. 4) have reported an increase in basal dehydroepiandrosterone sulfate (DEAS) in hyperprolactinemic, nonpregnant women, lending further support to this stimulatory mechanism, since this steroid represents the main substrate for estrogen synthesis by the placenta. However, recent work by del Pozo et al. (1978, 1980a) has shown that substantial reductions in maternal and fetal PRL throughout pregnancy have failed to modify DEAS concentrations in either organism. Furthermore, adrenal tissue examination of one dead fetus revealed normal development of the androgenic zone despite a marked reduction in plasma PRL caused by bromocriptine administration to the mother (del Pozo et al. 1980a). Hence, the control of fetal production of adrenal androgenic substrates for estrogen synthesis is not specifically dependent upon PRL.

3.4 Nyctohemeral PRL Rhythm During Pregnancy and Effect of TRH and Dopaminergic Blockade

As in the normal nonpregnant individual, women exhibit during gestation a biphasic PRL rhythm during the night with nocturnal peak values recorded about 3–4 h after initiation of sleep (Boyar et al. 1975). Figure 3.3 shows the nocturnal PRL profiles in three such women in the gestational week 30, 32, and 38 respectively. There is a clear sleep-dependent elevation, proportionally similar to the profile recorded in nonpregnant women. Although Robyn et al. (1977) have reported a decrease in amplitude of the sleep-related PRL peak in the last trimester of pregnancy, as will be shown in the next chapter, the nocturnal PRL secretion is disturbed in situations of pathologic hyperprolactinemia.

In order to assess the influence of changes in endocrine milieu on the control of PRL release during pregnancy, the response of the pituitary lactotropes to a TRH challenge has been investigated (Bohnet et al. 1975; Bellmann 1977; Ylikorkala et al. 1979a). In all three studies a positive PRL response was found although the magnitude of the peak was proportionally smaller when compared with the PRL elevation elicited in nonpregnant women. Kletzky et al. (1980), by calculating the areas under the curve for plasma PRL increases following TRH found amplified responses during the first and second trimester. The slightly reduced peak values observed by Bellmann (1977) and Ylikorkala et al. (1979a) can be interpreted as the result of a mild tonic inhibiting effect of estrogen on the PIF system as suggested by Meites (1973). Indeed, the PRL increase induced by TRH became correspondingly greater in the postpartum as the basal levels fell (Bohnet et al. 1975) at a time when placental estrogens had been completely cleared from the blood. The plasma PRL increases recorded after dopamine receptor blockade with sulpiride (Freeman et al. 1976; Guitelman et al. 1978; Massara et al. 1980) reflect the integrity of the mechanisms regulating the function of the pregnant pituitary mediated by that neurotransmitter.

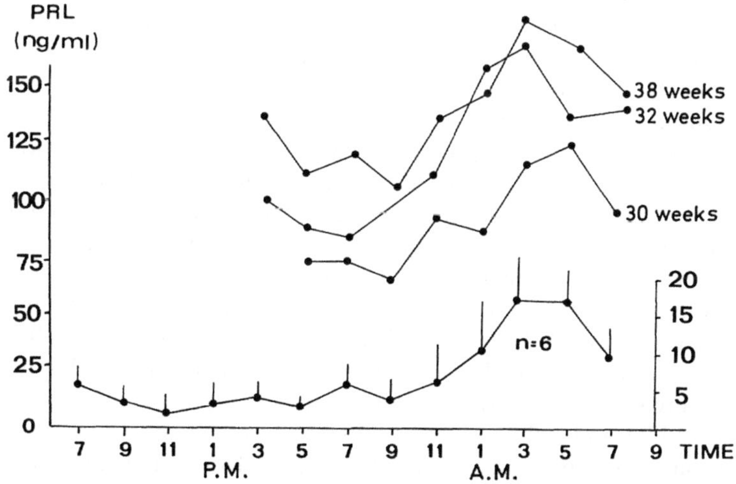

Fig. 3.3. Sleep-related PRL elevations are preserved in physiologic hyperprolactinemia. PRL profiles from six normal nonpregnant females are represented in the *lower part* of the figure

3.5 Mechanisms of Milk Secretion (Lactogenesis); Maintenance of Lactation (Lactopoiesis)

In some species, milk secretion can be detected in the mammary alveoli during pregnancy at a time when plasma prolactin has not reached lactational levels. It can be assumed that, during this period, the lactogenic effect is being induced by placental hormones and not originated in the pituitary of the mother animal. Indeed, mammary gland receptors bind both prolactin and placental lactogen (Shiu et al. 1973) lending further support to the postulated mechanism. Somewhat different are data collected in ruminants: secretory activity is quite marked during the second half of pregnancy, and suckling of cows (before calving) has been recommended as a routine procedure in dairy husbandry. In man, there is distension of the alveoli by colostrum towards the third trimester of pregnancy but no actual milk secretion occurs. Similar data have been collected in the rhesus monkey. However, it should be mentioned, that in the human, plasma prolactin concentrations are already in the lactational range before midpregnancy (Hwang et al. 1971; L'Hermite et al. 1972; Tyson et al. 1972). This means that, at this period of gestation, plasma PRL concentrations able to initiate lactogenesis can be measured. Indeed, interruption of pregnancy after the 20th week is regularly followed by milk letdown of variable intensity (Lehmann et al. 1979), whereas lactation is much less common when termination occurs before the 14th week (Pusterla and del Pozo, unpublished data). The mechanism of the secretory block has not been completely elucidated although the inhibitory action of placental steroids, presumably by interfering with the synthesis of peripheral lactogenic receptors has been documented. Thus, Kledzik et al. (1976) and Djiane and Durand (1977) were able to demonstrate experimentally that exposure of mammary tissue to elevated steroid concentrations would reduce PRL receptors. In humans, copious milk secretion does not appear until 2–5 days after child birth, this being about the time required by the body to completely eliminate the residual sex steroids still present in blood and storage tissues after expulsion of the placenta (del Pozo et al. 1976a; Hiba et al. 1977). Considerable differences as to the onset of postpartum lactation are found in other mammals, presumably due to differences in the interaction between sex steroids and prolactin. The sequence of hormonal changes leading to human lactation is depicted in Fig. 3.4. The disappearance of sex steroids between the 2nd and 3rd postpartum days coincides with the onset of milk secretion. At this moment average plasma PRL concentrations are elevated, although values are already below predelivery levels and a tendency to fall toward normal nonpregnant values can be recognized. Thus, prolactin in the absence of circulating sex steroids would freely act on the mammary epithelium to induce milk secretion. Subsequently, prolactin values gradually decline to reach nonpregnant levels during the 2nd or 3rd postpartum week. Despite this fall in lactogen concentration milk volumes continue to increase and stabilize in the 2nd week of lactation. This would impugn the requirement of prolactin for maintenance of milk secretion. However, Brun del Re et al. (1973) and del Pozo et al. (1975b) have demonstrated that, in the human, prolactin plasma values in the normal nonlactational range are required for the maintenance of lactation. This was recorded by inhibiting PRL secretion with bromocriptine at different times in the postpartum, and indicates that little PRL is

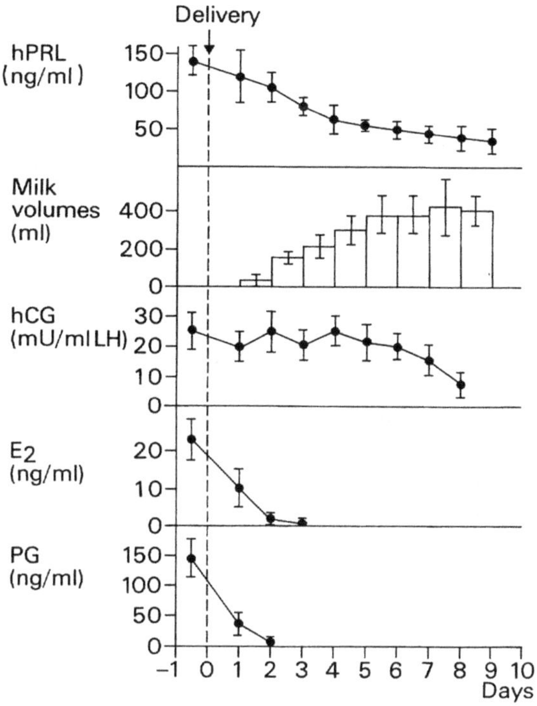

Fig. 3.4. Sequence of hormonal changes inducing postpartum lactation in the human. (del Pozo et al. 1976a)

required for maintenance of peripheral receptors once they have been induced, as supported by experimental data by Bohnet and Friesen (1976) and Aragona et al. (1977).

3.5.1 The Suckling Reflex: Effect on Prolactin and Milk Production

Much attention has been directed toward the effect of suckling by the infant in the maintenanance of milk secretion after it has been shown that each breast-feeding episode increases plasma PRL concentration in the mother (Hwang et al. 1972; Frantz et al. 1972; L'Hermite et al. 1972).In the past it was widely accepted that oxytocin released by the suckling stimulus would provoke a burst of PRL secreted by the anterior pituitary. Recently, Weitzman et al. (1980) showed that suction of the nipple is followed by an oxytocin peak at approximately 9 min whereas the PRL elevation appears after 20 min or longer, and that there is no causal relationship between these two events. In addition it has been shown that the administration of oxytocin to normal or postpartum women does not modify basal plasma PRL concentrations (del Pozo et al. 1980b).

Tyson et al. (1972; 1975) demonstrated that stimulation of basal PRL with TRH in the early postpartum would enhance lactational performance. These authors (Tyson et al. 1972a) found that the prolactin-stimulating action of suckling decreases in long-term lactation, so that a blunted response is elicited beyond 80 days. Despite this insensitivity, lactation can be maintained for weeks or months making doubtful the necessity of prolactin surges for lactopoiesis. Aono et al. (1977), however, found that milk yields correlated well with postsuckling PRL increases in the early postpartum. Furthermore, they showed that administration of the PRL-stimulating drug sulpiride could augment puerperal lactation (Aono et al. 1979). The latter could be only partially confirmed by Guzman et al. (1979) by using another dopamine antagonist, metoclopramide. This drug can induce sustained PRL elevations when administered chronically, and these investigators reported that postpartum women exhibiting rapidly decreasing plasma PRL also presented a decline in milk production, and that metoclopramide treatment would restore lactational performance. It can be assumed that a fall in circulating PRL would understimulate the synthesis of peripheral receptors and that this inadequate mechanism was corrected by enhancing PRL secretion (defective up-regulation).

More recent work by Conti et al. (1979) and Howie et al. (1980) rather supports a dependence of milk production on the effectiveness of the neural pathway mediating the suckling effect. Thus, treatment with a weak dopamine agonist led to almost complete suppression of the postsuckling PRL peak in some women but no change in milk volumes was recorded (Fig. 3.5), nor was the quality of milk modified by treatment. Since basal PRL had remained in the lactational range it was assumed that basal PRL was primarily responsible for maintance of lactation. Howie et al. (1980) arrived to the same conclusion using a different methodological approach: postsuctional PRL peak was absent in puerperae using a vacuum pump for milk

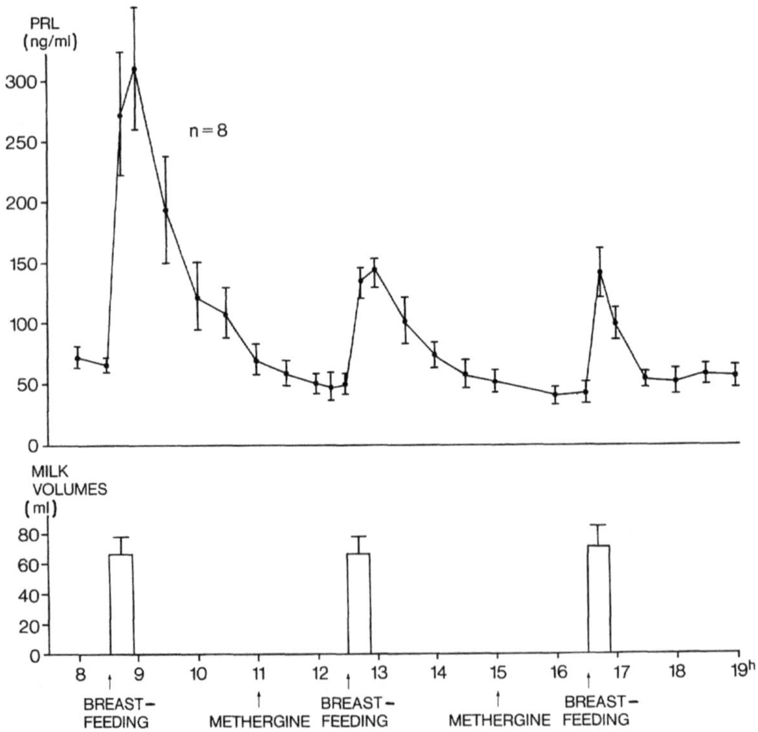

Fig. 3.5. Lack of effect of reduction of the postsuckling PRL peak by methergine on milk volumes during established lactation. (Conti et al. 1979)

extraction avoiding nipple stimulation. Milk amounts produced were not different from that recorded in women breast-feeding normally and exhibiting significant PRL responses to suckling. Delvoye et al. (1978a) found that full lactating African women failed to exhibit any suckling-induced PRL peak when investigated during the 1st postpartum year, also suggesting that elevations above basal secretion rates are not crucial for the maintenanance of milk secretion. Tyson et al. (1977) have conducted a series of experiments in normally lactating women. The application of an anesthetic ointment to the right nipple blunted the PRL response to suckling but a significant increment in plasma PRL was noted when the infant sucked the unanesthezized nipple. Also complete severance of the nipple nerve supply abolished nursing-induced PRL release despite normal lactational performance in a woman who had undergone mammoplasty and reimplantation of both nipples prior to pregnancy. It can be concluded that, although the PRL surge after breast-feeding is in causal relationship with nipple stimulation, the lactogen excess has little or no bearing on the maintance of milk volumes in established lactation. This may explain the failure of hyperprolactinemia induced by exogenous TRH to increase milk production in full lactating women (Zarate et al. 1976), whereas enhancement of PRL secretion may prove useful in lactational deficiency although the exact role of this lactogen needs clarification (Tyson et al. 1976).

On the basis of work with animal mammary tissue it was shown that exposure to elevated PRL concentrations led to a progressive reduction in total receptor population (Djiane et al. 1979; Kelly et al. 1980). Thus, a down-regulation mechanism may play a crucial role in the control of PRL-binding sites and subsequently in the production of milk. Receptor exposure to elevated PRL concentrations would be without clinical effect in cases in which lactogen production was adequate. It should therefore be interesting to investigate the biologic properties of the circulating PRL in situations of hypogalactia in the presence of plasma PRL in the normal lactational range (45–250 ng/ml according to the author).

These findings suggest that prolactin is not only a regulatory hormone but an indispensable factor in the induction and maintenance of milk secretion. The question regarding the circulating concentration of PRL required to start lactogenesis is still unanswered. In studies conducted in the author's laboratory using a long-acting bromocriptine preparation to suppress PRL release it was shown that a constant fall below 25 ng/ml would be required to inhibit permanently lactogenesis. Figure 3.6 summarizes the result of these studies: a moderate PRL fall, sometimes but not continuously below the 20 ng line, is followed by milk secretion whereas prolonged PRL inhibition maintains agalactia. These findings would support the previous statement that a fall in circulating PRL below a certain limit would fail to maintain up-regulation of peripheral receptors and impair milk production. It should be borne in mind, however, that plasma PRL determinations do not provide any information as to the biologic properties of the circulating hormone nor is the response of tissue receptors characterized in some manner.

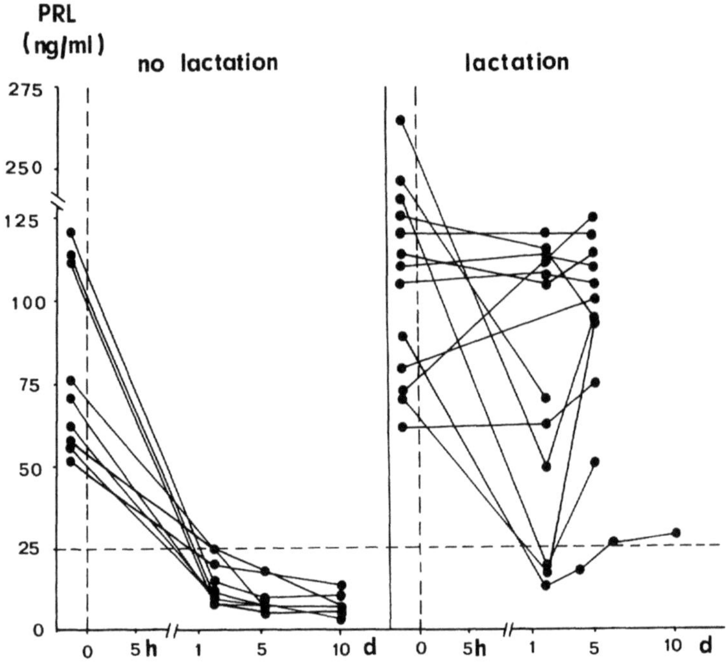

Fig. 3.6. Occurrence of milk letdown in women undergoing inadequate suppression of PRL secretion in comparison with puerperae exhibiting plasma PRL values consistently lower than 20 ng/ml.

3.5.2 Role of Neurotransmitters in the Initiation and Maintenance of Lactation

As shown along this synopsis, there is general consensus about the crucial role of dopamine in the control of PRL release and maintenance of adequate lactotrope cell function, probably through interplay with steroid receptors at the level of the pituitary gland and suprasellar structures. It is well known that prolonged exposure of experimental animals to dopamine-blocking agents elevates pituitary and circulating PRL and that this effect is accompanied by mammary tissue proliferation (Sulman 1970). Prolonged administration of such drugs to nonpregnant women has been followed by galactorrhea in some instances (Apostolakis et al. 1972), and the paramount role of PRL in this particular situation has been documented (Beumont et al. 1974). Also the suppressing effect that dopaminergic agonists exert on PRL secretion, especially compounds of the ergot series, speak for a principal role of this neurotransmitter in the control of release of this lactogen. Since this latter action will be the subject of detailed analysis elsewhere in this review, it will not be commented on further at this stage.

The part other neurotransmitters may play in lactogenesis is more difficult to understand. There are also discrepancies between results of animal experimentation and clinical observations. Serotonin, for instance, has been shown in the previous section to stimulate PRL secretion, and this effect has also been recorded in rats with the use of quipazine, a serotonin agonist (Clemens et al. 1977). Also suckling-mediated PRL secretion has been found to be serotoninmodulated in these animals (Kordon et al. 1973/74; Rowland et al. 1978).

Knowing that breast stimulation and suckling markedly releases PRL in women (see previous section) studies have been conducted in order to elucidate the role of serotonin in human lactation. Lancranjan et al. (1979) observed the effect of two serotonin antagonists on postpartum PRL secretion in an acute trial. The authors recorded a highly significant decrease in plasma PRL after administration of methysergide, an ergot derivative, whereas pizotifen, a nonergot compound, was without effect. The authors concluded that the inhibitory effect of the former was mediated by a dopaminergic mechanism inherent to its ergot structure and not to the antiserotonin component. Crosignani et al. (1979) made the same observation also by using other ergot and nonergot compounds. Thus, the role of serotonin in the modulation of PRL secretion cannot be clarified until pure antagonists are made available.

3.6 Impact of Lactation on Fertility

3.6.1 The Hypothalamic-Pituitary-Ovarian Axis in the Early Postpartum

Although a process of restoration of endocrine mechanisms leading to reinitiation of ovulatory cycles is started with the expulsion of the placenta, there are enough chronological differences to differentiate the early and the late puerperium. For the purpose of clarity a succint review will be made on that basis.

In the previous sections the endocrine status of puerperal women was described as a situation of PRL dominance with a quiescent pituitary-gonadal axis. With the advent of sensitive radioimmunologic techniques, the pituitary and ovarian responsiveness to circulating and exogenous gonadotropins has been studied in the postpartum. In a trial to induce ovarian response with a preparation of menopausal FSH/LH, urinary estrogens failed to increase (Zarate et al. 1972). Also ovarian refractoriness to circulating gonadotropins has been reported (Tyson et al. 1972; Reyes et al. 1972; Keettel and Bradbury 1961) and the pituitary gland has been found to be resistant to exogenous GnRH in the puerperium (Lemaire et al. 1974; Canales et al. 1974; Nakano et al. 1974; Jeppsson et al. 1974). In the light of new knowledge concerning the kinetics of PRL in the galactorrhea syndromes, PRL-induced puerperal pituitary and ovarian resistance may be assumed (McNeilly 1974; Rolland et al. 1975). In order to clarify this issue, del Pozo et al. (1975a) compared a group of puerperal women with another group of patients with idiopathic hyperprolactinemic anovulation. Results of pituitary and ovarian stimulation under basal conditions and after PRL inhibition (Figs. 3.7, 3.8) indicated that this lactogen is not implicated in the pituitary and ovarian resistance to stimulation found in the early postpartum. In these women, inhibition of PRL secretion down to nonpregnant levels failed to abolish the resistance of the pituitary and ovaries to GnRH challenge or to exogenous gonadotropins, whereas in nonpuerperal hyperprolactinemia both organs were responsive. These findings support the nonexistence of a PRL-mediated block in the early postpartum. It rather seems that the long-term exposure of the hypothalamic-pituitary-ovarian axis to high concentrations of circulating estrogens would temporarily eradicate or

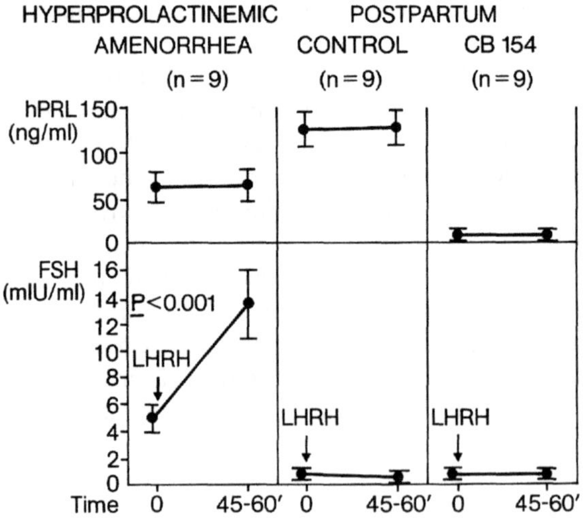

Fig. 3.7. Significant FSH increments ($P<0.001$) after LHRH administration were elicited in the group with hyperprolactinemic amenorrhea, whereas no effect was recorded in the postpartum women. In the latter, prolactin suppression with CB 154 failed to modify the response pattern to LHRH (mean ± SEM). (del Pozo et al. 1975 a)

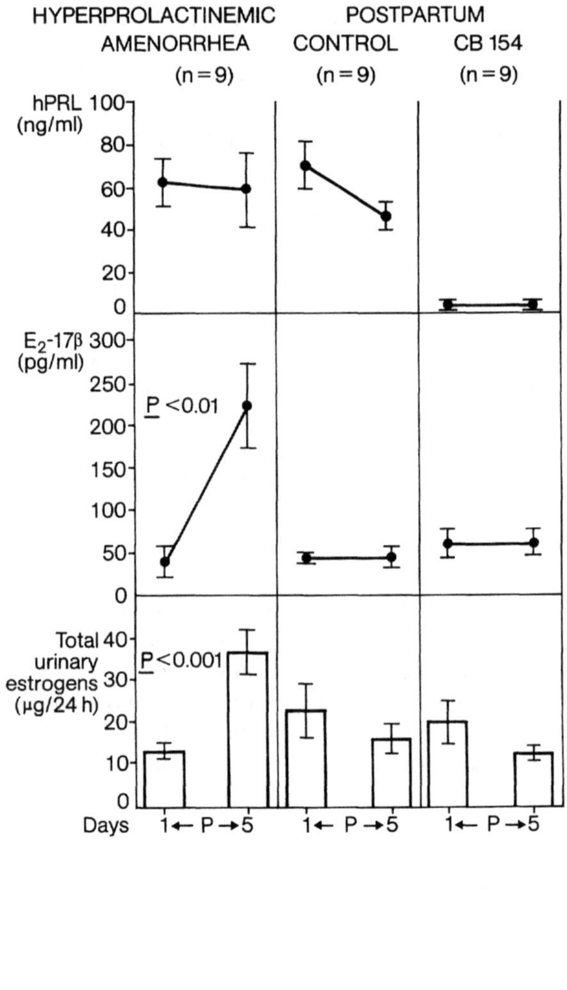

Fig. 3.8. Treatment with Pergonal 500 (P) induced adequate ovarian response in the women with pathologic hyperprolactinemia as manifested by increases in plasma E_2-17β ($P<0.01$) and total urinary estrogens ($P<0.001$). No response was observed in the postpartum women despite a fall in plasma hPRL to the low-normal range (mean ± SEM). (del Pozo et al. 1975 a)

Table 3.3. Onset of menses after delivery (in days) in women on no treatment and in puerperae receiving continuously bromocriptine, a PRL inhibitor. Differences between bromocriptine treatment and the other group are significant at the $P<0.01$ level

	No treatment (physical measures)	Bromocriptine[a] (5 mg daily)	Lactating women[a]
n	20	10	10
Mean $\pm$SE	45.2$\pm$2.4	35.1$\pm$2.0	68.6$\pm$7.0
Range	32–70	27–48	40–96

[a] Data adapted from Rolland et al. (1975)

inactivate GnRH and gonadotropin receptors. Indeed, pituitary and ovarian response to physiologic or pharmacologic stimulation is restored about 2 weeks postdelivery: probably the time required for the de novo synthesis of receptor material. Tolis et al. (1974) have shown by repeated GnRH injections that the LH response to this stimulus has been restored by the 3rd postpartum week in nonnursing mothers whereas lactating women exhibited a blunted LH profile. Nevertheless, PRL may indeed play a role in the pituitary refractoriness of the very early puerperium since restoration of FSH-secreting activity and ovulation can be accelerated by PRL suppression with bromocriptine (LeMaire et al. 1974; Rolland et al. 1975) and comparative studies between nonlactating women on no medicamentous therapy and bromocriptine-treated puerperae have shown a significantly longer latency period before onset of cyclic activity for the former group. These differences are depicted in Table 3.3: time lag until first postpartum bleeding in the women on continuous bromocriptine treatment is shorter than in untreated puerperae. As already commented on, lactating women presented a further delay in resuming menses. It can be speculated that PRL suppression would accelerate the return of cyclic activity by deblocking central or ovarian receptors, presumably gonadotropin-binding sites.

3.6.2 Lactation and Fertility in Late Puerperium

Demographic and social studies clearly show that long-term breast-feeding is associated with prolongation of postpartum amenorrhea and ovarian quiescence. Indeed, long-term lactation is a common contraceptive practice in some cultural groups, especially in developing countries. This effect has also been shown in other mammals. Thus, regular milking may delay the onset of estrus in dairy cattle (Wiltbank and Cook 1958; Morrow et al. 1969) and Meites (1974) demonstrated that the suckling stimulus induces natural puerperal infertility in the rat. In primates, Tutin (1975) has shown that in the chimpanzee lactational amenorrhea can last for several years. In humans, however, civilization, nursing habits, environmental factors, and the availability of substitutes for women's milk have practically abolished the preventive effect of natural lactation on conception in western societies. The relationship between lactation and postpartum fertility has been the subject of intensive research. One way to measure the impact of lactation fertility in noncontracepting populations is to compare pregnancy intervals of women who breast-fed their babies with those of women who did not nurse. If

Table 3.4. Median interval in days of lactational amenorrhea in caucasian women of Boston, USA, in relation to the duration of lactation (Salber et al. 1966)

Duration of lactation (days)	Distribution (%)	Median duration of amenorrhea (days)
0	77.9	55
1– 29	4.6	56
30– 59	3.6	68
60– 89	2.6	85
90–119	2.3	112
120–149	2.8	124
150–179	1.2	155
180–209	1.7	163
210–239	1.5	136
240–269	0.5	184
270–299	0.8	154
300–329	0.4	183

populations practicing sexual abstinence after parturition are discarded, the practice of nursing increases the intervals between pregnancies by about 15%–33% (van Ginneken 1975). Also birth or pregnancy intervals in nursing and noncontracepting women among poorly developed countries were on the average 5 to 10 months longer than those of women who had fetal or infant deaths and did not nurse (Potter et al. 1965; Knodel 1969; Jain 1969). Perez et al. (1972) found that the intensity and length of nursing affected the date in which the first postpartum ovulation occurred. Thus, intervals in days of lactational amenorrhea in relation to the duration of lactation were studied by Salber et al. (1966) in caucasian women of Boston. Although there was a clear tendency toward prolonged postpartum amenorrhea (Table 3.4) in women breast-feeding for longer periods, the anticonceptional effect was relative: menses resumed after 85 days in women lactating up to 3 months and the amenorrhea interval seemed to stabilize at an average of 162 days for lactational periods of 150–329 days. Work by Delvoye et al. (1978) conducted in Zaire has confirmed the higher incidence of pregnancies when lactational periods are shorter, lending further support to the birth control effect of breast-feeding. It was noticed, however, that industrialization around larger cities in such countries has led to differences in postpartum fertility patterns between urban and rural areas. It is obvious that modernization of the countryside in the old continent has balanced the situation regarding the impact of pregnancy on lactational infertility. Thus, when nomadic Kalahari hunters settled in agricultural communities and previous nonsupplemented breast-feeding was switched to combined bottle-feeding and grain meal, a reduction in the birth interval of more than 25% was recorded (Kolata 1974). Also considerable differences were observed among African women in rural and urban areas: it was found that in the former, infants were breast-fed on demand, whereas in the latter a more strict timetable was followed (Bonte et al. 1974). In Table 3.5 the duration of postpartum amenorrhea in European women living in industrialized areas as compared with rural populations is presented. A brief lactation due mainly to social and cultural pressures is reflected in extreme shortening of postpartum amenorrhea. The expected increase in pregnancy rate, however, is missing due to contraceptive measures. In contrast,

Table 3.5. Mean duration of lactation and postpartum amenorrhea in different countries. Partially adapted from Van Ginneken (1975) and Ebrahim (1976)

Country	Sample	Lactation (months)	Amenorrhea (months)
Switzerland	50	2.3[a]	2.1
South America (urban)	1100	7.6	3.9
Venezuela			
Colombia			
Africa (rural)	900	18	11
Ruanda			
Asia (rural)			
India	200	19	12
South Korea	4900	24	15
Bangladesh	200	25	16.5

[a] Probably unrealistic because of limitation of pregnancy leave among the working population.

traditionally rural women exhibit more prolonged lactation periods and accordingly longer postpartum amenorrhea.

Among nonnursing women, on the other hand, postpartum amenorrhea is under 8 weeks duration. Salber et al. (1966) found a median duration of 55 days in 1712 women. Tietze (1961) reported this interval to be between 42 and 56 days, and Berman et al. (1972) estimated a time lag of 52 days. In the author series, menses returned on average after 45 days in a group of 32 nonnursing puerperae treated with physical measures only.

From the preceding section it becomes obvious that differences in nursing habits between western societies and more primitive ethnic groups are just ones of grade.

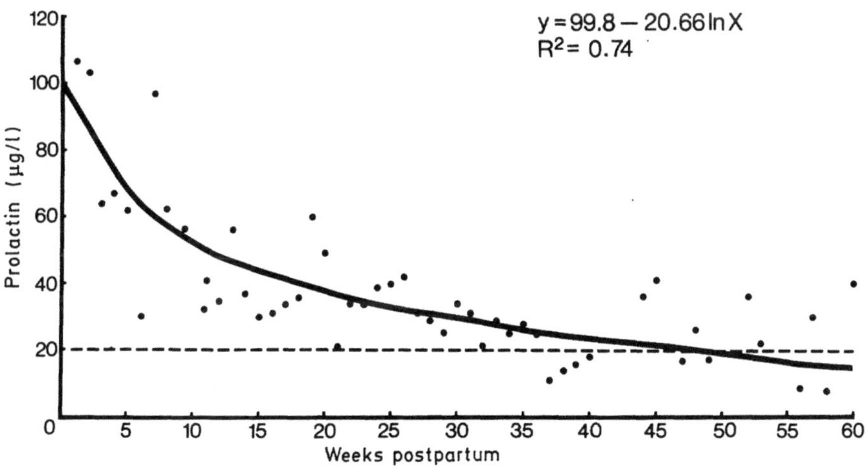

Fig. 3.9. This picture demonstrates the effect of prolonged nursing on plasma PRL in 18 breast-feeding mothers in Canberra, Australia. Mean duration of amenorrhea was 39.2 ± 3.6 weeks (mean ± SEM), and hyperprolactinemia was recorded beyond 40 weeks postpartum. (Courtesy of Drs. B. A. Gross and C. J. Eastman)

In Australia, for instance, a nation enjoying advanced industrial development, a countrywide campaign monitored by professional and nonprofessional groups has been operating now for years in order to promote natural lactation (Fig. 3.9). Gross et al. (1980) collected data on lactational performances, plasma PRL and duration of postpartum amenorrhea in a cross-cultural study involving Australian and Philippino women of urban and rural areas. Basal plasma PRL was elevated in amenorrheic lactating women up to 2 months postpartum, and was significantly higher than in mestruating lactating women. There was a correlation between basal PRL and the number of breast-feeds per day. It was concluded that in two different cultural, socioeconomic and ethnic populations, with similar breast-feeding behaviour, similar patterns of PRL secretion occur. The determinants of PRL secretion and menstrual status for late puerperium will be reviewed in the next section.

3.6.3 Mechanisms of Lactational Infertility: Effect of Suckling

The mechanism involved in lactational infertility has been the subject of numerous studies. The observation that, in the postpartum, women undergo biologic changes reminiscent of the female postpuberal period has long been recognized: after restoration of menstrual flow, subfertility through anovulation or inappropriate luteal function are common findings. Berman et al. (1972) demonstrated that the number of anovulatory cycles after menstruation recurs is greater for nursing than for nonnursing women. Recent work has brought prolactin to light as a paramount factor in lactational amenorrhea. Delvoye et al. (1978a) and Konner and Worthman (1980) found that the maternal gonadal function was apparently suppressed by a PRL-mediated effect of breast stimulation. Also a number of reports are consistent with the occurrence of hyperprolactinemia being more prolonged in lactating women when compared with nonnursing mothers (Rolland et al. 1975; Delvoye et al. 1978b; Bunner et al. 1978; Duchen and McNeilly 1980), and a close parallelism between the degree of hyperprolactinemia and the absence of menses has been found (Delvoye et al. 1978a).

Recent inrestigation has been focused on the endocrine mechanism maintaining lactational amenorrhea. Progress in the physiology of nonpuerperal hyperprolactinemia has helped to understand this mechanism. In this situation, central estrogen receptor blockade or resistance to estrogenic stimulation has been demonstrated by lack of positive LH response to exogenous estrogens. Thus, Baird et al. (1979) have shown that nursing women lack this positive estrogen effect during the first 100 days of lactation. Although a certain follicular development can take place, ovulation and corpus luteum formation are missing.

Concerning the hypothalamic control of LH secretion there is some uncertainty as to the factors linked to lactation involved in maintenance of postpartum amenorrhea. GnRH activity governs the episodic secretion of LH and there is experimental and clinical evidence to show that this pusatile secretion is necessary for the induction of ovulation through endogenous mechanisms. Since this pattern can be absent in pathologic hyperprolactinemia one could assume that suckling-induced PRL bursts or maintenance of supranormal plasma levels prolong lactational amenorrhea by interfering with hypothalamic mechanisms. Indeed, hyperprolactinemia has been shown to suppress the release of LH in lactating monkeys (Maneckjee et al. 1976). Data in humans, however, are far from clear. Tyson et al. (1978) reported a decrease in plasma PRL and an increase in LH after weaning but no correlation between these parameters and suckling had been established. Recently, Aznar et al. (to be published) found no differences in integrated LH profiles between full lactating mothers in the 4th to 5th postdelivery week and women in whom lactation had been prevented by administration of bromocriptine in the immediate postpartum. Analysis of the LH pulsatile pattern during breast-feeding failed to show an inhibitory action by endogenous PRL fluctuations. Also Soria et al. (1976) had arrived to the same conclusion after studying plasma gonadotropin changes during lactational episodes at different times in puerperium.

Maintenance of hyperprolactinemia seems to depend directly on adequacy and frequency of suckling stimulus. Delvoye et al. (1978a) found that PRL is kept elevated by repeating the suckling stimulus at least six times daily and Konner and Worthman (1980) observed persistent inhibition of gonadal function in women nursing their infants several times an hour. In the series reported by Gross et al. (1980) the frequency of suckling was a major determinant in the maintenance of lactational amenorrhea. Data collected by McNeilly et al. (1980) suggest that the PRL concentration is also related to the total time of the suckling stimulus over a 24-h period. Thus, even at a stage of lactation when each suckling episode may not release a PRL peak, the stimulus is adequate to maintain hyperprolactinemia and inhibit central mechanisms required for resumption of normal menstrual cyclicity. Indeed, return of estrous in suckled sows is delayed in comparison with nonmilked animals (Elsaesser and Parvizi 1980). There is also experimental evidence obtained in sheep to confirm this central mechanism of suckling (Kann and Martinet 1975; Kann et al. 1977). Interruption of neural pathways would reduce basal PRL secretion and abolish the suckling-induced PRL surge. In addition, there is central insensitivity to the positive feedback effect of estrogens during physiologic lactation (Wright et al. 1980; Plant et al. 1980) or after rendering sheep hyperprolactinemic with TRH (Kann et al. 1977). The latter effect has also been observed in nonpregnant women on whom plasma PRL was constantly elevated through the administration of dopamine-blocking agents (Robyn et al. 1976, 1977; L'Hermite et al. 1978). However, data recorded in bovine experimentation do not support the above observations: return to estrous in suckled cows is delayed compared with that of cows milked only twice a day despite lack of difference in blood PRL in these two situations, suggesting a dominance of the neural suckling stimulus over the hormonal controls (Peters et al. 1979). Therefore, it cannot be established with certainty whether it is the neural suckling stimulus alone, the elevated circulating PRL, central resistance to regulatory mechanisms, or a combination of these factors which is important for the maintenance of lactational infertility (McNeilly 1980). Finally it should be borne in mind that nutritional and socioeconomic factors may have a bearing on the prolongation of postpartum amenorrhea (Bonte et al. 1974; Whitehead et al. 1978; Delgado et al. 1979).

3.7 Psychosomatic Aspects of Human Lactation

Although prolactin seems to be essential for milk secretion, adequate breast-feeding depends upon a constellation of factors, some of them organic in nature and others of psychogenic origin. Active ejection of milk depends on the interplay of emotions and neurogenic mechanisms connected with suckling and leading not only to a PRL surge but to the secretion of oxytocin. Actually, PRL is not directly involved in the mechanisms governing the excretion of the produced milk. Also the old theory of oxytocin-mediated PRL secretion can be discarded in the light of data showing that the administration of this neuropeptide to women has failed to modify plasma PRL concentrations (del Pozo et al. 1980b) and the lack of a relationship between PRL and oxytocin kinetics after nipple stimulation in nursing mothers (Weitzman et al. 1980).

Great importance has been attributed to the psychogenic components of breast-feeding and to its influence in the induction of milk letdown and maintenance of lactation. A number of variables are related to the mother's feelings toward the infant. Physical contact and the presence of the baby (Hartemann et al. 1962; de Chateau and Wiberg 1977), auditory and visual stimulation, nursing by instinct and not by the clock (Robinson 1951) and frequent nipple stimulation (Egli et al. 1961; Hartemann and Richon 1962) are, besides socioeconomic and cultural aspects, important factors in the maintenance of lactation. Contrary to this, spaced nursing, loss of interest in breast-feeding, and also social embarrassment may contribute to reduce milk production and thus anticipate weaning. Absence of milk ejection due to psychogenic overlay can be overcome with exogenous oxytocin (Weiderman et al. 1963).

A detailed analysis of these interesting aspects of human lactation is beyond the scope of this monograph and the reader is referred to the excellent review by Newton and Newton (1967).

3.8 Postpartum Hypoprolactinemia

A situation of selective PRL deficiency in the postpartum has not been described to date. This condition is commonly associated with deficient secretion of other pituitary hormones due to preexistent disease or to pituitary apoplexy at delivery as reported by Sheehan and Murdoch (1938). Lactotrope cell reactivity has been tested in the latter by Tyson et al. (1975) and Alger et al. (1979), and although detectable basal PRL concentrations were found, these were significantly lower than levels exhibited by normal puerperae. In addition, PRL responses to stimulation with TRH or to dopaminergic blockade were missing. It is interesting to note that PRL was not completely absent from plasma in the cases described and it can be assumed that a cluster of functional cells was able to maintain a certain secretory activity though lacking reserve capacity. The possibility of a hypothalamic localization of the lesion obstructing the blood supply to the pituitary as has been suggested (Sheehan and Standfield 1961) seems improbable since, as pointed out by Alger et al. (1979), liberation of pituitary lactotropes from hypothalamic connections would have enhanced the secretory activity of these cells. Presumably, the low PRL levels recorded in peripheral blood already reflect maximal lactotrope secretory activity.

3.9 Suppression of Puerperal Lactation

New knowledge concerning the hormonal background governing the onset and maintenance of milk secretion in humans has led to considerable improvement in the control of puerperal lactation. A review of the suppression of lactation by physical measures, the use of sex steroids, and by specific PRL inhibitors, will be presented in the following sections.

3.9.1 Suppression of Milk Secretion by Physical Measures

Natural suppression of lactation depends on exclusion of neurohormonal stimulation of breast secretion and a series of physical measures directed to alleviate clinical symptoms caused by milk letdown such as mammary engorgement and mastodynia. Breast binders, cathartics, fluid restriction and local application of plasters have been widely prescribed in the past.

The results of natural lactation inhibition are quite variable. According to Elder and Hakim (1974), breast engorgement is recorded in about 50% of nonlactating postpartum women on no treatment, and 15% experienced marked discomfort. In addition to the above measure, analgesic medication can give some relief of the symptoms. In the last 2 decades, natural suppression of lactation has been replaced by the use of drugs, commonly estrogens, and more recently prolactin antagonists. These compounds act by blocking the effect of this lactogen at mammary tissue level or by inhibiting its release from the pituitary. Its clinical application will be rewiewed in the following sections.

3.9.2 Inhibition of Lactation with Estrogens

It has been generally accepted that the declining levels of estrogens in the early puerperium facilitate the action of PRL on mammary tissue, while high plasma estrogen concentrations prevent milk secretion during pregnancy. It is this property that is used to suppress lactation in the postpartum by artificially prolonging the endocrine situation before delivery. After initial favorable results at the end of the 3rd decade and in the 1940s (Kellar and Sutherland 1939; Barnes 1942; Jeffcoate et al. 1948), estrogenic compounds have been regularly used either alone or in association with diuretics (Hodge 1967) or in estrogen-androgen/progesterone combinations (Morris et al. 1970). More recently there has been a tendency to use long-acting injectable preparations, and semisynthetic or nonsteroidal compounds which have a prolonged action due to storage in body fat tissue or by slow conversion into active substances (Morris et al. 1970; Player 1971; Hepp and Ellringmann 1972).

The combined effect of female and male steroids does not seem to differ from their use separately. Usually, estradiol and testosterone derivatives in retarded form are used. The mode of action of androgens in lactation is not clearly understood. They may potentiate the blocking effect of estrogens on PRL at mammary tissue level.

Since oral contraceptives containing estrogen can be prescribed during the puerperium, the possible effect of such compounds on lactation has been investigated. Although there is a predominant view that estrogens may have adverse effects on milk secretion the intensity of this inhibitory effect may depend on the type and amount of steroid used and the moment of administration in the postpartum (Chopra 1972). It seems that lactation is little or not compromised at all if contraceptive therapy is started after full lactation has been established. A negative effect at an earlier stage in puerperium could be understood as being the restoration of the predelivery conditions, knowing that sex steroids will block the effect of prolactin on mammary tissue. This subject has been reviewed by Thomson et al. (1975). It should be mentioned in this context that progestogens do not impair lactation (Zañartu et al. 1966; Martinez-Manautou et al. 1967; Kamal et al. 1969) and that these steroids may contribute to prolong milk secretion (Zañartu, personal communication). Although mechanical contraception has been claimed to enhance lactation this has not been supported by substantial data (Thomson et al. 1975). Administration of estrogens in the postpartum has been shown to enhance the activity of factor IX, essential for the development of thromboembolism (Daniel et al. 1967; Hakim et al. 1969). In a survey of 9324 deliveries, puerperal thromboembolism was found in 44. There were influences by the age and parity of the women but even when allowance was made for these factors there was still a significantly higher incidence in the patients who had lactation suppressed. It has therefore been recommended that the use of estrogens be avoided in the high risk group consisting of women over 30 with a previous history of thrombosis, varicose veins or other predisposing factors (Elder and Hakim 1974).

3.9.3 Specific Prolactin Suppression with Dopamine Agonists

The recent introduction of an ergot derivative, bromocriptine (Parlodel®) as a specific inhibitor of PRL secretion has contributed to the elucidation of the mechanisms involved in physiologic lactation. The endocrine and pharmacologic profiles of this and others compounds characterized as dopamine agonists have been extensively reviewed in Chap. 2. The first successful suppression of postpartum lactation with bromocriptine was reported by Varga et al. (1972). Shortly thereafter, del Pozo et al. (1972) and Brun del Re et al. (1973) demonstrated that the absence of milk letdown in the treated women was preceded by a dramatic fall in circulating PRL (Fig.3.10). Kinetics of plasma lactogen after administration of a first dose of 2.5 mg revealed an average PRL drop to 83% of the starting basal plasma value (Brun del Re et al. 1973). It is interesting to note that Crosignani et al. (1977) have found persistence of the PRL response to a TRH challenge during bromocriptine treatment in the early puerperium. Later on, the effectiveness of this compound in suppressing postpartum milk secretion has been amply confirmed (Rolland and Schellekens 1973; Rolland et al. 1975; Cooke et al. 1976; Weinstein et al. 1976; Nilsen et al. 1976; Gerstner et al. 1977; MacLeod et al. 1977). Bromocriptine also effectively suppressed lactation once it had been initiated and blunted the PRL surge that normally occurs in response to suckling (Fig.3.11) (Brun del Re et al. 1973; del Pozo et al. 1975a). Comparison studies with exogenous sexual steroids have yielded a higher success rate for bromocriptine (Weinstein et al. 1976; Schmidt-Gollwitzer et al. 1977; Villaverde-Fernandez et al. 1977; Clavero-Salvador and Villarreal-Alcaraz 1977). The better results obtained with the ergot derivative can be explained through the specific PRL inhibition exerted by the compound whereas estrogens

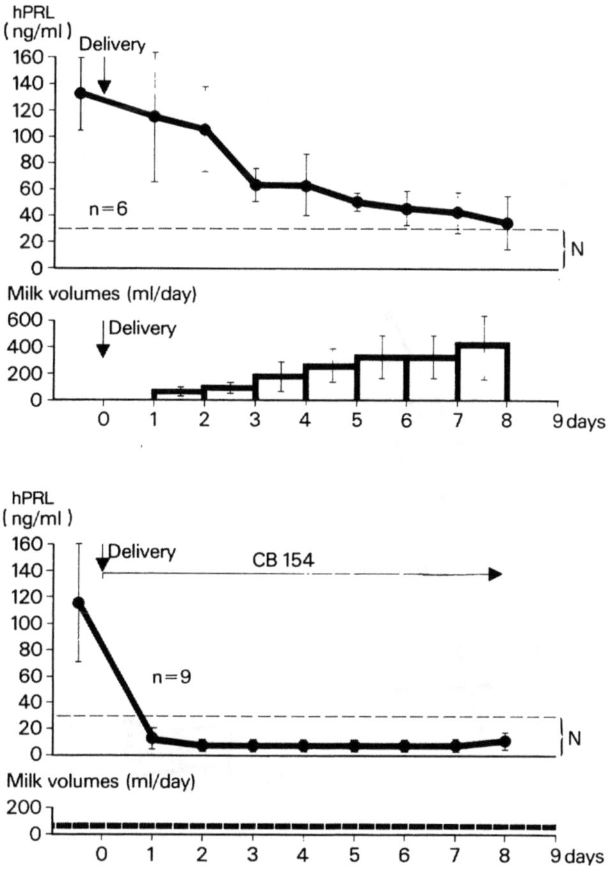

Fig. 3.10. Human PRL measurements and milk volumes obtained in six normally breast-feeding women. There is a gradual decrease in serum PRL concentrations in the postpartum period and a concomitant rise in milk secretion. Bromocriptine (CB 154) treatment reduced prolactin to nonpregnant levels in another group of nine puerperae. No milk was produced (mean ± SEM). (Brun del Re et al. 1973)

have been found to enhance PRL secretion in the postpartum (del Pozo et al. 1975b; Weinstein et al. 1976; Schmidt-Gollwitzer et al. 1977).

Therapy with bromocriptine has also proven effective in patients requiring rapid release of mammary engorgement such as in puerperal mastitis. Peters and Breckwoldt (1977) observed that signs and symptoms of early mastitis could be easily controlled by treatment with bromocriptine in 52 puerperae. Additional therapy with antibiotics was necessary only in two cases. It was concluded that suppression of milk secretion would prevent the extension of the infectious process by depriving bacteria of culture substrate. These authors have also found this treatment effective in the management of nonpuerperal mastitis (Peters and Breckwoldt 1978). Furthermore, Canales et al. (1977a) were able to inhibit and reinitiate lactation in women with premature infants by administration of bromocriptine in the immediate postpartum and provoking milk letdown with TRH at a later date. Brun del Re (unpublished data) used the same procedure in the

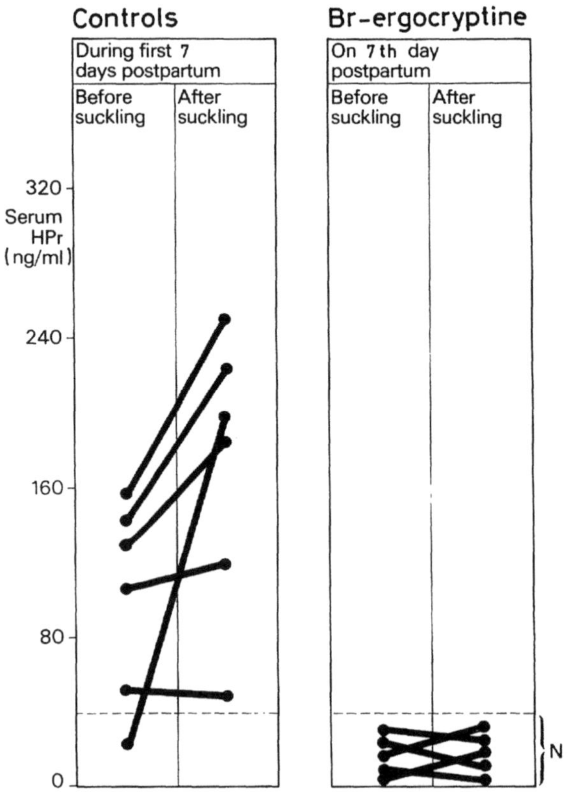

Fig. 3.11. Human prolactin (PRL) peak at *left* in response to breast-feeding in six normally lactating mothers. At *right*, therapy with Bromocriptine (CB 154) depressed basal PRL values to the normal nonpregnant range. The postsuction PRL peak has been abolished. (Brun del Re et al. 1973)

treatment of mastitis: lactation was suppressed with bromocriptine for a few days in order to accelerate the resolution of the inflammatory process. If restoration of milk secretion was desired, discontinuation of medication and the suckling stimulus by the infant were in general succesful in reinitiating lactogenesis.

Pusterla et al. (1977) have tested the effect of bromocriptine in women undergoing therapeutic abortion between pregnancy weeks 8 and 14. Spontaneous milk secretion was recorded in 19% of control women on no therapy whereas only 2% receiving bromocriptine exhibited milk leakage. Thus, this agent also represents a useful tool in the preventive treatment of postabortion lactation. More recently other ergot derivatives have become available for clinical evaluation. Thus, Delitala et al. (1977) and Crosignani et al. (1978) have reported prevention of milk letdown with metergoline and also adequate suppression when the drug was administered once lactation had started. However, it is noteworthy that the fall in plasma PRL recorded with these drugs in the postpartum barely reached 50% of the basal value. In the case of lisuride maleate, Schmidt-Gollwitzer et al. (1977) found the effect was not different from placebo, whereas Hardt et al. (1979) reported better results with higher doses.

3.9.4 Lactation Suppression with Nonergot Drugs

Recent experimental findings have suggested an antigalactogenic effect of estrogen antagonists, and Zuckerman and Carmel (1973) and Weinstein et al. (1976) reported a suppressive action of clomiphene on physiologic lactation. Also Masala et al. (1977) were able to blunt the postsuctional PRL surge by pretreatment with this compound. In contrast, Thorner et al. (1974a) saw no modification of basal and stimulated PRL secretion by clomiphene in normal subjects, nor was it effective in the treatment of nonpuerperal hyperprolactinemia (Thorner et al. 1974b). Furthermore, Canales et al. (1977b) reported no effect of this drug in the postpartum: it neither inhibited milk secretion nor had a suppressive action on serum PRL.

In the assumption that pyridoxine, in the form of pyridoxal phosphate, serves as a coenzyme in the synthesis of dopamine, its use as a lactation inhibitor was advocated by Foukas (1973). Later studies, however, have failed to support such an action. This vitamin was ineffective in the control of physiologic lactation nor had its administration been followed by a decrease in plasma PRL concentration (del Pozo et al. 1975c; Canales et al. 1976; del Pozo and Brun del Re 1979). Piribedil, a piperazine with strong dopaminergic properties was found to inhibit PRL secretion in postpartum women (Müller et al. 1977) whereas its effect was inconsistent in nonpuerperal galactorrhea. Also a 5-day course of Piribedil caused a significant decrease in the PRL response to mechanical breast stimulation in the puerperium (Alagna et al. 1979). Despite basal plasma PRL levels still in the lactational range the authors reported inhibition of puerperal milk secretion. It can therefore be concluded that, despite claims of efficacy of some steroid and nonergot compounds in the suppression of physiologic lactation, at the present time the data are unconvincing.

3.10 References

Abraham GE, Odell WD, Swerdloff RS, Hopper K (1972) Simultaneous radioimmunoassay of plasma FSH, LH, progesterone, 17-hydroxyprogesterone, and estradiol-17β. J Clin Endocrinol Metab 34:312–318

Alagna S, Lodico G, Devilla L, Masala A, Delitala G, Rovasio PP, Stoppelli I (1979) Piribedil inhibition of prolactin release induced by mechanical breast emptying in puerperal women. Clin Endocrinol (Oxf) 11:301–305

Alexander DP, Britton HG, Buttle HL, Nixon DA (1973) Impermeability of the sheep placenta towards endogenous prolactin. Res Vet Sci 14:271–272

Alger M, Caballero M, Canales ES, Fonseca ME, Forsbach G, Zarate A (1979) Estudio de la capacidad de secrecion de prolactina en el sindrome de Sheehan mediante la prueba de metoclopramida. Rev Invest Clin 31:211–215

Amenomori Y, Meites J (1970) Effects of hypothalamic extract on serum prolactin levels during the estrus cycle and lactation. Proc Soc Exp Biol Med 134:492–495

Andersen AN, Hertz J, Kjer JJ, Eskildsen PC, Larsen PS, Svenstrup B, Nielsen J, Arends J (1980) Ovarian and placental hormones during prolactin suppression and stimulation in early human pregnancy. Clin Endocrinol (Oxf) 13:151–155

Aono T, Shioji T, Shoda T, Kurachi K (1977) The initiation of human lactation and prolactin response to suckling. J Clin Endocrinol Metab 44:1101–1106

Aono T, Shioji T, Aki T, Hirota K, Nomura A, Kurachi K (1979) Augmentation of puerperal lactation by oral administration of sulpiride. J Clin Endocrinol Metab 48:478–482

Apostolakis M, Kapetanakis S, Lazos G, Madena-Pyrgaki A (1972) Plasma prolactin activity in patients with galactorrhea after treatment with psychotropic drugs. In: Wolstenholme GEW, Knight J (eds). Lactogenic hormones. Livingstone, Edingburgh London, pp 349–359

Aragona C, Bohnet HG, Friesen HG (1977) Localization of prolactin binding in prostate and testis: the role of serum prolactin concentration on the testicular LH receptor. Acta Endocrinol (Copenh) 84:402–409

Aubert ML, Grumbach MM, Kaplan SL (1975) The ontogenesis of human fetal hormones. III. Prolactin. J Clin Invest 56:155–164

Aznar A, Gerber L, del Pozo E (to be published) Lack of effect of suckling on LH pulsatility during lactation.

Badawi M, Perez-Lopez FR, Robyn C (1973) Human prolactin and chorionic somatomammotrophin (HCS) in amniotic fluid. Acta Endocrinol [Suppl] (Copenh) 177:237

Baird DT, McNeilly AS, Sawers RS, Sharpe RM (1979) Failure of estrogen-induced discharge of luteinizing hormone in lactating women. J Clin Endocrinol Metab 49:500–506

Baker BL, Yu Y (1977) An immunocytochemical study of human pituitary mammotropes from fetal life to old age. Am J Anat 148:217–240

Barberia JM, Abu-Fadil S, Kletzky OA, Nakamura RM, Mishell DR (1975) Serum prolactin patterns in early human gestation. Am J Obstet Gynecol 121:1107–1110

Barnes J (1942) Inhibition of lactation by synthetic oestrogenic substances with special reference to dienoestrol. Br Med J 1:601–604

Bassi F, Giusti G, Borsi L, Cattaneo S, Giannotti P, Forti G, Pazzagli M, Vigiani C, Serio M (1977) Plasma androgens in women with hyperprolactinemic amenorrhoea. Clin Endorinol (Oxf) 6:5–10

Bellmann O (1977). Prolaktin in der Schwangerschaft und im Wochenbett – zu Sekretion, Regulation und Funktion. Gynaekologe 10:66–83

Berman ML, Hanson K, Hellman IL (1972) Effect of breast-feeding on postpartum menstruation, ovulation, and pregnancy in Alaskan Eskimos. Am J Obstet Gynecol 114:524–534

Beumont PJV, Gelder MG, Friesen HG, Harri GW, MacKinnon PCB, Mandelbrote BM, Wiles DH (1974) Effects of phenothiazines on endocrine function. I. Patients with inappropriate lactation and amenorrhea. Br J Psychiatry 124:413–419

Bigazzi M, Ronga R, Lancranjan I, Ferraro S, Branconi F, Buzzoni P, Martorana G, Scarselli GF, del Pozo E (1979 a) A pregnancy in an acromegalic woman during bromocriptine treatment: effects on growth hormone and prolactin in the maternal, fetal, and amniotic compartments. J Clin Endocrinol Metab 48:9–12

Bigazzi M, Pollicino F, Nardi E (1979 b) Is human decidua a specialized endocrine organ? J Clin Endocrinol Metab 49:847–850

Biswas S, Rodeck CH (1976) Plasma prolactin levels during pregnancy. Br J Obstet Gynecol 83:683–687

Bohnet HG, Friesen HG (1976) Effect of prolactin and growth hormone on prolactin and LH receptors in the dwarf mouse. J Reprod Fertil 48:307–311

Bohnet HG, Heuberger LW, Dahlen HG, Schneider HPG (1975) TRH-Stimulation als Nachweisversuch für Induktion und Involution prolaktinsezernierender Hypophysenzellen in Schwangerschaft und Wochenbett sowie bei pathologischer Hyperprolaktinämie. Endokrinologie 66:173–183

Bohnet HG, Naber NG, del Pozo E, Schneider HPG (1978) Effects of synthetic gestagens on serum prolactin and growth hormone secretion in amenorrheic patients. Arch Gynecol 226:233–240

Bonte M, Akingeneye E, Gashakamba M, Nbarusto E, Nolens M (1974) Influence of the socio-economic level on the conception rate during lactation. Int J Fertil 19:97–102

Boyar RM, Finkelstein JW, Kapen S, Hellman L (1975) Twenty-four hour prolactin (PRL) secretory patterns during pregnancy. J Clin Endocrinol Metab 40:1117–1120

Bruce JO, Ramirez VD (1970) Site of action of the inhibitory effect of estrogen upon lactation. Neuroendocrinology 6:19–29

Brun del Re R, del Pozo E, de Grandi P, Friesen H, Hinselman M, Wyss H (1973) Prolactin inhibition and suppression of puerperal lactation by a Br-ergocryptine (CB 154). Obstet Gynecol 41:884–890

Bunner DL, van der Laan EF, van der Laan WP (1978) Prolactin levels in nursing mothers. Am J Obstet Gynecol 131:250–252

Canales ES, Zarate A, Garrido J, Carlos Leon J, Soria A, Schally W (1974) Study on the recovery of pituitary FSH function during puerperium using synthetic LRH. J Clin Endocrinol Metab 38:1140–1142

Canales ES, Soria J, Zarate A, Mason M, Molina M (1976) The influence of pyridoxine on prolactin secretion and milk production in women. Br J Obstet Gynecol 83:387–388

Canales ES, Lasso P, Murrieta S, Fonseca E, Soria J, Zarate A (1977 a) Feasibility of suppressing and reinitiating lactation in women with premature infants. Am J Obstet Gynecol 128:695–697
Canales ES, Lasso P, Soria J, Zarate A (1977 b) Effect of clomiphene on prolactin secretion and lactation in puerperal women. Br J Obstet Gynecol 84:758–759
Chochinov RH, Ketupanya A, Mariz IK, Underwood LE, Daughaday WH (1976) Amniotic fluid reactivity detected by somatomedin C radioreceptor assay: correlation with growth hormone, prolactin and fetal renal maturation. J Clin Endocrinol Metab 42:983–986
Chopra JG (1972) Effect of steroid contraceptives on lactation. Am J Clin Nutr 25:1202–1214
Clavero-Salvador PA, Villarreal-Alcáraz A (1977) Inhibición de la lactancia con estrógenos, vitamina B6 y CB-154. Acta Ginecol 30:339–349
Clemens JA, Sawyer BD, Cerimele B (1977) Further evidence that serotonin is a neurotransmitter involved in the control of prolactin secretion. Endocrinology 100:692–698
Clements JA, Reyes FI, Winter JSD, Faiman C (1977) Studies on human sexual development. IV. Fetal pituitary and serum, and amniotic fluid concentrations of prolactin. J Clin Endocrinol Metab 44:408–413
Conti A, del Pozo E, Derrer F, Litschgi M, Wyss H (1979) Significance of the post-suckling prolactin surge in the maintenance of puerperal lactation. IXth world congress of gynecology and obstetrics, Tokyo, Abstr no 690, p 302
Cooke I, Jenkins A, Foley M, Obiekwe B, Lenton E, McNeilly A, Preston E, Parsons J, Millar D, Kennedy G (1976) The treatment of puerperal lactation with bromocriptine. Postgrad Med J 52:75–80
Cowie AT, Tindal JS (1971) The physiology of lactation. Arnold, London
Crosignani PG, Lombroso GC, Trojsi L, Caccamo A, Reschini E, Peracchi M (1977) Prolactin response to TRH during inhibition of puerperal lactation with ergolinic drugs. Horm Metab Res 9: 163–164
Crosignani PG, Lombroso GC, Caccamo A, Reschini E, Peracchi M (1978) Suppression of puerperal lactation by metergoline. Obstet Gynecol 51:113–115
Crosignani PG, Lombroso GC, Mattei A, Caccamo A, Trojsi L (1979) Effect of three serotonin antagonists on plasma prolactin response to suckling in puerperal women. J Clin Endocrinol Metab 48:335–337
Daniel DG, Campbell H, Turnbull AC (1967) Puerperal thromboembolism and suppression of lactation. Lancet 2:287–289
De Chateau P, Wiberg B (1977) Long-term effect on mother-infant behaviour of extra contact during the first hour post-partum I. First observations at 36 hours. Acta Paediatr Scand 66:137–143
Delgado H, Brineman E, Lechtig A, Bongaarts J, Martorell R, Klein RF (1979) Effect of maternal nutritional status and infant supplementation during lactation on postpartum amenorrhea. Am J Obstet Gynecol 135:303–307
Delitala G, Masala A, Alagna S, Devilla L, Lodico G, Lotti G (1977) Metergoline in the inhibition of puerperal lactation. Br Med J 1:744–746
Delvoye P, Badawi M, Demaegd M, Robyn C (1978 a) Long lasting lactation is associated with hyperprolactinemia and amenorrhea. In: Robyn C, Harter M (eds), Progress in prolactin physiology and pathology. Elsevier/North-Holland Biomedical Press, Amsterdam, pp 213–232
Delvoye P, Demaegd M, Uwayitu-Nyampeta, Robyn C (1978 b) Serum prolactin, gonadotropins, and estradiol in menstruating and amenorrheic mothers during two years' lactation. Am J Obstet Gynecol 130:635–639
Djiane J, Durand P (1977) Prolactin-progesterone antagonism in self regulation of prolactin receptors in the mammary gland. Nature 266:641–643
Djiane J, Clauser H, Kelly PA (1979) Rapid downregulation of prolactin receptors in mammary gland and liver. Biochem Biophys Res Commun 90:1371–1378
Duchen MR, McNeilly AS (1980) Hyperprolactinemia and long-term lactational amenorrhea. Clin Endocrinol (Oxf) 12:621–627
Ebrahim GJ (1976) Cross-cultural aspects of breastfeeding. Ciba Found Symp 45:195–204
Egli GE, Egli NS, Newton M (1961) Influence of number of breast feedings on milk production. Pediatrics 27:314–317
Elder MG, Hakim CA (1974) The puerperium. In: Hawkins DF (ed). Obstetric therapeutics. Clinical pharmacology and therapeutics in obstetric practice, Baillière Tindall, London pp 510–553
Elsaesser F, Parvizi N (1980) Partial recovery of the stimulatory oestrogen feedback action on LH release during late lactation in the pig. J Reprod Fertil 59:63–67

Erdheim J, Stumme E (1909) Über Schwangerschaftsveränderung der Hypophyse. Beitr Pathol Anat 46:1–132

Fang VS, Kim MH (1975) Study on maternal, fetal, and amniotic human prolactin at term. J Clin Endocrinol Metab 41:1030–1034

Forsbach G, Ayala A, Soria J, Arcovedo F, Mason M, Canales ES, Zarate A (1976) Secrecion de prolactin en recién nacidos normales y en anencéfalos. Arch Invest Med 7:85–90

Foukas MD (1973) An antilactogenic effect of pyridoxine. J Obstet Gynaecol Br Commonw 80:718–720

Frame LT, Wiley L, Rogol AD (1979) Indirect immunofluorescent localization of prolactin to the cytoplasm of decidua and trophoblast cells in human placental membranes at term. J Clin Endocrinol Metab 49:435–437

Frantz AG, Kleinstein DL, Noel GL (1972) Studies on prolactin in man. Recent Prog Horm Res 28:527–590

Freeman R, Lev-Gur M, Boyar RM, Schulman H (1976) Studies of maternal plasma prolactin and amniotic fluid prolactin. Effects of chlorpromazine and prostaglandins F2 alpha. Obstet Gynecol 47:282–286

Friesen H, Hwang P, Guyda H, Tolis G, Tyson J, Myers R (1972) A radioimmunoassay for human prolactin. In: Boyns AR, Griffiths K (eds) Prolactin and carcinogenesis. Alpha Omega Alpha, Cardiff, pp 64–80

Gaede P, Trolle D, Pedersen H (1978) Extremely low placental lactogen hormone (hPL) values in an otherwise uneventful pregnancy preceding delivery of a normal baby. Acta Obstet Gynecol Scand 57:203–209

Geiger W (1974) Radioimmunologische Bestimmungen von HCG, HCS, STH und TSH. Fortschr Geburtsh Gynaekol 52:1–120

Gerstner G, Mick R, Spona J (1977) Serum prolactin levels during inhibition of lactation with 2-bromo-α-ergocryptin (CB 154). Wien Klin Wochenschr 89/19:664–667

Gitlin D, Biasucci A (1969) Ontogenesis of immunoreactive growth hormone, follicle-stimulating hormone, thyroid-stimulating hormone, chorionic prolactin and chorionic gonadotropin in the human conceptus. J Clin Endocrinol Metab 29:926–935

Giusti G, Bassi F, Forti G, Giannotti P, Calabresi E, Pazzagli M, Fiorelli G, Mannelli M, Misciglia N, Serio M (1978) Effects of prolactin on androgen secretion by the human adrenal cortex. In: Robyn C, Harter M (eds). Progress in prolactin physiology and pathology. Elsevier/North-Holland Biomedical Press, Amsterdam, pp 293–303

Gluckman PD, Ballard PL, Laplan SL, Liggins GC, Grumbach MM (1978) Prolactin in umbilical cord blood and the respiratory distress syndrome. J Paediatr 93:1011–1014

Golander A, Hurley T, Barrett J, Handwerger S (1979) Synthesis of prolactin by human decidua in vitro. J Endocrinol 82:263–267

Goluboff LG, Ezrin C (1969) Effect of pregnancy on the somatotroph and the prolactin cell of the human adenohypophysis. J Clin Endocrinol 29:1533–1538

Griffith RW, Turkalj I, Braun P (1978) Outcome of pregnancy in mothers given bromocriptine. Br J Clin Pharmacol 5:227–231

Griffith RW, Turkalj I, Braun P (1979) Pituitary tumours during pregnancy in mothers treated with bromocriptin. Br J Clin Pharmacol 7:393–396

Gross BA, Haynes SP, Eastman CJ, Balderrama-Guzman V, del Casillo LV (1980) A cross-cultural comparison of prolactin secretion in long-term lactation. Prog Reprod Biol 6:179–186

Guitelman A, Aparicio NJ, Mancini A, Debeljuk L (1978) Release of prolactin during pregnancy: effect of sulpiride. Fertil Steril 30:42–44

Guzman V, Toscano G, Canales ES, Zarate A (1979) Improvement of defective lactation by using oral metoclopramide. Acta Obstet Gynecol Scand 58:53–55

Hakim CA, Elder MG, Hawkins DF (1969) Plasma factor IX levels in patients given hexoestrol or stilboestrol to suppress lactation. Br Med J 4:82–84

Hamosh M, Hamosh P (1977) The effect of prolactin on the lecithin content of fetal rabbit lung. J Clin Invest 59:1002–1005

Hardt W, Schmidt-Gollwitzer M, Horowski R (1979) Suppression of lactation with lisuride. Gynecol Invest 10:95–105

Hart IC (1974) The relationship between lactation and the release of prolactin and growth hormone in the goat. J Reprod Fertil 39:485–499

Hartemann J, Richon J (1962) Maternal nursing and reduction of number of infant feeding times. Bull Fed Gynecol Obstet Fr 14:773–775

Hartemann J, Ribon M, Dellestable P (1962) Effects of attraction of mother and child on maintenance of nursing. Bull Fed Gynecol Obstet Fr 14:770–773
Hauth JC, Parker CR, MacDonald PC, Porter JC, Johnston JM (1978) A role of fetal prolactin in lung maturation. J Obstet Gynecol 51:81–88
Healy DL, Müller HK, Burger HG (1977) Immunofluorescence shows localisation of prolactin to human amnion. Nature 265:642–643
Hepp H, Ellringmann R (1972) Hemmung der Laktation mit Estrovis 400 und Ablacton. Geburtshilfe Frauenheilkd 32:471–476
Hiba J, del Pozo E, Genazzani A, Pusterla E, Lancranjan I, Sidiropoulos D, Gunti J (1977) Hormonal mechanisms of milk secretion in the newborn. J Clin Endocrinol Metab 44:973–976
Hodge C (1967) Suppression of lactation by stilboestrol. Lancet 2:286–287
Howie PW, McNeilly AS, McArdle T, Smart L, Houston M (1980) The relationship between suckling-induced prolactin response and lactogenesis. J Clin Endocrinol Metab 50:670–673
Hwang P, Guyda H, Friesen H (1971) A radioimmunoassay for human prolactin. Proc Natl Acad Sci USA 68:1902–1906
Hwang P, Guyda H, Friesen H (1972) Purification of human prolactin. J Biol Chem 247:1955–1958
Jacobs LS, Mariz IK, Daughaday WH (1972) A mixed heterologous radioimmunoassay for human prolactin. J Clin Endocrinol 34:484–490
Jaffe RB, Ho Yuen B, Keye WR, Midgley AR (1973) Physiologic and pathologic profiles of circulating human prolactin. Am J Obstet Gynecol 117:757–773
Jain AD (1969) Pregnancy outcome and the time required for next conception. Popul Stud 23:421–533
Jeffcoate TNA, Lister UM, Hargreaves B, Roberts M (1948) Ethinyl oestradiol Br Med J 2:809–812
Jeppsson S, Rannevik G, Kullander S (1974) Studies on the decreased gonadotropin response after administration of LH/FSH-releasing hormone during pregnancy and the puerperium. Am J Obstet Gynecol 120:1029–1034
Josimovich JB (ed) Wiley, New York, Lactogenetic hormones, fetal nutrition and lactation (1974) (Problems in reproduction, vol 2)
Josimovich JB, Archer DF (1977) The role of lactogenic hormones in the pregnant woman and the fetus. Am J Obstet Gynecol 129:777–780
Josimovich JB, Merisko K, Boccella L (1977) Amniotic prolactin control over amniotic and fetal extracellular fluid water and electrolytes in the rhesus monkey. Endocrinology 100:564–570
Kamal I, Hefnawi F, Ghoneim M, Talaat M, Younis N, Tagui A, Abdalla M (1969) Clinical, biochemical, and experimental studies on lactation. I. Lactation pattern in Egyptian women. Am J Obstet Gynecol 105:314–323
Kann G, Denamur R (1974) Possible role of prolactin during the oestrus cycle and gestation in the ewe. J Reprod Fertil 39:473–483
Kann G, Martinet J (1975) Prolactin levels and duration of postpartum anestrus in lactating ewes. Nature 257:63–69
Kann G, Martinet J, Schirar A (1977) Modifications of gonadotrophin secretion during natural and artifical hyperprolactinaemia in the ewe. In: Crosignani PG, Robyn C (eds). Prolactin and human reproduction. Academic Press, London, pp 47–59
Kaplan SL, Grumbach MM (1974) Lactogenic hormones, fetal nutrition and lactation. In: 2. Josimovich JW, Reynolds M, Cobo E (eds) Problems in reproduction, ser 2. Wiley, New York, pp 183–197
Kaplan SL, Grumbach MM, Shepard TH (1972) The ontogenesis of human fetal hormones. I. Growth hormone and insulin. J Clin Invest 51:3080–3093
Kaplan SL, Grumbach MM, Aubert ML (1976) The ontogenesis of pituitary hormones and hypothalamic factors in the human fetus: maturation of the central nervous system regulation of anterior pituitary function. Recent Prog Horm Res 32:161–233
Karg H, Schams D (1974) Prolactin release in cattle. J Reprod Fertil 39:463–472
Keettel WC, Bradbury JT (1961) Endocrine studies of lactation amenorrhea. Am J Obstet Gynecol 82:995–1002
Kellar RL, Sutherland JK (1939) Clinical experiences with new synthetic oestrogen-stilboestrol (Diethylstilboestrol); report to Therapeutic Trials Committee of Medical Research Council. J Obstet Gynecol Br Emp 46:1–7 (1939)
Kelly PA, Posner BI, Friesen HG (1975) Effects of hypophysectomy, ovariectomy, and cycloheximide on specific binding sites for lactogenic hormones in rat liver. Endocrinology 97:1408
Kelly PA, Tsushima T, Shiu RPC, Friesen HG (1976) Lactogenic and growth hormone-like activities in pregnancy determined by radioreceptor assays. Endocrinology 99:765–774

Kelly PA, Djiane J, de Léan A (1980) Interaction of prolactin with its receptor: dissociation and downregulation. In: MacLeod RM, Scapagnini U (eds). Central peripheral regulation of prolactin function. Raven, New York, pp 173–188

King KC, Adam PAJ, Schwartz R, Teramo K (1971) Human placental transfer of human growth hormones-I^{125}. Pediatrics 48:534–539

Kivinen S, Ylikorkala O, Puukka M (1979) Prolactin response to thyrotropin-releasing hormone in normal and complicated late pregnancies. Obstet Gynecol 54:695–698

Kledzik GS, Bradley CJ, Marshall S, Campbell GA, Meites J (1976) Effect of high doses of estrogen on prolactin binding activity and growth of carcinogen-induced mammary cancers in rats. Cancer Res 36:3265–3268

Kletzky OA, Marrs RP, Howard WF, McCormick W, Mishell DR (1980) Prolactin synthesis and release during pregnancy and puerperium. Am J Obstet Gynecol 136:545–550

Knodel J (1968) Infant mortality and fertility in three Bavarian villages; an analysis of family histories form the 19th century. Popul Stud 22:297–318

Kolata GB (1974) Kung hunter-gatherers: feminism, diet and birth control. Science 185:932–934

Konner M, Worthman C (1980) Nursing frequency, gonadal function, and birth spacing among Kung hunter-gatherers. Science 207:788–791

Kordon C, Blake CA, Terkel J, Sawyer CH (1973/74) Participation of the serotonin-containing neurons in the suckling-induced rise in plasma prolactin levels in lactating rats. Neuroendocrinology 13:213–223

Lancranjan I, del Pozo E, Picciolini E, D'Antona N, Genazzani AR (1979) Effect of two serotonin antagonists on prolactin release induced by breast stimulation in postpartum women. Horm Res 10:14–19

Laron Z, Mannheimer S, Guttmann S (1965) Plasma disappearance of various ^{131}I-labelled growth hormone preparations in man and rabbit. Nature 207:298–299

Laron Z, Pertzelan A, Mannheimer S, Goldman J, Guttmann S (1966) Lack of placental transfer of human growth hormone. Acta Endocrinol (Copenh) 53:687–692

Lehmann WD, Musch K, Wolf AS (1979) Influence of bromocriptine on plasma levels of prolactin and steroid hormones in the 20th week of pregnancy. J Endocrinol Invest 2:251–255

LeMaire WJ, Shapiro AG, Riggall F, Yang NST (1974) Temporary pituitary insensitivity to stimulation by synthetic LRF during the postpartum period. J Clin Endocrinol Metab 38:916–918

L'Hermite M, Delvoye P, Nokin J, Vekemans M, Robyn C (1972) Human prolactin secretion as studied by radioimmunoassay. In: Boyns AR, Griffiths K (eds). Prolactin and carcinogenesis. Alpha Omega Alpha, Cardiff, pp 81–97

L'Hermite M, Delogne-Desnoeck J, Michaux-Duchene A, Robyn C (1978) Alteration of feedback mechanism of estrogen on gonadotropin by sulpiride-induced hyperprolactinemia. J Clin Endocrinol Metab 47:1132–1136

MacLeod SC, Scott J, Lord L, Brodie G, Perlin I, Simpson AA (1977) Prevention and suppression of postpartum lactation with 2-bromo-alpha-ergocryptine (CB-154). Clin Endocrinol [Suppl] (Oxf) 6:65s–70s

Mancuso A, Baviera G, Quattrone D, Alessandrini G (1979) Prolattina durante la gravidanza ed il parto (1979). In: Montanari GD, Volpe A (eds) La prolattina in gynecologia. Piccin Editore, Padua, 89–100.

Maneckjee R, Srinath BR, Moudgal NR (1976) Prolactin suppresses release of luteinizing hormone during lactation in the monkey. Nature 262:507–508

Masala A, Delitala G, Lodico G, Alagna S, Devilla L (1977) Effect of clomiphene on release of prolactin induced by mechanical breast emptying in women post partum. J Endocrinol 74:501–502

Massara F, Camanni F, Belforte L, Campagnoli C, Novi RF, Molinatti GM (1980) Effect of sulpiride on plasma prolactin levels in women with puerperal or pathological hyperprolactinemia. Acta Endocrinol (Copenh) 95:145–150

Martinez-Manautou J, Giner-Velásquez J, Cortés-Gallegos V, Aznar R, Rojas B, Gutiérrez-Najar A, Rudel HW (1967) Daily progestogen for contraception: a clinical study. Br Med J 2:730–732

Matsuzaki F, Irie M, Shizume K (1971) Growth hormone in human fetal pituitary glands and cord blood. J Clin Endocrinol Metab 33:908–911

McNeilly AS (1974) Prolactin and human reproduction. Br J Hosp Med 12:57–62

McNeilly AS (1980) Prolactin and the control of gonadotrophin secretion in the female. J Reprod Fertil 48:537–549

Meites J (1973) Control of prolactin secretion in animals. In: Pasteels JL, Robyn C (eds). Human prolactin. Excerpta Medica/American Elsevier, New York, pp 105–118

Meites J (1974) Neuroendocrinology of lactation. J Invest Dermatol 63:119–124

Morishige WK, Rotchild I (1974) Temporal aspects of the regulation of corpus luteum function by luteinizing hormone, prolactin and placental luteotrophin during the first half of pregnancy in the rat. Endocrinology 95:260–274

Morris JA, Creasy RK, Hohe PT (1970) Inhibition of puerperal lactation. Double-blind comparison of chlorothianesene, testosterone enanthate with estradiol valerate and placebo. Obstet Gynecol 36:107–114

Morrow DA, Roberts SJ, McEntee K (1969) A review of postpartum ovarian activity and involution of the uterus and cervix in cattle. Cornell Vet 59:173–190

Müller EE, Genazzani AR, Murru S, Fioretti P (1977) Effect of piribedil on plasma prolactin levels in women with puerperal or pathological hyperprolactinaemia. Acta Endocrinol (Copenh) 86:33–41

Nakano R, Kayashima F, Mori A, Kotsuje S, Hashiba N, Pojo S (1974) Gonadotropin response to luteinizing hormone releasing factor (LRF) in puerperal women. Acta Obstet Gynecol Scand 53:303–307

Newton N, Newton M (1967) Psychologic aspects of lactation. N Engl J Med 277:1179–1188

Nielsen PV, Pedersen H, Kampmann EM (1979) Absence of human placental lactogen in an otherwise uneventful pregnancy. Am J Obstet Gynecol 135:322–326

Nilsen PA, Meling A-B, Abildgaard U (1976) Study of the suppression of lactation and the influence on blood clotting with bromocriptine (CB 154, Parlodel): a double blind comparison with diethylstilboestrol. Acta Obstet Gynecol Scand 55:39–44

Pasteels JL (1963) Recherches morphologiques et expérimentales sur la sécrétion de prolactine. Arch Biol (Liege) 74:439–553

Pasteels JL, Gausset P, Danguy A, Ectors F (1972) Immunofluerescent studies on prolactin and the pituitary. In: Boyns AR, Griffiths K (eds). Prolactin and carcinogenesis. Alpha Omega Alpha, Cardiff, pp 128–136

Pencharz RI, Long JA (1933) Hypophysectomy in pregnant rat. Am J Anat 53:117–139

Pepperell RJ, McBain JC, Winstone SM, Smith MA, Brown JB (1977) Corpus luteum function in early pregnancy following ovulation induction with bromocriptine. Br J Obstet Gynecol 84:898–903

Perez AL, Vela P, Masnick GS, Potter RG (1972) First ovulation after childbirth: the effect of breast-feeding. Am J Obstet Gynecol 114:1041–1047

Peters AR, Vyvoda S, Lamming GE (1979) Comparison of plasma prolactin levels in the milked and suckled cow. J Endocrinol 83:27P–28P

Peters F, Breckwoldt M (1977) Neue Aspekte bei der Behandlung der puerperalen Mastitis. Dtsch Med Wochenschr 102:1754–1758

Peters F, Breckwoldt M (1978) Nonpuerperale Mastitis bei hormonell stimulierter Mamma. Geburtshilfe Frauenheilkd 38:754–757

Plant TM, Schallenberger E, Hess DL, McCormack JT, Dufy-Barbé L, Knobil E (1980) Influence of suckling on gonadotropin secretion in the female rhesus monkey (macaca mulata). Biol Reprod 23:760–766

Player HP (1971) Inhibition of lactation by oestrogens. A comparison of quinestrol (Estrovis) with hexoestrol. Clin Trials J 8:25–27

Potter RG, New ML, Wyon JB, Gordon JE (1965) Applications of field studies to research on the physiology of human reproduction: lactation and its effects upon birth intervals in eleven Punjab villages, India. J Chronic Dis 18:1125–1140

Pozo E del, Brun del Re R (1979) Letter to the editor. N Engl J Med 301:107

Pozo E del, Brun del Re R, Varga L, Friesen H (1972) The inhibition of prolactin secretion in man by CB-154 (2-Br-α-ergocryptine). J Clin Endocrinol Metab 35:768–771

Pozo E del, Varga L, Schulz KD, Künzig HJ, Marbach P, Lopez del Campo G, Eppenberger U (1975 a) Pituitary and ovarian response patterns to stimulation in the postpartum and in galactorrhea-amenorrhea. Obstet Gynecol 46:539–543

Pozo E del, Goldstein M, Friesen H, Brun del Re R, Eppenberger U (1975 b) Lack of action of prolactin suppression on the regulation of the human menstrual cycle. Am J Obstet Gynecol 123:719–723

Pozo E del, Brun del Re R, Hinselmann M, Wyss H (1975 b) Wirkung von Methergin, Oxytocin, Pyridoxin (Vitamin B_6) und Quinestrol auf die Prolaktin- und Milchsekretion im Puerperium. Arch Gynaekol 219:469–470

Pozo E del, Flückiger E, Lancranjan I (1976 a) Endogenous control of prolactin release and its modification by drugs. In: Charro Salgado A, Fernández Durango R, López del Camp G (eds). Basic applications and clinical uses of hypothalamic hormones. Proceedings of the first international symposium on basic applications and clinical uses of hypothalamic hormones, Madrid 5–7 May, 1975. Excerpta Medica, Amsterdam pp 137–150

Pozo E del, Hiba J, Lancranjan I, Honacker M, Pusterla E (1976 b) Materno-fetal correlations: the role of lactogenic hormones. Fifth International Congress of Endocrinology, Hamburg, July 18–24, Abstr no 614

Pozo E del, Hiba J, Lancranjan I, Kunzig HJ (1977) Prolactin measurements throughout the life cycle: endocrine correlations. In: Crosignani PG, Robyn C (eds). Prolactin and human reproduction. Academic Press, London, pp 61–69

Pozo E del, Schulz KD, Wyss H, Lancranjan I, Braun P (1978) Effect of prolactin on the mechanism of ovulation and on pregancy: some recent findings. In: Robyn C, Harter M (eds). Progress in prolactin physiology and pathology. Elsevier/North-Holland Biomedical Press, Amsterdam, pp 281–292

Pozo E del, Bigazzi M, Calaf J (1980 a) Induced human gestational hypoprolactinemia: Lack of action on fetal adrenal androgen synthesis. J Clin Endocrinol Metab 51:936–938

Pozo E del, Kleinstein J, Brun del Re R, Derrer F, Martin-Perez J (1980 b) Failure of oxytocin and lysine-vasopressin to stimulate prolactin release in humans. Horm Metab Res 12:26–28

Pusterla E, Gunti J, Kuchen J, del Pozo E, Hiba J, Alijnovic A, Gigon U, Haldemann R (1977) Sécrétion et congestion mammaires dans les interruptions artificielles de grossesse du premier trimestre: premières expériences de traitement prophylactique avec le CB 154. In: Bricaire H (ed). La bromocriptine. Colloque de Paris, 5–6 November 1976. Sandoz editions, pp 347–351

Rakoff JS, Yen SSC (1978) Progesterone induced acute release of prolactin in estrogen primed ovariectomized women. J Clin Endocrinol Metab 47:918–921

Ranta T, Lehtrovirta P, Stenman U-H, Laatikainen T, Seppälä M (1980) Acute changes in serum prolactin concentration have no effect on the secretion of progesterone, estradiol, or chorionic gonadotropin during early pregnancy. J Clin Endocrinol Metab 51:544–547

Raymond V, Beaulieu M, Labrie F, Boissier J (1978) Potent antidopaminergic activity of estradiol at the pituitary level on prolactin release. Science 200:1173–1175

Redman CWG, Bonnar J, Beilin LJ, McNeilly AS (1975) Prolactin in hypertensive pregnancy. Br Med J 1:304–306

Resko JA, Pleom JG, Stadelman HL (1975) Estrogens in fetal and maternal plasma of the rhesus monkey. Endocrinology 97:425–430

Reyes FI, Winter JSD, Faiman C (1972) Pituitary-ovarian interrelationships during the puerperium. Am J Obstet Gynecol 114:589–594

Reyes FI, Winter JSD, Faiman C, Hobson WC (1975) Serial serum levels of gonadotrophins, prolactin and sex steroids in the nonpregnant chimpanzee. Endocrinology 97:1447–1455

Riddick DH, Kusmik WF (1977) Decidua: a possible source of amniotic fluid prolactin. Am J Obstet Gynecol 127:187–190

Rigg LA, Yen SSC (1977) Multiphasic prolactin secretion during parturition in human subjects. Am J Obstet Gynecol 128:215–218

Rimoin DL, Holzman GB, Merimee TJ, Rabinowitz D, Barnes AC, Tyson JEA, McKusick VA (1968) Lactation in the absence of human growth hormone. J Clin Endocrinol 28:1183

Rivera EM (1972) Influence of hormones on enzyme activity in mouse mammary gland in vitro. J Dairy Sci 55:1308–1316

Robinson M (1951) Infant morbidity and mortality: study of 3266 infants. Lancet 1:788–794

Robyn C, Vekemans M, Delvoye P, Joostens-Defleur V, Caufriez A, L'Hermite M (1976) Prolactin and fertility control in women. In: Pecile A, Müller EE (eds). Growth hormone and related peptides. Excerpta Medica, Amsterdam, pp 396–406

Robyn C, Delvoye P, van Exter C, Vekemans M, Caufriez A, de Nayer P, Delogne-Desnoeck J, L'Hermite M (1977) Physiological and pharmacological factors influencing prolactin secretion and their relation to human reproduction. In: Crosignani PG, Robyn C (eds). Prolactin and human reproduction. Academic Press, London New York San Francisco, pp 71–96

Rolland R, Schellekens L (1973) A new approach to the inhibition of puerperal lactation. J Obstet Gynaecol Br Commonw 80:945–951

Rolland R, de Jong FH, Schellekens LA, Lequin RM (1975) The role of prolactin in the restoration of ovarian function during the early post-partum period in the human female. I. A study during physiologic lactation. Clin Endocrinol (Oxf) 4:15–25

Rowland D, Steele M, Moltz H (1978) Serotoninergic mediation of the suckling-induced release in the lactating rat. Neuroendocrinology 26:8–14

Sadovsky E, Weinstein D, Ben-David M, Polishuk WZ (1977) Serum prolactin in normal and pathologic pregnancy. Obstet Gynecol 50:559–561

Salber EJ, Feinleib M, MacMahon B (1965) The duration of postpartum amenorrhoea. Am J Epidemiol 82:347–358

Schenker JG, Ben-David M, Polishuk WZ (1975) Prolactin in normal pregnancy: relationship of maternal, fetal and amniotic fluid levels. Am J Obstet Gynecol 123:834–838

Schmidt-Gollwitzer M, Hardt W, Schmidt-Gollwitzer K, Michel E, Nevinny-Stickel J (1977) Vergleichsstudie über den Einfluß verschiedener Sexualsteroide, 2-Br-Alpha-Ergokryptin und Lisuridhydrogenmaleat auf die postpartalen Prolaktin-Serumkonzentrationen und die Laktation. Geburtshilfe Frauenheilkd 37:500–508

Schulz K-D, Geiger W, del Pozo E, Kunzig HJ (1978) Pattern of sexual steroids, prolactin, and gonadotropic hormones during prolactin inhibition in normally cycling women. Am J Obstet Gynecol 132:561–566

Seron-Ferré M, Monroe SE, Hess D, Parer JT, Jaffe RB (1979) Prolactin concentrations in the monkey fetus during the last third of gestation. Endocrinology 104:1243–1246

Sheehan HL, Murdoch R (1938) Postpartum necrosis of the anterior pituitary: pathological and clinical aspects. Br J Obstet Gynaecol Br Emp 45:456–489

Sheehan HL, Standfield JP (1961) The pathogenesis of postpartum necrosis of the anterior lobe of the pituitary gland. Acta Endocrinol (Copenh) 37:479–510

Shiu RP, Kelly PA, Friesen HG (1973) Radioreceptor assay for prolactin and other lactogenic hormones. Science 180:968–971

Shupnik MA, Baxter LA, French LR, Gorski J (1979) In vivo effects of estrogen on ovine pituitaries: prolactin and growth hormone biosynthesis and messenger ribonucleic acid translation. Endocrinology 104:729–735

Siler-Khodr TM, Morgenstern LL, Greenwood FC (1974) Hormone synthesis and release from human fetal adenohypophyses in vitro. J Clin Endocrinol Metab 39:891–905

Soria J, Zárate A, Canales ES, Villalobos H (1976) Effect of suckling on serum follicle-stimulating hormone and luteinizing hormone in nursing women. Neuroendocrinology 20:43–46

Soria J, Canales ES, Forsbach G, Karchmer S, Guzman V, Zarate A (1977) Relationship of maternal, fetal, and amniotic fluid prolactin levels. Ann Endocrinol (Paris 38:55–59

Spellacy WN, Carlson KL, Birk SA (1966) Dynamics of human placental lactogen. Am J Obstet Gynecol 96:1164–1173

Spellacy WN, Buhi WC, Bradieu B, Bradley D, Holsinger KK (1973) Maternal, fetal and amniotic fluid levels of glucose, insulin and growth hormone, Obstet. Gynecol 41:323–331

Spellacy WN, Buhi WC, Schram JD, Birk SA, McCreary BS (1971) Control of human chorionic somatomammotropin levels during pregnancy. Obstet Gynecol 37:567–573

Stock RJ, Josimovich JB, Kosor B, Klopper A, Wilson GR (1971) The effect of chorionic gonadotrophin and chorionic somatomammotrophin on steroidogenesis in the corpus luteum. J Obstet Gynaecol Br Commonw 78:549–553

Sulman FG (1970) Hypothalamic control of lactation. Heinemann, London

Thomas CMG, Corbey RS, Rolland R (1977) Assessment of unconjugated oestradiol and progesterone serum levels throughout pregnancy in normal women and in women with hyperprolactinaemia, who conceived after bromocriptine treatment. Acta Endocrinol (Copenh) 86:405–414

Thomson AM, Hytten FE, Black AE (1975) Lactation and reproduction. Bull WHO 52:337–349

Thorner MO, Besser GM, Hagen C, McNeilly AS (1974a) The relationship between prolactin and gonadotrophins effect of clomiphene administration in normal men. J Endocrinol 63:43P–44P

Thorner MO, McNeilly AS, Hagan C, Besser GM (1974b) Long-term treatment of galactorrhoea and hypogonadism with bromocriptine. Br Med J 2:419–422

Tietze C (1961) The effects of breastfeeding on the rate of conception. Proceedings of the international population conference, London, International Union for the Scientific Study of Population, 1961, vol 2, pp 129–136

Tobon H, Josimovich JB, Salazar H (1972) The ultrastructure of the mammary gland during prolactin-induced lactogenesis in the rabbit. Endocrinology 90:1569–1577

Tolis G, Guyda H, Pillorger R, Friesen HG (1974) Breast-feeding: Effects of the hypothalamic pituitary gonadal axis. Endocrinol Res Commun 1:293–303

Tulchinsky D, Hobel CJ, Yeager E, Marshall JR (1972) Plasma estrone, estradiol, estriol, progesterone, and 17-hydroxyprogesterone in human pregnancy. I. Normal pregnancy. Am J Obstet Gynecol 112:1095–1100

Tutin CEG (1975) Sexual behaviour and mating patterns in a community of wild chimpanzees (Pantroglodytes schweinfurthii). Ph D dissertation, University of Edinburgh

Tyson JE, Friesen HG (1972) Human lactational and ovarian response to endogenous prolactin release. Science 77:897–900

Tyson JE, Hwang P, Guyda H, Friesen HG (1972 a) Studies of prolactin secretion in human pregnancy. Am J Obstet-Gynecol 113:14–20

Tyson JE, Khojandi M, Huth J, Andreassen B (1975) The influence of prolactin secretion on human lactation. J Clin Endocrinol Metab 40:764–773

Tyson JE, Perez A, Zanartu J (1976) Human lactational response to oral thyrotropin releasing hormone. J Clin Endocrinol Metab 43:760–768

Tyson JE (1977) Nursing and prolactin secretion: principal determinants in the mediation of puerperal infertility. In: Crosignani PG and Robyn C (eds) Prolactin and human reproduction. Academic Press, London, pp 97–108

Tyson JE, Carter JN, Andreassen B, Huth J, Smith B (1978) Nursing-mediated prolactin and luteinizing hormone secretion during puerperal lactation. Fertil Steril 30:154–162

Van Ginneken JK (1975) Prolonged breastfeeding as a birth spacing method. J Trop Pediatr 21: 59–66

Varga L, Lutterbeck PM, Pryor JS, Wenner R, Erb H (1972) Suppression of puerperal lactation with an ergot alkaloid: a double-blind study. Br Med J 2:743–744

Vician L, Shupnik MA, Gorski J (1979) Effects of estrogen on primary ovine pituitary cell cultures: stimulation of prolactin secretion, synthesis, and preprolactin messenger ribonucleic acid activity. Endocrinology 104:736–743

Villaverde-Fernández S, Fernández Pérez-Lastra JC, Peneda-Cuesta E, Suárez F, Barangana F, Junceda-Avello E (1977) Estudio comparativo de la bromoergocriptina metilergonovina y clorotrianisene en la inhibición de la prolactina sérica puerperal. Acta Ginecol 30:99–106

Wiederman J, Freund M, Stone ML (1963) Human breast and uterus: comparison of sensitivity to oxytocin during gestation. Obstet Gynecol 21:272

Weinstein D, Ben-David M, Polishuk WZ (1976) Serum prolactin and the suppression of lactation. Br J Obstet Gynaecol 83:679–682

Weiss RR, Frantz AG, Macri JN, Robins J, Merker JG (1977) Prolactin and human placental lactogen changes in maternal serum and amniotic fluid in midtrimester induced abortions. Am J Obstet Gynecol 129:9–13

Weitzman RE, Leake RD, Rubin RT, Fisher DA (1980) The effect of nursing on neurohypophyseal hormone and prolactin secretion in human subjects. J Clin Endocrinol Metab 51:836–839

Whitehead RG, Gutton M, Müller E, Rowland MGM, Prentice AM, Paul A (1978) Factors influencing lactation performance in rural gambian mothers. Lancet 2:178–181

Wiltbank JN, Cook AC (1958) The comparative reproductive performance of nursed and milked cows. J Anim Sci 17:640–648

Winters AJ, Colston C, MacDonald PC, Porter JC (1975) Fetal plasma prolactin levels. J Clin Endocrinol Metab 41:626–629

Wright PJ, Geytenbeek PE, Clarke IJ, Findlay JK (1980) Pituitary responsiveness to LH-RH the occurrence of oestradiol-17 -induced LH-positive feedback and the resumption of oestrous cycles in ewes post partum. J Reprod Fertil 60:171–176

Ylikorkala O, Kauppila A, Haapalahti J (1976) Follicle stimulating hormone, thyrotrophin, human growth hormone and prolactin in hyperemesis gravidarum. Br J Obstet Gynaecol 83:529–533

Ylikorkala O, Kivinen S, Reinilä M (1979 a) Serial prolactin and thyrotropin responses to thyrotropin-releasing hormone throughout normal human pregnancy. J Clin Endocrinol Metab 48:288–292

Ylikorkala O, Huhtaniemi I, Tuimala R, Seppälä M (1979 b) Subnormal postconceptional levels of prolactin do not interfere with the early events of human pregnancy. Fertil Steril 32:286–288

Ylikorkala O, Kivinen S, Rönnberg L, Viinikka L (1980 a) Sulpiride treatment during early human pregnancy effect on the levels of prolactin, sex steroids, and placental lactogen. J Clin Endocrinol Metab 51:155–157

Ylikorkala O, Kivinen S, Rönnberg L (1980 b) Bromocriptine treatment during early human pregnancy: effect on the levels of prolactin, sex steroids and placental lactogen. Acta Endocrinol (Copenh) 95:412–415

Yokoyama A, Shinde Y, Ota K (1969) Endocrine control of changes in lactose content of the mammary gland in rats shortly before and after parturition. In: Reynolds M, Folley SJ (eds). Lactogenesis: The initiation of milk secretion at parturition. University of Pennsylvania Press, pp 65–71

Yoshinaga K, Hawkins RA, Stocker JF (1969) Estrogen secretion by the rat ovary in vivo during the estrous cycle and pregnancy. Endocrinology 85:103–112

Zañartu J, Rice-Wray E, Goldzieher JW (1966) Fertility control with long-acting injectable steroids. A preliminary report. Obstet Gynecol 28:513–515

Zarate A, Canales ES, Soria J, Ruiz F, MacGregor C (1972) Ovarian refractoriness during lactation in women: effect of gonadotropin stimulation. Am J Obstet Gynecol 112:1130–1132

Zarate A, Villalobos H, Canales ES, Soria J, Arcovedo F, MacGregor C (1976) The effect of oral administration of thyrotropin-releasing hormone on lactation. J Clin Endocrinol Metab 43:301–305

Zuckerman H, Carmel S (1973) The inhibition of lactation by clomiphen. J Obstet Gynaecol Br Commonw 80:822–823

4 Nontumoral Hyperprolactinemia

4.1 Tests of Lactotrope Cell Function

4.1.1 Dopamine (DA) Antagonists and Effect of TRH

In the first chapter the dopaminergic mechanisms controlling PRL release have been amply analyzed. Most PRL-stimulating agents act by blocking DA synthesis or DA receptors (e. g., neuroleptics), depleting central catecholamine stores (e. g., reserpine) (Turkington 1972d; Lee et al. 1976) or interfering with some step of its synthesis (e. g., α-methyl-dopa or 3-I-L tyrosine) (Steiner et al. 1976; Smythe et al. 1975). Figure 4.1 presents the stimulatory action on PRL secretion of TRH and several DA-blocking agents in current clinical use. The administration of some of these drugs has been proposed as a clinical test of pituitary PRL reserve.

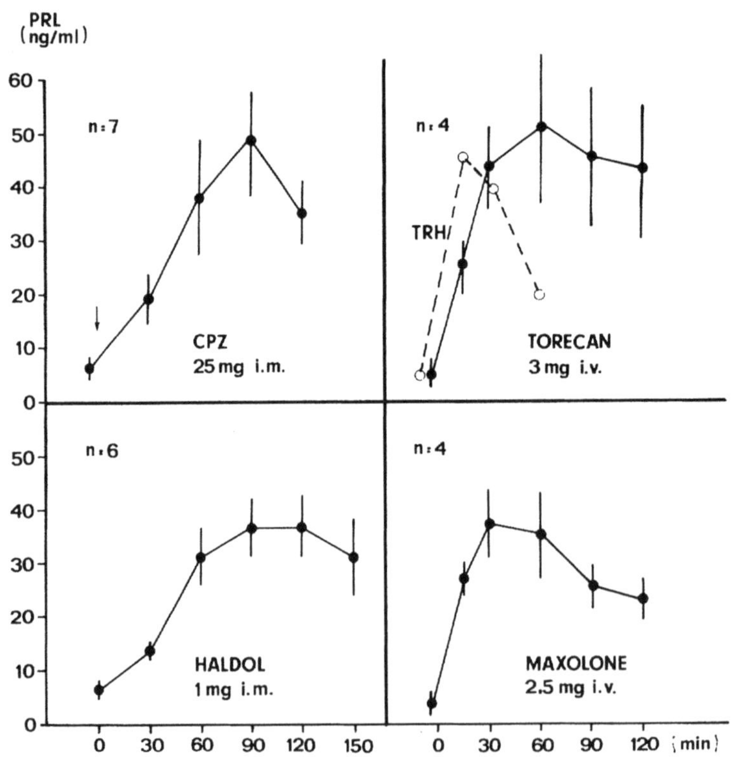

Fig. 4.1. Stimulation of PRL by different DA antagonists. The PRL response to TRH has been extrapolated for comparative purposes. *CPZ*, Chlorpromazine; *Maxolone*, metoclopramide; *Torecan*, thyethylperazine; *Haldol*, haloperidol

A single intravenous injection of 50–400 µg of TRH elevates PRL with peak values recorded within 15–20 min. (Bowers et al. 1971; Jacobs et al. 1971, 1973; Friesen et al. 1972). When TRH is administered as a continuous infusion plasma PRL increases progressively to reach a persistent plateau (Noel et al. 1974a; Rabello et al. 1974). Several authors (McNeilly et al. 1974a; McCallum et al. 1976; Judd et al. 1976; Sowers et al. 1976; Healy and Burger 1977) found that the administration of 2.5–10 mg metoclopramide to normal adult individuals would induce a burst of PRL with plasma values peaking 30 min after intravenous administration. In our experience, the injection of 2.5 mg produces similar effects to higher doses avoiding undesirable side effects such as dizziness and nausea. Intramuscular administration of 25 mg chlorpromazine has disclosed a plasma PRL elevation after a time lag of approximately 90 min (Frantz et al. 1972a; Turkington 1972a; Tolis et al. 1973; Langer et al. 1977), whereas the action of another neuroleptic drug sulpiride is more immediate (L'Hermite et al. 1972, 1978a; Thorner et al. 1974a; Mancini et al. 1976). Domperidone is a DA antagonist which does not cross the blood-brain barrier. By acting at the level of the median eminence or on the pituitary itself, this compound induces a prompt increase in plasma PRL after intramuscular administration (Pourmand et al. 1980; Delitala et al. 1980).

The site of action of some of these compounds is difficult to determine with the exception of domperidone since DA receptors have been localized in the hypothalamus and also in the pituitary gland (Benakis and Stefan 1977; Labrie et al. 1978). There is also evidence that pituitary dopamine contents may reflect changes in hypothalamic catecholamine turnover rendering difficult the interpretation of central drug actions. As recently mentioned, some compounds such as metoclopramide and also sulpiride have a more rapid onset of action than phenotiazines suggesting at least partially a direct effect on the pituitary lactotropes. In addition, the PRL-releasing effect of Torecan®, a phenotiazine derivative, when injected intravenously is somewhat delayed if compared with that of TRH and sulpiride, possibly reflecting for this compound an intermediary mechanism. TRH itself does not seem to play a relevant role in the physiologic control of PRL after Gautvik et al. (1973) were able to dissociate the response in TSH and PRL after breast stimulation by suckling and administration of TRH, indicating that the PRL elevation following breast-feeding is not induced by the release of endogenous TRH. Klindt et al. (1979) arrived at the same conclusion after using a continuous TRH infusion pump for 5 days in ovariectomized ewes. The stimulating effect of TRH on PRL is several fold above that recorded for TSH, and is assumed to take place via direct action on the pituitary (Jacobs et al. 1973) since TRH is also active on the isolated pituitary (Tashjian et al. 1971). The administration of TRH to two women after section of the pituitary stalk for the treatment of breast cancer enhanced PRL secretion lending further support to a primary pituitary site of action (Lister et al. 1974).

The response to these stimulating agents can be modified by a number of factors. Thus, peak PRL responses to TRH are lower in men and in the menopause than in normally cycling women, due to the lack of tonic effect of estrogens (Jacobs et al. 1973). Accordingly, the administration of pure estrogen or oral contraceptives can augment basal PRL secretion (Yen et al. 1974) and amplify the PRL response to neuroleptics or TRH (Buckman and Peake 1973; Carlson et al. 1973; Reymond and Lemarchand-Beraud 1976). Estrogen modulation of the TRH receptor may be an important factor in the regulation of TRH effect (Gershengorn et al. 1979).

4.1.2 Dopaminergic Drugs

Since PRL secretion in the human is under dopaminergic inhibitory control administration of L-dopa and dopamine agonists such as apomorphine and bromocriptine has led to suppression of PRL secretion in normal and hyperprolactinemic subjects (Frantz et al. 1972b; Turkington 1972b). However, the magnitude of the response to acute L-dopa administration seems to be variable among hyperprolactinemic subjects (Friesen et al. 1973) although in most cases a significant fall in plasma lactogen is recorded. The DA agonist bromocriptine and other ergot derivatives induce a prompt reduction in plasma PRL in normal and hyperprolactinemic individuals (Besser et al. 1972; del Pozo et al. 1972, 1974; Friesen et al. 1973). The use of these agents will be discussed in detail in the treatment section (p. 125).

Recently, Müller et al. (1978, 1980) have reported that nomifensine, an antidepressant drug which inhibits dopamine reuptake (Hunt et al. 1974), was able to suppress plasma PRL levels in functional hyperprolactinemia with no effect on prolactinoma patients. These authors have also shown that estrogen-induced pituitary tumors in rats may develop resistance to nomifensine after long-term treatment as judged by measurements of PRL in plasma (Müller et al. 1980). Camanni et al. (1980) found no PRL response to this drug in subjects with elevated plasma PRL and intact sella turcica, stating that these patients may harbor prolactinomas during the early stage of treatment. However, these findings are at variance with results obtained by Scanlon et al. (1977), Ferrari et al. (1979), Moriondo et al. (1980), and Crosignani et al. (1980). Thus, the usefulness of a nomifensine test in differentiating between hyperprolactinemic patients with a radiologically invisible microadenoma and those with functional hyperprolactinemia requires further evaluation. It should be kept in mind that evidence for a common central defect in hyperprolactinemic patients with and without radiologic signs of prolactinoma has been provided (Crosignani et al. 1980).

4.1.3 Insulin Hypoglycemia

Administration of insulin to normal subjects leads not only to stimulation of growth hormone and ACTH secretion but also to transient hyperprolactinemia (Frantz et al. 1972b; Wilson et al. 1972; del Pozo et al. 1977a). This effect, however, is less marked than that observed on other "stress" hormones and has little clinical applications.

4.1.4 Water Load Test

Buckman et al. (1973) proposed the use of a water-loading test to differentiate between functional and tumoral hyperprolactinemia. Although Friesen et al. (1973) were able to demonstrate a correlation between the degree of hydration, estimated as the fall in plasma osmolality, and a reduction in PRL, a consistent response which would justify the clinical use of the test has not been substantiated (Adler et al. 1975; Archer and Josimovich 1975; Fossati et al. 1976a; L'Hermite et al. 1978a).

4.2 Factors Influencing Basal Prolactin Secretion in the Clinical Practice

4.2.1 Chronobiologic Fluctuations

After integration of PRL values obtained from continuous blood sampling throughout the day, Dagget and Nabarro (1979) concluded that single PRL estimations taken under resting conditions are adequate for routine diagnostic use. There are, however, a number of factors which may influence the interpretation of single plasma measurements. Thus, circulating PRL is subjected to random fluctuations (Ehara et al. 1973; McNeilly et al. 1974b; McNeilly and Chard 1974; Leighton et al. 1976) and although this matter has not been completely solved, evidence has been provided that average plasma PRL concentrations around ovulation may be elevated in comparison with the other phases of the cycle (Robyn et al. 1973; Buckman et al. 1980). These differences and the somewhat higher plasma PRL concentrations recorded in normal women in comparison with men are obviously due to the priming effect of endogenous estrogens (del Pozo et al. 1977b). Accordingly female subjects exhibit significantly higher sleep-related PRL elevations than males (L'Hermite et al. 1972). Also blood sampling performed in the early morning hours may still be influenced by the physiologic elevations present during sleep (Parker et al. 1973). In this context it is interesting to note that the nocturnal increment in plasma PRL is reduced in hypercortisolism (Krieger et al. 1976). Some authors have reported varying PRL responses to TRH in the different stages of the menstrual cycle (Sanchez-Franco et al. 1973; Reymond and Lemarchand-Beraud 1976) but others found no such changes (McNeilly and Hagen 1974; Sawin et al. 1978). Finally, it should be mentioned that basal PRL levels can be subjected to seasonal variations in women, although geographic and racial factors may be implicated (Haus et al. 1980). Circannual fluctuations have not been detected in men (Gala et al. 1977; Robyn et al. 1977).

4.2.2 Effect of Stress and Other Factors

Surgical procedures and the effect of anesthesia have been reported to increase PRL secretion (Noel et al. 1972; Frantz 1973; Friesen et al. 1972). Later on it has been shown that during epidural block basal PRL remains unchanged, in contrast with the effect of general anesthesia (Hagen et al. 1980). It is the authors' experience that this hormone is less sensitive to stress situations than growth hormone and corticotropine as judged by the response patterns to standard stimuli such as hypoglycemia and exercise. It is interesting to mention a report by Brisson et al. (1980) in which postadolescent girls with a history of sport training exhibited higher PRL responses to exercise than untrained controls. The authors proposed a PRL-mediated mechanism in the pubertal delay observed in young female athletes.

Although the effect of venous punction may be of minor relevance for routine clinical purposes (Koninckx 1978), a fall in plasma PRL can be observed if cannulation is prolonged for several hours. Breast stimulation has been reported to increase plasma PRL in nonlactating women (Kolodny et al. 1972; Noel et al. 1974b) but inconsistency of response was reported by Archer and Josimovich (1975), and Jarrell et al. (1980) found that routine clinical examination including pelvic digital examination and breast palpation did not modify basal plasma PRL. In the authors' experience, breast stimulation during sexual forplay was unable to consistently enhance PRL secretion (del Pozo, unpublished data).

4.3 Physiopathology of Functional Hyperprolactinemia

4.3.1 Effect of Hyperprolactinemia in Women

4.3.1.1 Mechanism of Cyclic Disturbances Induced by Hyperprolactinemia

In women, galactorrhea may be found in the presence of ovulatory cycles and normoprolactinemia, but increasing plasma concentrations of this lactogen are accompanied by cyclic disturbances ranging from irregular menstrual bleeding or inappropriate luteal function to anovulatory periods and cessation of menses (Fig. 4.2). The mechanisms by which PRL exerts its anovulatory effect are still far from clear. In some instances a variety of menstrual disturbances such as anovulatory periods, oligomenorrhea and menometorrhagia can be found in the presence of moderate hyperprolactinemia (del Pozo et al. 1974; Jaszmann and Sternthal 1974) indicating that a certain periodic activity can subsist in hyperprolactinemic states. In the authors' experience, values above 70 ng/ml are almost inevitably associated with amenorrhea. Thus, initial interference with ovarian function followed by a suppressive effect at the level of the cycling center seems a likely mechanism.

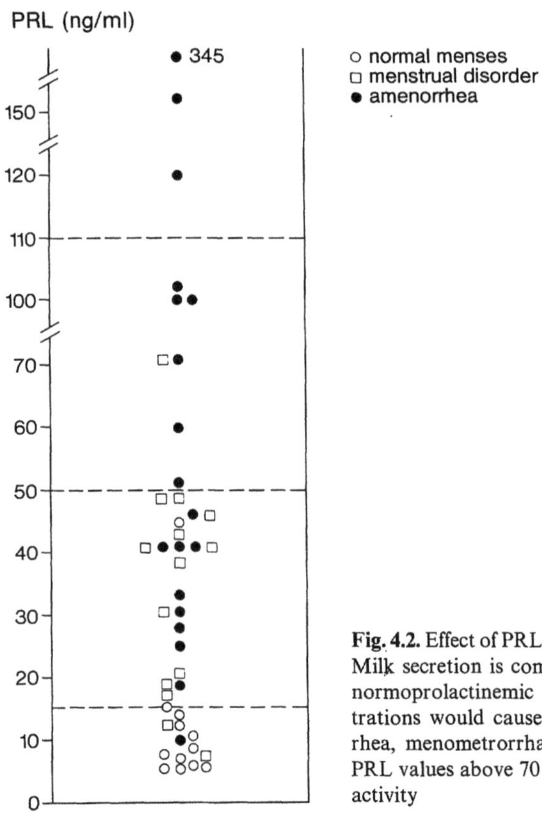

Fig. 4.2. Effect of PRL on cyclic function in galactorrheic patients. Milk secretion is compatible with normal ovulatory periods in normoprolactinemic women. Increasing plasma PRL concentrations would cause menstrual disorders, mainly oligomenorrhea, menometrorrhagia and deficient luteal function. Plasma PRL values above 70 ng/ml are usually incompatible with cyclic activity

4.3.1.2 Effect of Hyperprolactinemia on the Ovary

There is human and experimental evidence for an action of PRL on the corpus luteum. This lactogenic hormone has been found not to be necessary for the maintenance of luteal function in experimental animals (Döhler and Wuttke 1974; Flückiger and del Pozo 1978). Studies by McNatty et al. (1974) have helped considerably in understanding the regulatory role of PRL on progesterone synthesis by the corpus luteum. In vitro work by these authors has shown that synthesis of progesterone by human granulosa cells is dependent on the PRL concentration in the culture medium. Similar data have recently been reported by Mohini et al. (1980) using goat ovarian tissue. The inhibitory effect of this lactogen on human luteal function has been documented by Delvoye et al. (1974) and Robyn et al. (1976) in experiments conducted in normally menstruating women (Fig. 4.3). Elevation of plasma PRL concentrations induced with sulpiride, a neuroleptic drug with PRL-releasing properties, caused a shortening of the hyperthermic phase and a reduction in progesterone by the corpus luteum. Also L'Hermite et al. (1979) were able to

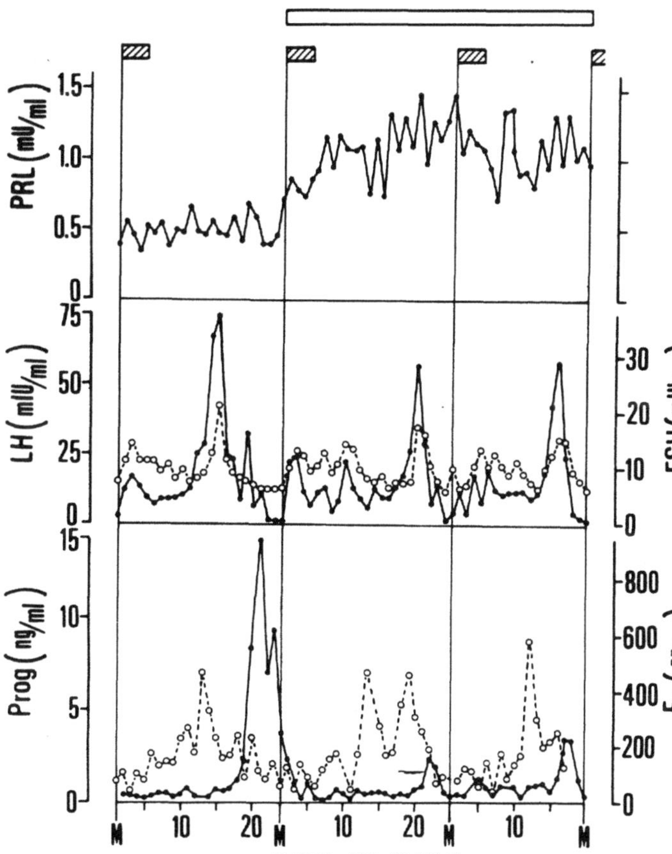

Fig. 4.3. Effect of PRL elevation with sulpiride on the hormonal sequence of the female cycle. A reduction in progesterone secretion and a displacement of the LH peak to later in the luteal phase are conspicuous. (Robyn et al. 1976)

reproduce this effect by treating normal women with tiapride, a benzamide derivative with dopamine antagonistic action. After Wenner (1975) had suggested a relationship between infertility due to a short postovulatory phase and hyperprolactinemia, del Pozo et al. (1976) reported that short luteal phase and low progesterone secretion may be associated with elevated plasma PRL concentrations in some infertile women. Seppälä et al. (1976b), were able to follow the sequence of events leading first to inappropriate luteal function and later to amenorrhea when bromocriptine treatment was discontinued in two women with a previous history of galactorrhea-amenorrhea. Also Corenblum et al. (1976) observed restoration of menses but short hyperthermic phase in two women with galactorrhea-amenorrhea syndrome who were treated with clomiphene, alone or combined with mestranol. Furthermore, insufficient PRL suppression in a case of galactorrhea-amenorrhea allowed restoration of menses but luteal progestogen production was obviously deficient. Only complete normalization of PRL brought about a prolongation of the hyperthermic phase into the fertile range (del Pozo et al. 1976) (Fig. 4.4).

In this context it should be mentioned that there is no homogeneity among hyperprolactinemic patients exhibiting different degrees of luteal inadequacy. This was demonstrated in an analysis of 44 cases of infertility due to deficient postovulatory mechanisms (Wyss and Huber, unpublished), and agrees with data which have appeared in the literature showing a lack of correlation between plasma PRL and midluteal progesterone measurements in regulary menstruating women (Sarris et al. 1978). Although other investigators found lower midluteal progesterone values and higher PRL in infertile women with ovulatory periods (Ranta et al. 1979), there was no correlation between the levels of these hormones. In the authors' experience, single midcycle progesterone measurements do not provide accurate information on the quality of the functioning corpus luteum since the induction of an adequate endometrium requires proper chronological setting and steroid exposure for a certain time. If the patients' selection is made on the basis of

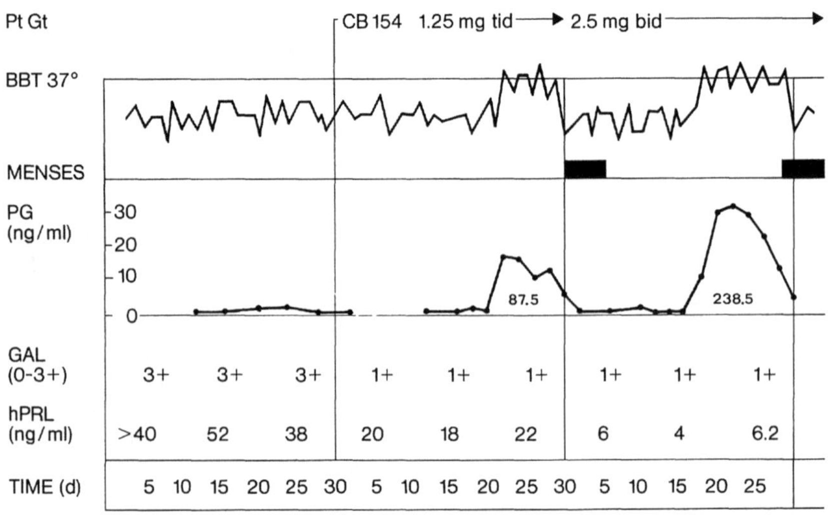

Fig. 4.4. Hyperprolactinemic short luteal phase: effect of incomplete PRL normalization (explanation in text). Integrated PG (progesterone) profiles are expressed as a luteal index (Normal > 100)

short luteal phase and hyperprolactinemia, normalization of plasma PRL is likely to reestablish adequate luteal function (see below). Moreover, ovarian sensitivity to PRL is extremely variable. Thus, moderately elevated PRL can have no effect or cause luteal inadequacy in some cases but induce amenorrhea in others.

The relationship between PRL and ovarian estrogen synthesis in the galactorrhea syndromes is not completely understood. As judged by the near normal plasma estrogen levels exhibited by some of these patients, a certain degree of follicular maturation can be assumed. In Fig. 4.5 taken from Kandeel et al. (1979), basal PRL values are plotted against the sum of estradiol and estrone increments following clomiphene treatment. There is a negative correlation between estrogen production and PRL. Data by McNatty et al. (personal communication) collected from human granulosa cell cultures, indicate that exposure to increasing PRL concentrations in the medium does not modify estrogen synthesis by the cultured cells, whereas progesterone production is impaired. Indeed, ovarian response to exogenous gonadotropin stimulation is maintained in hyperprolactinemic amenorrhea (del Pozo et al. 1975d; Archer and Josimovich 1976; Kemmann et al. 1977; Aono et al. 1978a) and in normally cycling women in whom hyperprolactinemia was induced by sulpiride (Aono et al. 1978b). More recently, Caro and Woolf (1980) were able to show that the estradiol elevation that follows the gonadotropin surge induced by exogenous LHRH was not different in normal and hyperprolactinemic women. The

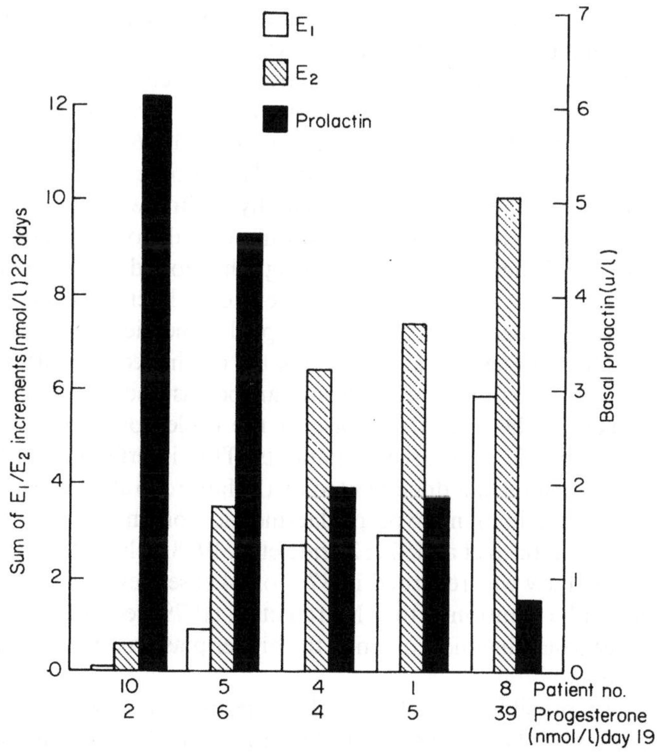

Fig. 4.5. Negative correlation between increasing plasma PRL values and estrogen concentrations. (Kandeel et al. 1979)

finding that hypoestrogenism of hyperprolactinemia may be accompanied by bone mineral loss requires further evaluation (Klibanski et al. 1980; Kollin et al. 1981).

4.3.1.3 Effect of Hyperprolactinemia on Gonadotropin Release

Excessive PRL secretion is frequently associated with amenorrhea, and delayed puberty or primary amenorrhea may result when hyperprolactinemia has been present since before the time of menarche (Jürgensen et al. 1976; Bergh et al. 1977; Kleinberg et al. 1977; Kemmann and Jones 1979). That this effect of PRL is exerted at suprasellar level is demonstrated by the normal to exaggerated LH response found after LHRH administration in such patients (Zarate et al. 1973b; del Pozo et al. 1974; Thorner et al. 1974a; Archer et al. 1976; Healy et al. 1977; Forsbach et al. 1977). However, reduced gonodotropin reserve has been described in hyperprolactinemic women after prolonged LHRH stimulation (Caro and Woolf 1980).

Bohnet et al. (1974) were the first to show that hyperprolactinemia induced with TRH would disturb LH pulsatility in normal women, knowing that a fluctuating gonadotropin pattern is essential for ovulation to occur. Indeed, idiopathic hyperprolactinemia is characterized by absence of rhythmicity in the secretion of LH (Bohnet et al. 1976). In another report, however, Boyar et al. (1974) restricted the absence of LH pulsatility to hyperprolactinemia associated with pituitary adenomas after normal LH surges were recorded in some patients with functional hyperprolactinemia, although a trend towards dampened gonadotropin release during episodes of active PRL hypersecretion was discernible. No particular LH fluctuating pattern has been found in male hyperprolactinemics with secondary hypogonadism (Thorner and Besser 1977).

The basic cyclic disturbance in hyperprolactinemic anovulation seems to involve blockade of estrogenic receptors at the level of the hypothalamus. Several investigators (Glass et al. 1975; Aono et al. 1976; Schaison et al. 1977; Forsbach et al. 1977) have reported a lack of LH response to an estrogen challenge in these patients in comparison with normally cycling women. Furthermore, L'Hermite et al. (1978b) were able to show that in normal women rendered hyperprolactinemic with sulpiride, a neuroleptic drug, the gonadotropin response to an estrogen challenge was abolished. As will be shown later, PRL normalization has restored positive estrogen feedback in hyperprolactinemic patients (Aono et al. 1979a). Thus, receptor interference at the level of the cycling center with abolition of the positive estrogenic feedback mechanism can be assumed. Figure 4.6 shows this effect in women in the follicular phase of the cycle compared with normo- and hyperprolactinemic amenorrheic patients. This interference mechanism on estrogenic receptors is further documented by the failure to elicit an LH surge after clomiphene stimulation in many hyperprolactinemic women (Thorner et al. 1974b; Healy et al. 1977; Forsbach et al. 1977; Zarate et al. 1978; Glass et al. 1979). Indeed, clomiphene insensitivity was recorded in 87% of 74 cases reviewed by the author. However, it should be mentioned that March et al. (1979) recorded a high rate of ovulation in hyperprolactinemic women in the higher plasma estrogen range found among these patients.

The possible role of central progestogenic receptors in PRL-mediated anovulation has been the subject of recent work. After it was shown that progesterone can induce the release of LH and FSH in estrogen-primed menopausal women (Odell and Swerdloff 1968; Nillius and Wide 1971), Rakoff and Yen (1978)

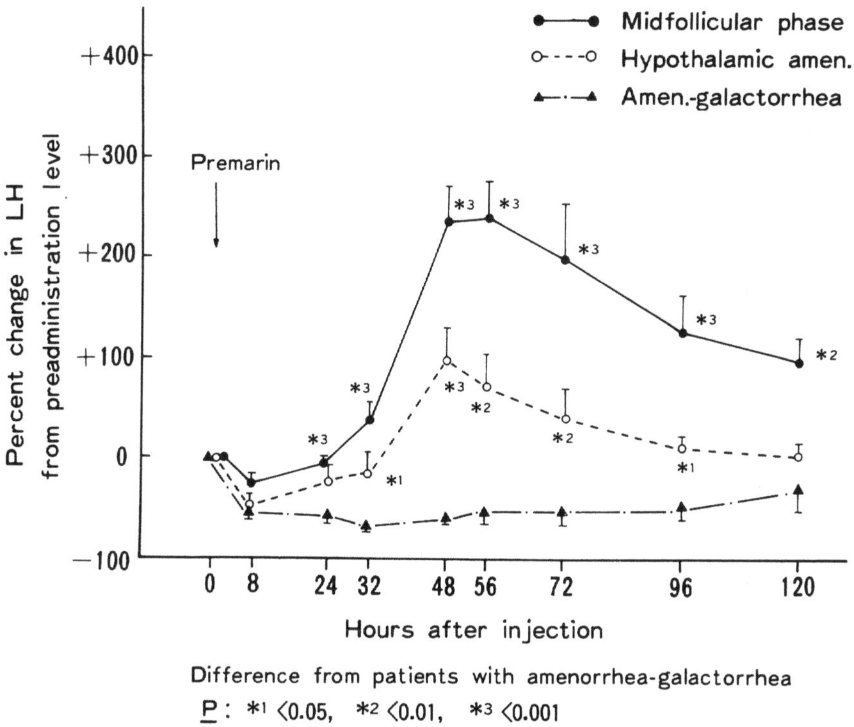

Fig. 4.6. Absence of positive estrogen feedback in hyperprolactinemic amenorrhea. (Aono et al. 1976 b)

conducted a study in hypogonadal women including four cases of hyperprolactinemia: the sequential administration of estradiol benzoate and progesterone failed to modify basal gonadotropin secretion. The authors interpreted this finding as indicative of hypothalamic dysfunction. Interestingly enough, they were able to dissociate the positive feedback of estrogens on endogenous LHRH, which was abolished, from the amplifying effect of sex steroids on the LH response to exogenous LHRH, which was preserved. More recently, Travaglini et al. (1979) demonstrated that the administration of progesterone to estradiol-primed hyperprolactinemic women was able to induce an LH surge. Thus, it can be speculated that in situations of inappropriate PRL secretion the hypothalamic resistance to the effect of estrogen may be mediated also by progestogenic receptors, presumably in connection with dopaminergic mechanisms (Rakoff and Yen 1978). However, results of dopamine receptor stimulation on LH release in hyperprolactinemia are controversial. Whereas Lachelin et al. (1977b) reported a significant fall in plasma LH after acute administration of bromocriptine based on percentage decreases, Evans et al. (1980) found no effect after evaluating cumulative basal values. In this context it should be mentioned that prolonged treatment with the dopamine agonist bromocriptine has failed to modify cyclic gonadotropin secretion in normal women (del Pozo et al. 1975a; Schulz et al. 1978), in contrast with data collected by Lachelin et al. (1977) in an acute trial. Thus, results of work on modification of dopaminergic control of LH release in humans should be interpreted cautiously unless a biologic effect is clearly demonstrated in a chronic trial.

4.3.1.4 Adrenal Function in Hyperprolactinemia

Earlier reports by Bigazzi et al. (1974) Bassi et al. (1977) and Carter et al. (1977) indicated excessive production of Δ_5-androgens (androstenediol, dehydroepiandrosterone-sulfate) in hyperprolactinemic women with long-standing amenorrhea, while Δ_4-androgens (androstenedione and testosterone) were found to be within the normal range. These findings have been subsequently confirmed (Vermeulen et al. 1977; Vermeulen and Ando 1978). Since dehydroepiandrosterone-sulfate is almost exclusively synthesized in the adrenals (Kalliala et al. 1970), whereas androstenedione and testosterone come partially from the ovary as well, it was proposed that hyperprolactinemia would stimulate the adrenal Δ_5 synthetic pathway without affecting ovarian androgenic production. Indeed, suppression of PRL secretion with bromocriptine has led to normalization of plasma adrenal androgens (Bassi et al. 1977; Carter et al. 1977). It should be pointed out, however, that an effect of PRL on adrenal sex steroid synthesis is restricted to pathologic hyperprolactinemia (Giusti et al. 1978) since no such effect has been demonstrated for instance in gestational women (del Pozo et al. 1980). However, it should be mentioned that Turpin et al. (1979) have found no difference in plasma concentrations of androgenic precursors in hyperprolactinemic women when compared with normal controls.

4.3.1.5 Hyperprolactinemia and the Polycystic Ovary Syndrome (PCO)

Elevated PRL has been found in patients with PCO (Duignan 1976; Falaschi et al. 1977). On the other hand, the presence of sclerocystic ovaries has been reported in some instances of classic galactorrhea-amenhorrhea (Lavric 1969; Thorner et al. 1974a). Although there is evidence for a distorted relationship between estrogen feedback and the hypothalamic control of pituitary gonadotropes and lactotropes in both conditions, the mechanisms involved deserve closer attention since estrogens are usually low in galactorrhea-amenorrhea and elevated in PCO. Thus, increased sensitivity to clomiphene and adequate LH response to exogenous estrogen in PCO reflects intergrity of central estrogenic-mediated mechanisms. As discussed previously, these are blocked in galactorrhea-amenorrhea. In the former the estrogen pool is augmented: absolutely by enhanced peripheral steroid synthesis (estrone), and relatively through decreased circulating steroid-binding globulin. Thus, it can be assumed that pituitary priming by estrogens can augment PRL secretion in PCO. However, the validity of this proposed mechanism is not free of criticism since here excessive pituitary estrogenization is only relative: estrone is a metabolite of lower estrogenic activity than estradiol and, accordingly, basal plasma PRL levels could not be correlated with estrone concentrations in PCO patients (Falaschi, unpublished). Kandeel et al. (1977) found no PRL stimulation by this steroid when administered to normally cycling women, whereas estradiol showed an immediate effect. On the other hand, van der Gugten et al. (1970) found adequate lactotrope cell priming by estrone in the experimental animal.

Also the effect of androgens on PRL secretion in PCO has not been clearly established. Acute administration of testosterone has failed to modify PRL levels in normal volunteers despite a significant fall in plasma LH (Martin-Perez et al. 1979), this being in agreement with data published by D'Agata et al. (1979). However, prolonged exposure of the pituitary gland to exogenous testosterone has reportedly stimulated PRL secretion in rats (de las Heras and Negro-Vilar 1979). Thus, enhancement of PRL release after chronic stimulation by androgens, presumably through aromatization into estrogens, remains another possibility in PCO. Further clinical and differential diagnostic features of PCO are presented on p. 120.

4.3.1.6 Prolactin and Premature Menopause

Thorner et al. (1974a) and Darragh et al. (1976) have recognized the presence of high plasma or urinary gonadotropins in some middle-aged women with galactorrhea-amenorrhea mimicking premature menopause. Ovulatory cycles resumed upon normalization of PRL secretion with bromocriptine, and at least in one patient pregnancy ensued. The cause of the elevated gonadotropin secretion in this situation is far from clear but one explanation could be central dopamine deficiency. This has been proposed for the disturbance in gonadotropin secretion typical of other conditions unrelated to ovarian failure (Yen 1980). In this context it should be mentioned that in the menopause, characteristically elevated plasma LH canot be reduced by dopaminergic stimulation with bromocriptine (Gerber and del Pozo, to be published).

4.3.1.7 Diabetogenic Effect of Prolactin

Diabetogenic properties have been attributed to PRL (Harter et al. 1978) probably due to similarity regarding structure and action with growth hormone. Berle (1973) found no changes in blood glucose after intravenous administration of human PRL to normal women but Tourniaire et al. (1974) observed reduced glucose tolerance in hyperprolactinemia accompanying pituitary tumors. Landgraf et al. (1977) confirmed these findings and found an elevated and prolonged insulin response to glucose challenge. Normalization of the excessive glucose and insulin reaction was restored after PRL suppression with bromocriptine. Steininger and Sulm (1978) subjected six hyperprolactinemic women to an oral glucose tolerance test and found a normal profile in five. In one case exhibiting a diabetogenic type of reaction, treatment with bromocriptine normalized glucose levels but in another case the hyperglycemic type of response was exaggerated by treatment despite a 93% fall in circulating PRL. Thus, although glucose utilization can be impaired in hyperprolactinemia a wide range of variability can be observed among patients. Recently, Johnston et al. (1980) described a state of hyperinsulinemia in hyperprolactinemic women. There were increased responses to meals in blood lactate, piruvate and alanine, as well as a negative correlation between mean insulin and 3-hydroxybutirate, all this reflecting a state of relative hyperinsulinism.

4.3.1.8 The Psychogenic Component of Hyperprolactinemia

With the advent of new laboratory sensitive techniques to measure human plasma PRL and with the confirmation of the causative role of this lactogen in the galactorrhea-amenorrhea syndromes, little attention has been paid to the possible etiologic implication of psychogenic factors. A relationship between dopaminergic pathways, PRL, and mental illness has been well established. Neuroleptic drugs enhance PRL secretion and improve the mental condition of schizophrenic subjects, and there is a good correlation between the endocrine effect and the antipsychotic action. A number of other compounds interfering with some step of catecholamine metabolism such as reserpine and α-methyl-dopa, may cause depression and at the same time elevate plasma PRL. There are, however, clinical observations suggesting that plasma PRL concentrations are not directly connected with the underlying physiopathology of mental disease. Thus, after Meltzer and Fang (1976) recognized that a certain relationship existed between the diminishing psychotic behavior and rising PRL secretion, it was recently demonstrated that normalization of PRL secretion with bromocriptine administered concomitantly with neuroleptic drugs did not interfere with the antipsychotic effect of the latter (del Pozo, to be published). Thus, plasma PRL changes may be useful as a therapeutic indicator but they do not reflect subtile changes of dopamine metabolism in localized areas of the brain. Indeed, Beumont et al. (1974a, b) and Kolakowska et al. (1975) have reviewed the endocrine effects of antischizophrenic therapy and found that Parkinsonian symptoms and tardive diskinesia were relatively frequent, whereas basal plasma PRL is normal in Parkinson's disease, a condition characterized by focal dopamine deficiency (Polleri et al. 1977).

4.3.2 Effect of Hyperprolactinemia in Male Subjects

Although PRL circulates in male blood in appreciable concentrations its physiologic role has not been clarified. The lack of lactational requirements does not preclude that under adequate priming the male mammary gland will respond to a PRL challenge with milk production. Thus, Wyss and del Pozo (unpublished data) found that PRL stimulation with TRH was able to induce milk secretion in a male individual pretreated with estrogens. Certainly, the chronic ingestion of dopamine antagonists or estrogens may lead to sustained hyperprolactinemia, and the same effect can be expected in male subjects on chronic estrogen therapy of prostatic cancer or transsexualism (Frantz 1973; del Pozo, to be published). Although not yet described, hyperprolactinemia can be predicted in male subjects harboring an estrogen-producing adrenal tumor since the pituitary lactotropes seem very sensitive to the amplifying effect of estrogens. Recently, hyperprolactinemia was described in a male subject with a testicular tumor secreting estrogen. Removal of the source of estradiol was followed by normalization of plasma PRL (Perez et al. 1980).

A possible relationship between PRL and pubertal gynecomastia seems questionable. Normal plasma PRL has been reported in this situation (Turkington 1972c) although mean diurnal and sleep values may be higher than in control subjects (Large et al. 1980). The self-limiting character of the condition in adolescent subjects makes the study of a causal relationship difficult.

Testicular androgen synthesis is usually reduced in hyperprolactinemic males. The central inhibitory effect of PRL can be localized at suprasellar level since a positive LH response to exogenous LHRH has been recorded in idiopathic (Franks et al. 1978; Carter et al. 1978) or provoked hyperprolactinemia (Falaschi et al. 1978). The degree of response, however, can be reduced in the presence of prolactinomas (Nagulesparen et al. 1978). A disturbance of endogenous control of LHRH can be assumed on the basis of animal experimental data but the primary central mechanism causing gonadal insufficiency remains uncertain. Most male patients with hyperprolactinemia present with normal basal plasma LH (Thorner et al. 1974) but no reports are available on the biologic efficacy of the latter hormone nor have its secretory rhythms been studied. Similar to hyperprolactinemic women, the male

PRL response to a TRH challenge is blunted (see below), probably due to a primary hypothalamic DA disturbance or by PRL itself via a short loop mechanism. It is interesting to note that Thorner and Besser (1977) found no changes in the plasma gonadotropin hourly profiles in hyperprolactinemic males after normalisation of PRL secretion as compared with untreated profiles. This questions the role of episodic secretion in the maintenance of male gonadal function (see previous section).

Despite experimental evidence of a direct action of PRL at the level of the gonad, this effect has not been clearly demonstrated in man. Hypoandrogenic males with elevated plasma PRL usually exhibit adequate sex steroid responses to exogenous chorionic gonadotropin (Friesen et al. 1973; Fossati et al. 1976; Franks et al. 1978; Carter et al. 1978). Low response rates can be considered normal in these subjects if basal plasma testosterone levels reflect the functional capacity of the Leydig cells. Therefore, absolute androgen increases after stimulation with chorionic gonadotropin in hyperprolactinemia should be considered cautiously when compared with normal subjects, since these exhibit higher basal androgen production and accordingly higher reserve capacity.

A direct effect of PRL on spermatogenesis seems unlikely and limited experience by the authors in male hyperprolactinemia rather indicates androgen dependency in defective sperm maturation (del Pozo, to be published). There is no clear evidence for a role of PRL in oligospermia accompanied by normal testosterone and lactogen levels. This is supported by the poor response following PRL inhibition (see treatment section).

4.4 Control of Lactotrope Cell Function in Pathologic Hyperprolactinemia

4.4.1 Effect of Dopaminergic Blockade and Stimulation with TRH

The effect of dopaminergic blockade in hyperprolactinemia has been the subject of a series of publications, mainly with the purpose of discriminating between functional and tumoral processes. In hypothalamic disease without radiologic evidence of pituitary lesions there is a reduction in the amplitude of response to chlorpromazine, haloperidol or metoclopramide (del Pozo et al. 1974; Malarkey 1975; Archer and Josimovich 1975; Faglia et al. 1977; Healy et al. 1977; Yen 1978). It is noteworthy that the response becomes more attenuated with increasing basal PRL levels, indicating that in mild to moderate hyperprolactinemia residual dopaminergic activity can still be present. Usually plasma PRL profiles after TRH challenge parallel those of dopaminergic blockade (del Pozo et al. 1974; Snyder et al. 1974; Franks and Jacobs 1977; van Look et al. 1977; Healy et al. 1977; Wajchenberg et al. 1979). As will be discussed in the chapter dealing with prolactinomas, tests elaborated up until now do not seem to facilitate the differentiation between tumorous and functional hyperprolactinemias. It should be noted that responses to TRH stimulation and dopaminergic blockade in estrogen-mediated hyperprolactinemias such as in pregnancy, polycystic ovary syndrome or following estrogen treatment, are usually normal or exaggerated (Guitelman et al. 1978; Ylikorkala et al. 1979; Falaschi et al. 1980; Massara et al. 1980; del Pozo, to be

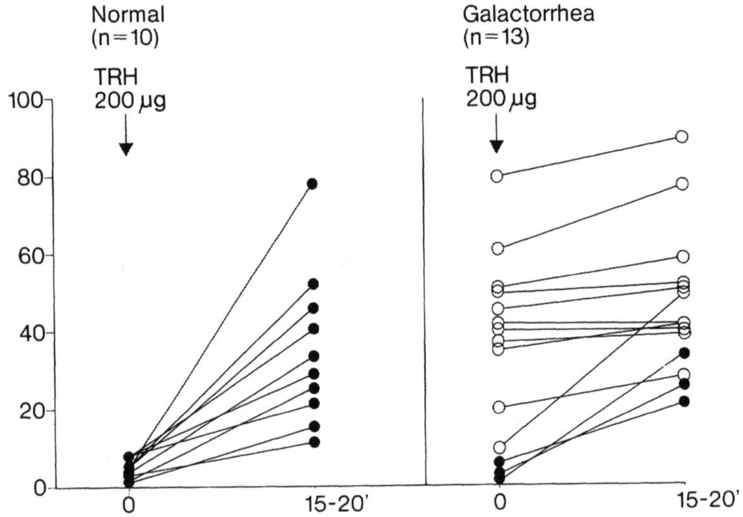

Fig. 4.7. Blunted or absent PRL response to a TRH challenge in hyperprolactinemia. It is noteworthy that the response becomes more atenuated with increasing basal PRL levels, indicating a certain degree of residual dopaminergic control in moderatw hyperprolactinemia. *Black dots* on *right panel* indicate regular menstrual periods

published). This is in contrast with the resistance to the stimulating agents recorded in classic galactorrhea-amenorrhea indicating that hyperprolactinemia due to estrogens does not alter the dopaminergic control of lactotrope cells (Figs. 4.7 and 4.8).

4.4.2 Circadian Rhythm of Prolactin in the Galactorrhea Syndromes

Prolactin secretion is subjected to a nyctohemeral rhythm with sleep, unrelated to daytime, as the main determinant of PRL elevations. Also frequent blood sampling shows episodic, although aperiodic, PRL bursts, occurring both during states of wakefulness and sleep (McNeilly et al. 1974b). The circadian rhythm is preserved in gestational hyperprolactinemia and in other situations of estrogen dominance. In pathologic states, blunting or absence of the nocturnal PRL elevation has been reported (Boyar et al. 1974; Malarkey 1975) although exceptions are possible (Kapen et al. 1975). Figure 4.9 presents the nyctohemeral PRL profile in a normal female as compared with postpartum hyperprolactinemia and a case of galactorrhea-amenorrhea. The sleep-related peak is missing in the latter, although the episodic day-time variation seems to be preserved.

The mechanisms involved in the abolition of the circadian PRL rhythm are far from clear. There is, however, indirect evidence for an endogenous blockade of dopaminergic mechanisms controlling PRL secretion. Thus, Malarkey (1975) reported not only blunting of the PRL response to chlorpromazine but also a reduction in the nocturnal elevation with progressively increasing basal PRL.

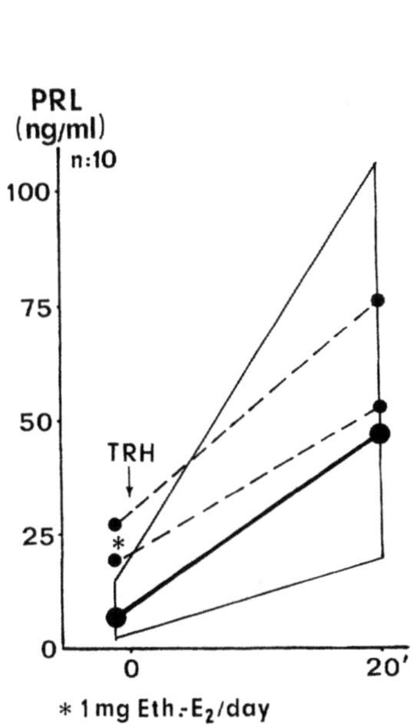

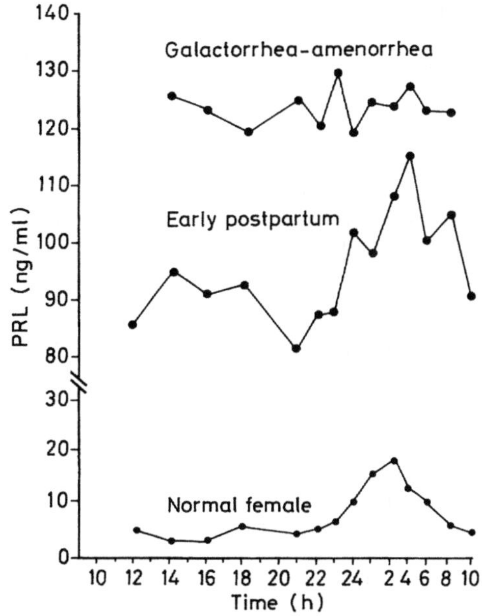

Fig. 4.8. **Fig. 4.9.**

Fig. 4.8. Normal PRL response to 200 μg TRH in two estrogen-primed male subjects despite moderate elevation of basal lactogen. *Continuous lines* represent mean values and range recorded in ten normal men

Fig. 4.9. Nyctohemeral PRL rhythm in a normal female as compared with a woman in the early postpartum and a case of galactorrhea-amenorrhea. The sleep-related PRL elevation is missing in the latter

4.5 Incidence of Hyperprolactinemia

Table 4.1 presents the incidence of hyperprolactinemia in a collective of 1969 amenorrheic women taken from 11 different sources. The frequency rate averages 24.2%, ranging from 11% to 47%. However, only 55,6% of all hyperprolactinemic cases exhibit galactorrhea, so that the typical biologic manifestation of elevated circulating lactogen is missing in a high percentage of cases. It should be pointed out that the symptoms leading the patient to seek medical advice, the patients' referral system, the technical support of a given clinical unit, and the physician's orientation toward fertility problems may represent important factors biasing the cumulative evaluation of hyperprolactinemic cases. Thus, Pepperell et al. (1977a) selected 37 anovulatory women on the basis of clomiphene resistance and found elevated PRL in 20. Clomiphene resistance, however, is frequent in hyperprolactinemia (see p. 110). Subjects with galactorrhea may be referred to an endocrinology unit, but if infertility or amenorrhea is the main feature they will be seen by a gynecologist.

As recorded in females, the incidence of hyperprolactinemia also varies in male individuals according to the selection procedures. Thus, this lactotrope has been found elevated in 2 of 65 hypogonadal males screened by the authors. However, and as will be shown in the next chapter, the incidence can increase to 65% in the

Table 4.1. Prevalence of hyperprolactinemia in amenorrhea

Reference	No. of Patients	Hyperprolactinemia (%)	Galactorrhea as % of hyperprolactinemia
Franks et al. (1975)	40	20	35
Seppälä et al. (1975)	34	47	53
Canales et al. (1976 a)	116	28	0
Bohnet et al. (1976)	127	13	53
Jürgensen et al. (1976)	305	11	–
Rjosk et al. (1976)	445	16	89
Bergh et al. (1977)	287	15	87
Seppälä et al. (1977)	123	33	38
Shearman and Fraser (1977)	90	39	83
Leyendecker et al. (1977)	50	24	66
Lunenfeld et al. (1980)	352	21	52
	Σ1969		
Mean and range		24.2 (11–47)	55.5 (0/35–89)

presence of pituitary tumors (Fossati et al. 1976b; Franks et al. 1978; Carter et al. 1978; Nagulesparen et al. 1978). On the other hand, oligospermia in the absence of endocrine abnormality exhibits a wide incidence of hyperprolactinemia, between 3.6% (da Rugna, unpublished), 9% and 40% (Roulier et al. 1976; Boucher et al. 1977).

4.6. Inappropriate Prolactin Secretion: A Classification Under Consideration of Etiologic Factors

4.6.1 Historical Background

The clinical picture of the galactorrhea-amenorrhea syndromes had been known for more than a century before their endocrine background was elucidated. In 1855 Chiari et al. described a syndrome of persistent postpartum lactation and amenorrhea, accompanied by uterine involution. The clinical symptoms and signs of this syndrome were clarified further by Frommel in 1882. A few more years elapsed before Argonz and del Castillo (1953) reported on a similar condition unrelated to pregnancy, and Forbes et al. (1954) described a syndrome of galactorrhea and amenorrhea with pituitary tumors. Thus, the presence of a radiologically normal sella and the absence of any neurologic lesions are generally thought to favor the diagnosis of hypothalamic dysfunction rather than of pituitary tumors. However, the reliability of these criteria is doubtful, since lesions which appear clinically hypothalamic may evolve to frank pituitary adenomas (Maas 1967; Young et al. 1967; Mahesh et al. 1969; Sandler and Gardner 1973; Gates et al. 1973). Lactotropic tumors involving the pituitary gland will be discussed in the next chapter.

4.6.2 Drugs as Causative Factors

With the increased availability of oral contraceptives, some women have developed secondary amenorrhea and galactorrhea during the use of these drugs or immediately after their withdrawal (Friedman and Goldfien 1969; Shearman and Turtle 1970; Rifkin et al. 1972; Shearman and Smith 1972; Hughes et al. 1972; Starup 1972; Tyson et al. 1975; Lavric 1975). Certainly, the prolonged intake of estrogens, in male subjects also, as observed in the treatment of prostatic carcinoma and in transsexuals, can lead to hyperprolactinemia (Frantz 1972b; del Pozo, to be published). Falaschi et al. (1978) found that prolonged administration of the dopamine antagonist metoclopramide can impair spermatogenesis in normal men, presumably secondary to drug-induced hyperprolactinemia. Also the prolonged use of neuroleptics, especially of the phenotiazine type, has frequently led, in women, to cessation of cyclic activity and to initiation of milk secretion (Apostolakis et al. 1972; Beumont et al. 1974c; de Rivera et al. 1976). There, is however, a prompt return to normality on discontinuation of the stimulatory agent, within a 2- to 8-week period (Turkington 1972b; Aono et al. 1978b). This is not always the case after use of oral contraceptives, although this clinical entity has better prognosis than the idiopathic type (see p. 125).

The effect of sex steroid antagonists has also been the subject of recent investigation. Although clomiphene, an estrogen receptor blocking agent, was first claimed to inhibit puerperal lactation (Zuckerman and Carmel 1973), Weinstein et al. (1976) found only partial suppression and Canales et al. (1976a) no action at all. Whereas acute administration of testosterone has no effect on basal plasma PRL (D'Agata et al. 1979; Martin-Perez et al. 1979), its antagonist cyproterone acetate has been found to elevate this lactogen in normal individuals (Gräf et al. 1978) in sexual offenders (Jeffcoate et al. 1980) and in precocious puberty (Fonzo et al. 1977). Its mechanism of action, however, remains obscure.

Morphine and methadone have been found to elevate plasma PRL levels (Tolis et al. 1975; Gold et al. 1977). This effect is believed to occur via dopaminergic pathways since it can be blocked by apomorphine, L-dopa and bromocriptine. In the clinical field, menstrual disturbances have been found in up to 72.5% of women taking heroin (Gaulden et al. 1964), and galactorrhea-amenorrhea has been described in such patients (Pelosi et al. 1974). Hallmarks of the syndrome were low plasma gonadotropins and signs of hypoestrogenism. Hypothalamic blockade by the opiate was proposed as the mechanism, this postulate being recently supported by del Pozo and Köbberling (1981) who showed abolition of LH pulsatile pattern after administration of a met-enkephalin to normal women. These substances possess specific affinity for brain morphine receptors, and the effect can be counteracted by simultaneous administration of naloxone, an opiate receptor blocking agent.

The PRL-releasing mechanism of some substances such as cimetidine (Carlson and Ippoliti 1977), a histamine$_2$ blocking agent, needs to be reevaluated since results of chronic administration may differ from the acute effect (delle Fave et al. 1977; Majumdar et al. 1978; Healy et al. 1979; Masala et al. 1980a).

4.6.3 Hyperprolactinemia Associated with Other Endocrinopathies

Galactorrhea and associated hyperprolactinemia have been recognized in a number of endocrinopathies such as hypothyroidism (Ross and Nusynowitz 1968; Lederer and Marchandise 1971; Boroditsky and Faiman 1973; Laurent et al. 1973; Fossati et al. 1975; Keye et al. 1976; Aono et al. 1977a), Cushing's syndrome (Mahesh et al. 1969; Berlinger et al. 1977) and acromegaly (Frantz et al. 1972a; Turkington 1972d; Franks et al. 1976; Aono et al. 1976a; Tolis et al. 1978b; Peillon et al. 1979). Hyperprolactinemia, not necessarily associated with galactorrhea, has been described in a variety of hormonal-mediated conditions such as adrenal hyperplasia (Boyar and Hellman 1974) and in stress situations (Frantz et al. 1972a; Noel et al. 1972).

Changes in PRL secretion in thyroid disease deserve closer attention. Plasma concentrations following a TRH challenge and estimation of the dischargeable stores were significantly higher in patients with primary hypothyroidism than in normal subjects (Foley et al. 1972; Refetoff et al. 1974; Yamaji . 1974; Wartofsky et al. 1976), and PRL response patterns can be normalized by substitutive treatment (Keye et al. 1976; Aono et al. 1977b; Tolis et al. 1978a). The opposite phenomenon has also been described in a woman with hyperthyroidism due to TSH hypersecretion and the syndrome of hyperprolactinemic amenorrhea (Benoit et al. 1980). Although galactorrhea has already been reported in Grave's disease (Zondek et al. 1951), TSH-induced thyrotoxicosis can be associated with pituitary tumors secreting both TSH and PRL (Horn et al. 1976; Duello and Halmi 1977). It seems that the thyroid status can also influence PRL release by common dopaminergic mechanisms (Feek et al. 1980).

Previous work by different investigators has demonstrated a relatively high incidence of hyperprolactinemia in patients with the syndrome of polycystic ovaries by standard diagnostic criteria (Duignan 1976; Falaschi et al. 1977). The coexistence of galactorrhea-amenorrhea and sclerocystic ovaries was recognized by Forbes et al. (1954) in their early description of inappropriate milk secretion in the presence of a pituitary tumor. Some clinical and biochemical features in several patients were reminiscent of the polycystic ovary syndrome. Subsequently, other confirmatory reports on this relationship have appeared (Zarate et al. 1968; Lavric 1969; Gambrell et al. 1971; Hughes et al. 1972; Thorner et al. 1974a; Seppälä and Hirvonen 1975). These patients exhibit adequate PRL response to TRH and dopaminergic blockade (Falaschi et al. 1980), and the hyperprolactinemia is only moderate. It is assumed that the increased circulating sex steriods characteristic of the polycystic ovary syndrome are responsible for the lactotrope cell stimulation. Thus, its well known that in situations of estrogen dominance, the lactotrope response to a TRH challenge or to dopaminergic blockade is augmented (Reymond and Lemarchand-Beraud 1976; Bellmann 1977; Guitelman et al. 1978) and exogenous estrogens can amplify the PRL response to dopaminergic blockade (Buckman and Peake 1973). The hypothalamic disturbance of hyperprolactinemic anovulation is lacking in this condition as judged by the positive response to clomiphene treatment and the persistence of a positive estrogen feedback on LH. Biochemical differentiation from classic galactorrhea-amenorrhea can be made on the basis of the preferential Δ_5 pathway for PRL stimulation of the adrenal cortex to produce dehydroepiandrosterone sulfate without effect on the Δ_4 (androstenedione)

Table 4.2. Endocrine profiles of polycystic ovary syndrome and hyperprolactinemia

Parameter	PCO	PRL↑	Parameter	PCO	PRL↑
FSH	↓	N/↑	E_2	N	↓
LH	↑↑↑	N	A-dione	↑	N
PRL	(↑)	↑↑	E_1	↑	↓
DEAS	↑	↑	SHBG	↓	N/(↓)
DEA	↑	(↑)	TRH/PRL	↑	↓
T(DHT)	↑	N	DAØ/PRL	↑	↓

Abbreviations: FSH/LH: Gonadotrophins; PRL: Prolactin; DEA (DEAS): Dehydroepiandrosterone (Sulphate); T (DHT): Testosterone (Dihydro-T); E_2: Estradiol; A-dione: Androstenedione; E_1: Estrone; SHBG: Sex steroid binding globulin; TRH/PRL: Prolactin response to TRH; DAØ/PRL: Prolactin response to Dopaminergic blockade

pathway favored by the ovary (Bassi et al. 1977; Giusti et al. 1978). In addition, low concentrations of steroid-binding globulin can enhance the biologic activity of some circulating steroids in the polycystic ovary syndrome whereas this globulin has been found only moderately reduced in galactorrhea-amenorrhea (del Pozo, unpublished data).

Hormonal profiles of the galactorrhea and PCO syndromes are sumarized in Table 4.2. In addition to the disturbances in steroid metabolism there are differences involving basal LH and its pulsatile pattern and also its response to LHRH. The possibility of simultaneous occurrence of both hyperprolactinemia and polycystic ovaries as two separate entities should also be kept in mind (Futterweit and Krieger 1979).

An interesting clinical observation is the implication of PRL in cases of premature menopause. Thorner et al. (1974a) and Darragh et al. (1976) have reported on middle-aged hyperprolactinemic women with anovulation and elevated basal gonadotropins consistent with early menopause. Suppressive PRL treatment, however, restored normal cyclic activity as demonstrated by the occurrence of normal ovulation and pregnancies. The causative mechanism of this condition remains unknown (see p. 113).

A considerable number of patients seeking medical advice can exhibit mild galactorrhea in the presence of regular mestrual periods and normoprolactinemia. This finding is fairly common among women with a history of previous pregnancy (Lavric 1975). Evaluation usually reveals normal hormone cyclic profiles with ovulatory periods and adequate response to TRH stimulation or dopaminergic blockade. PRL suppressive treatment is not required except in rare cases of more severe galactorrhea which may cause psychogenic stress to the patient.

Although the hormonal etiology of breast fibrosis is still the subject of controversy, a higher incidence of moderate hyperprolactinemia has been reported in this condition (Bischoff et al. 1980). This is in contrast with findings collected by Hiba and del Pozo (unpublished) in 72 cases of fibrocystic disease showing that basal PRL values were not different from normal age-matched controls. Finally, PRL deficiency has been described in a male subject and in two families with pseudohypoparathyroidism (Spitz et al. 1977; Carlson et al. 1977), whereas PTH has been found to elevate plasma PRL when infused into normal individuals (Isaac et al. 1978).

4.6.4 Association of Hyperprolactinemia with Nonendocrine Diseases

Although rare, the possibility of an ectopic source of PRL such as in hypernephroma or in lung cancer (Turkington 1971; Gropp et al. 1980) should also be considered. Elevated plasma PRL has also been associated with polyostotic fibrous dysplasia (Carr et al. 1979).

Prolactin has been found moderately elevated in 37% to 76% of patients with severe renal impairment (Nagel et al. 1973; Hagen et al. 1976; Czernichow et al. 1976; Sciarra et al. 1977; Weissel et al. 1979). The PRL response to TRH or dopaminergic blockade is blunted in these patients (Czernichow et al. 1976; Ramirez et al. 1977), and PRL is not normalized by dialysis but by kidney transplantation (Lim et al. 1979). In a recent study, Sievertsen et al. (1980) reported a decrease in metabolic PRL clearance in chronic renal failure, but resistance of lactotrope receptor to dopamine seems to be more important in the maintenance of elevated plasma PRL. This is in agreement with a previous report by Frohman et al. (1979).

Mild hyperprolactinemia has been described in advanced liver disease (Morgan et al. 1978; Wernze and Burghardt 1979). Indeed, a significantly higher percentage of PRL cells has been reported in pituitaries of cirrhotic patients at necropsy when compared with findings from subjects deceased from other causes (Jung and Russfield 1972). In this condition, a disturbance of estrogen metabolism is present and the PRL response to TRH stimulation in male cirrhotics is enhanced (Wernze and Schmitz 1977; Panerai et al. 1977). This is probably due to the priming effect of a relative excess of circulating sex steroids on the pituitary lactotropes (Zanoboni and Zanoboni-Muciaccia 1975), although other factors such as delayed metabolic disposal may also play a role (Panerai et al. 1977). Bauer et al. (1980) found that the hepatic elimination of PRL was not markedly affected by the severity of the liver involvement concluding that the kidney was the main site of PRL disposal. It is questionable whether the presence of gynecomastia in such patients is linked to the mild hyperprolactinemia recorded (Wernze and Schmitz 1977; Morgan et al. 1978). Interestingly, a blunted PRL response to TRH has been reported in hemocromatosis presumably due to iron deposition in the pituitary gland and surrounding tissues (Levy and Carlson 1978). Also, impaired GH and PRL responses to insulin have been reported in alcoholics after a period of abstinence (Chalmers et al. 1978).

Data have been controversial in the PRL screening of psychiatric diseases searching for a central dopaminergic component. Both elevated (Sachar 1976; Meltzer et al. 1974) and normal (Brambilla et al. 1976; Arana et al. 1977; Rotrosen et al. 1978) plasma levels have been reported in psychiatric populations. In addition, a disruption of the diurnal PRL secretory rhythm has been observed in depressive patients (Halbreich et al. 1979) whereas diurnal plasma levels have been reported within the normal range (Mendlewicz et al. 1977). In anorexia nervosa values are normal (Macaron et al. 1978; Isaacs et al. 1980) or low (Brauman 1973; Beumont et al. 1974c), the latter probably caused by the lack of tonic effect by estrogens although galactorrhea has been described in one case (Archer et al. 1974), and moderate hyperprolactinemia has been reported by others (Harrower et al. 1977). In contrast, impaired hypothalamic control of prolactin secretion has been reported in massively obese men and women (Kopelman et al. 1979). Finally, normal basal PRL has been measured in Parkinsonian subjects (Polleri et al. 1977; Hyyppä et al. 1978), a neurologic entity characterized by localized central dopamine deficiency. It is interesting to note that in this particular condition, and as observed in normoprolactinemic subjects, bromocriptine withdrawal can be accompanied by a PRL rebound phenomenon and galactorrhea has been reported in one case (Pentland and Sawers 1980).

4.6.5 Hyperprolactinemia Through Disturbances of Neural Pituitary Control

Milk secretion has also been associated with processes interfering with a normal hypothalamic control of the pituitary lactotropes, such as local sarcoidosis (Turkington and McIndoe 1972; Malarkey and Kataria 1974; Fromantin et al. 1975; Kleinberg et al. 1977), and histiocitosis X (Gates et al. 1973; Kleinberg et al. 1977). This mechanism might also have been operative in a case of galactorrhea associated with oncocitoma (Rollet et al. 1975) and in the elevated circulating PRL found in some patients with craneopharingiomas (Jenkins et al. 1976; Kapcala et al. 1980). The anatomic separation of the pituitary from its hypothalamic connection leads to hyperprolactinemia and galactorrhea. This has been reported after transsection of the pituitary stalk in the treatment of advanced breast cancer (Powell et al. 1966; Turkington et al. 1971). Also compression of the suprasellar structures by a tumoral mass, traumatic edema, or inflammatory brain diseases can produce functional hyperprolactinemia (Riese 1928; Kleinberg et al. 1977). Traumatic damage to the pituitary gland as a cause of hypoprolactinemia is rare event (Brown and Mcmillin 1977). Most cases with empty sella present with normal plasma PRL (Ridgeway et al. 1975; Schroeder et al. 1976) with occasional exceptions (Archer et al. 1978) probably caused by hypothalamic damage.

It has been shown that repeated breast stimulation can induce galactorrhea in normally cycling women (Tolis et al. 1974). This may take place through neural stimulation, similar to the milk secretion or moderate hyperprolactinemia reported after surgical procedures to the chest wall (Weir 1971), to the upper segments of the spinal chord (Boyd et al. 1978), or after breast self-stimulation (Boyd et al. 1978). However, it should be noted that ovulation is common among women exhibiting galactorrhea after self-manipulation or stimulation by partner (Tolis et al. 1974; Jones and Gentile 1975).

4.7 Clinical and Biochemical Effect of Inappropriate Prolactin Secretion

Elevated PRL secretion can lead in the nonpregant woman and in males to a series of clinical and biochemical effects as a result of central and peripheral actions already discussed in a previous section. These are sumarized in Table 4.3. The hallmark is subfertility in both sexes. In the female there are a number of intermediate situations characterized by the presence of cycles but inability to conceive or to maintain pregnancy. Prolactin-induced pubertal delay has been recognized (Jürgensen et al. 1976; Bergh et al. 1977; Kleinberg et al. 1977) and familial occurrence has also been described (Wider et al. 1969). A condition of hypoprolactinemia has not yet been described with the exception of the postpartum,

Table 4.3. Effects of inappropriately increased prolactin secretion

Females	
Clinical	Failure to enter menarche
	Galactorrhea with ovulatory cycles
	Short luteal phase
	Menometrorrhagia
	Anovulatory cycles
	Amenorrhea[a]
	Hirsutism
Biochemical	Elevated plasma PRL concentrations
	Lack of sleep-related PRL elevations
	Absence of LH pulsatility
	Lack of positive estrogen feedback on LH
	Low basal estrogens[b]
	Elevated androgenic adrenal steroids
	Positive pituitary and ovarian response to exogenous stimulation
	Clomiphene resistance
Males	
Clinical	Failure to enter puberty
	Galactorrhea[c]
	Oligoaspermia, subfertility
	Signs of androgen failure
Biochemical	Elevated plasma PRL concentrations
	Lack of PRL elevations during sleep
	Low androgen production
	Positive pituitary and testicular response to exogenous stimulation

[a] Not necessarily associated with galactorrhea.
[b] Subjected to variance: Normal basal estrogens with normal or even elevated FSH are not uncommon.
[c] Less frequent in males probably due to reduced mammary tissue mass.

but luteal inadequacy was observed following pharmacologically induced low PRL secretion (Schulz et al. 1975, 1978; Bohnet et al. 1977), probably as the result of lactogen oversuppression. It was assumed that the gonad was missing the stimulatory effect of PRL in maintaining a physiologic number of LH receptors. In the male, the antifertility effect of elevated circulating lactogen is less apparent clinically. Male potency can be preserved to a certain extent so that a common reason for seeking medical advice is barrenness. Clinical and biologic signs of excessive PRL production in the male are depicted in Table 4.3. They reflect a dual effect of this lactogen: disturbances of control mechanisms at the level of the central nervous system and possible interference with gonadal receptors. The clinical expression of this phenomenon is the failure to enter puberty in the case of preadolescent onset (Koenig et al. 1977), or the insidious development of a syndrome of relative hypogonadism and defective sperm maturation (Fossati et al. 1976b; Roulier et al. 1976; Boucher et al. 1977; Franks et al. 1978; Carter et al. 1978). As in the female subjects, the nyctohemeral rhythm of PRL is disturbed and the sleep-induced elevation is missing (del Pozo, unpublished data).

Biochemically there are a series of signs reflecting central and peripheral effects such as defective synthesis of progesterone by the corpus luteum, blockade of estrogen receptors and interference with LH pulsatility, the significance of which has already been discussed.

4.8 Natural History of Galactorrhea Syndromes

In 1972, a review of the literature up to 1970 revealed 110 cases of galactorrhea-amenorrhea (Hughes et al. 1972). With the advent of specific radioimmunoassays for PRL the diagnosis of PRL-dependent conditions has become relatively frequent in endocrine and gynecologic practice. Probably the extended use of galactogenic drugs such as neuroleptics and oral contraceptives containing estrogens has contributed to the increase in the number of cases diagnosed. If these possible etiologic factors are considered, there is some suggestion that patients in the group with persistent postpartum lactation beyond physiologic limits may have a more favorable prognosis than the idiopathic or prolactinoma types. The former have features in common with postpill galactorrhea-amenorrhea, and spontaneous resumption of menses can also occur here more frequently than in the other types (Rankin et al. 1969; Hughes et al. 1972; Starup 1972). It can be assumed that estrogens and also probably progestogens, endogenous in pregnancy and exogenous in postpill galactorrhea-amenorrhea, would disrupt the hypothalamic-pituitary axis resulting in innappropriate PRL secretion and blockade of the cycling center. The mechanisms involved seem extremely sensitive considering the concentration of circulating sex steroids in both conditions. Extreme hypothalamic lability in such patients would be supported by the observation that women with prior menstrual irregularities are more prone to developing postpill amenorrhea-galactorrhea (Gambrell et al. 1971).

4.9 Medical Therapy of Nontumorous Prolactin-Dependent Conditions

4.9.1 Dopaminergic Drugs: Bromocriptine

The medical management of galactorrhea-amenorrhea was unsatisfactory prior to the advent of drugs stimulating central dopaminergic mechanisms. Normalization of abnormal secretion of pituitary lactogen can be obtained by (a) discontinuation of medication enhancing PRL secretion, e.g., neuroleptic drugs, or replacement by less galactogenic medication, (b) renal transplantation in secondary hyperprolactinemia associated with uremia, and (c) administration of a dopamine agonist. Clomiphene citrate and gonadotropin preparations were used earlier when pregnancy was desired but no medical therapy was able to normalize excessive PRL secretion. After Malarkey et al. (1971) reported variable responses to the acute administration of L-dopa, Turkington (1972d) observed restoration of menses in 4 out of 14 galactorrhea women treated chronically with this drug. Later, Zarate et al. (1973a) reported resumption of menstrual cycles in 11 out of 14 patients on the same therapy and two women conceived during treatment. These results were further confirmed by Ayalon et al. (1974). The action of L-dopa is believed to occur via stimulation of dopamine synthesis and subsequent PIF activation.

Considerable therapeutic improvement was obtained with the introduction of bromocriptine, an ergot derivative, in the management of galactorrhea-amenorrhea. The characterization of this drug has been presented in the chapter concerned with the pharmacology of PRL. The compound is usually administered in doses of 1.25–2.5 mg two to three times daily. In rare instances a dosage increase to 10–15 mg daily is required in order to reestablish normal ovulatory mechanisms. The potent dopaminergic agonistic effect of the drug may induce hypotension, nausea and occasionally vomiting in sensitive subjects. This can be avoided by increasing dosage gradually after initiating therapy with 1.25 mg or less, the drug being taken after meals. In many instances intolerance symptoms will subside spontaneously after medication has been continued for several days. It should be mentioned that, at least in subjects treated with bromocriptine for nonhyperprolactinemic conditions, a PRL rebound effect has been observed, lasting for several weeks and slowly receding to normal (Groom et al. 1978; Pentland and Sawyers 1980). In the following section, the clinical effect of the drug will be illustrated in a series of case reports.

4.9.1.1 Galactorrhea-Amenorrhea Syndrome

After first reports by Lutterbeck et al. (1971) and Varga et al. (1973) the effectiveness of bromocriptine in hyperprolactinemic anovulation in galactorrhea-amenorrhea has been amply demonstrated (Besser et al. 1972; del Pozo et al. 1972, 1974; Jaszmann and Sternthal 1974; Rolland et al. 1974; Thorner et al. 1974a, 1975; Lloyd et al. 1975; Bohnet et al. 1976; Canales et al. 1976a; Fossati et al. 1976b; Jacobs et al. 1976; Rjosk et al. 1976; Pepperell et al. 1977a; Badano et al. 1977; Gomez et al. 1977; Friesen and Tolis 1977; Kleinberg et al. 1977). The sequence of clinical and biologic findings following bromocriptine treatment in this condition is illustrated in Fig. 4.10. A 26-year-old nulliparous woman reported amenorrhea of 12 months duration after discontinuation of an oral contraceptive. Plasma PRL was slightly elevated and urinary estrogens low.

Galactorrhea had not been noticed by the patient but milk could easily be obtained from both breasts on slight pressure. Onset of treatment with bromocriptine, 2.5 mg twice daily, was followed by rapid restoration of menstrual cycles and normal ovulatory mechanisms demonstrated through the biphasic pattern of basal body temperature, plasma progesterone elevations in the middle of the hyperthermic phase, and endometrial biopsy showing a secretory phase at the onset of vaginal bleeding. Urinary excretion of estrogen increased to within the normal range. It is interesting to note that ovulation took place shortly after initiation of therapy, an effect observed by other investigators (Bernal et al., unpublished data) suggesting that some follicles had already reached a certain stage of maturation

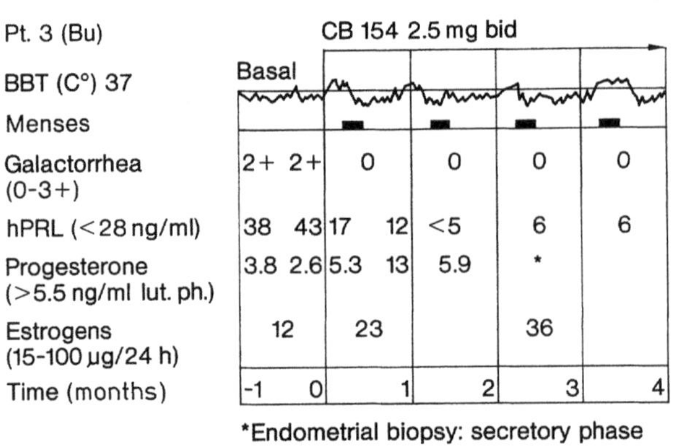

Fig. 4.10. Effect of bromocriptine treatment in a case of hyperprolactinemic amenorrhea (explanation in text). (del Pozo et al. 1974)

despite hyperprolactinemia. Thus, Seki and Seki (1974) reported a case in which ovulation and pregnancy followed a 10-day course of bromocriptine. Strauch et al. (1977a,b) and Pepperell et al. (1977b) have followed the hormonal kinetics in the early stage of treatment and found that ovulation is frequent in the first cycle of bromocriptine. In some cases, anovulation persisted due to incomplete PRL suppression but follicle rupture occurred after dosage increase. Resumption of plasma LH fluctuations has been recorded after bromocriptine administration (Bohnet et al. 1976), and restoration of prositive estrogenic feedback on LH has also been observed (Aono et al. 1979b). Since in galactorrhea-amenorrhea circulating gonadotropins are usually within the normal follicular range, the importance of restoration of LH pulsatility and estrogenic feedback becomes obvious. It is noteworthy that Polatti et al. (1978) restricted administration of the drug to the period prior and around the ovulatory temperature surge and found results equivalent to continuous treatment.

Acting predominantly at the level of the pituitary gland, the favorable effect of bromocriptine takes place independently of the cause of hyperprolactinemia. However, it should be stated that in medicamentous-induced galactorrhea, for example in psychiatric patients treated with neuroleptics, drug interaction effects may require higher daily doses of bromocriptine in order to overcome the dompaminergic blockade (del Pozo and Lancranjan 1978; del Pozo, to be published). Restoration of cyclic activity with bromocriptine in this particular situation has been reported by Beumont et al. (1975).

The mechanism by which hyperprolactinemia can simulate premature menopause has been discussed previously (see p. 113). Thorner et al. (1974a) and Darragh et al. (1976) recorded clinical and hormonal changes compatible with premature menopause in hyperprolactinemic women. Treatment with bromocriptine led to restoration of ovulatory cycles. Thus, the possibility of a PRL-mediated effect should be considered before the diagnosis of early menopause can be established with certainty in a woman with amenorrhea-galactorrhea, low estrogens and elevated plasma FSH.

4.9.1.2 Hyperprolactinemia and Inappropriate Luteal Phase

As described in the section dedicated to physiopathology, elevated PRL levels can be associated with luteal inadequacy (Seppälä et al. 1976b; del Pozo et al. 1976; Fredricsson 1977) and treatment with bromocriptine can restore ovulatory cycles (del Pozo et al. 1979). The following case illustrated this effect (Fig. 4.11). A 30-year-old woman had failed to conceive again after a previous normal pregnancy and delivery. When first seen at the endocrine clinic the patient was found to have galactorrhea and elevated resting plasma PRL. Menses were regular and an ovulatory LH peak was disclosed, but the basal body temperature record revealed a luteal phase of 8 days duration. Treatment with bromocriptine readily stopped milk secretion and brought plasma PRL to within the normal range. The hyperthermic phase of the cycle was prolonged to 12 days and the patient conceived in the second cycle of treatment. It should be mentioned, however, that excessive PRL suppression in previously hyperprolactinemic women may lead to reappearance of shortened luteal phases (Bohnet et al. 1977).

The role of PRL in the etiology of normoprolactinemic luteal inadequacy remains controversial. Whereas Diedrich et al. (1978) and Mühlenstedt et al. (1979) reported

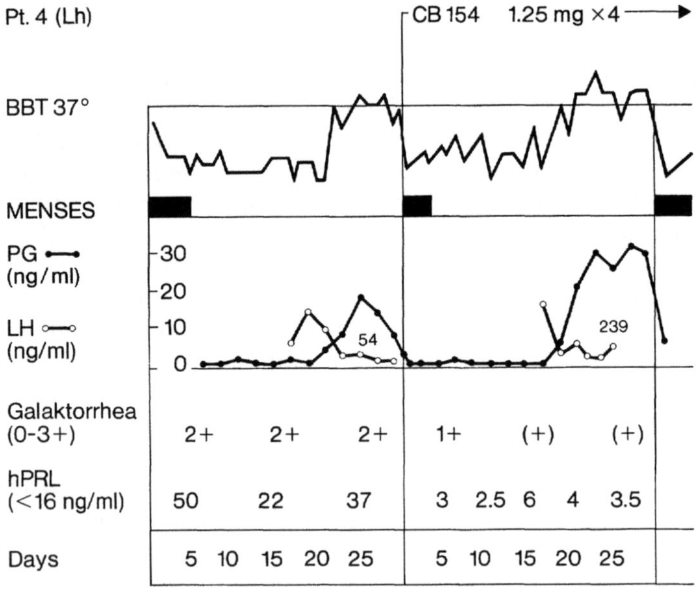

Fig. 4.11. Normalization of plasma PRL restored luteal function (normal luteal index > 100) and fertility in a woman with galactorrhea and regular menses (del Pozo et al. 1979)

a favorable effect of bromocriptine treatment in this condition, Saunders et al. (1979) found no action.

4.9.1.3 Hyperprolactinemic Anovulation Without Galactorrhea

As previously mentioned, the absence of milk secretion in hyperprolactinemic amenorrhea is not uncommon. The following case report will remind the clinician that the absence of galactorrhea does not exclude hyperprolactinemia as the cause (Fig. 4.12). A 25-year-old woman was first seen because of a history of oligomenorrhea and cessation of menses 2 years prior to the consultation. The physical examination was unremarkable and no milk could be obtained from her breasts by

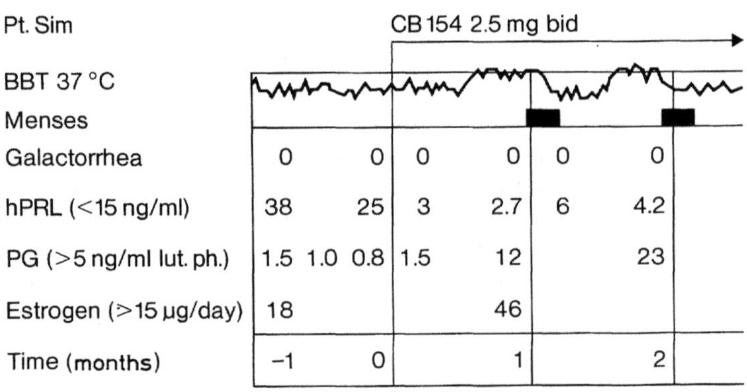

Fig. 4.12. Hyperprolactinemia may induce amenorrhea in the absence of galactorrhea. PRL suppression with bromocriptine restores cyclic activity and ovulation. (del Pozo et al. 1974)

manual expression. The biochemical picture presented moderate hyperprolactinemia, low estrogen, and somewhat elevated plasma dehydroepiandrosterone. Treatment with bromocriptine, 5 mg daily in divided doses, promptly normalized the hormonal picture and reestablished ovulation. A careful analysis of the discordant patterns of hyperprolactinemia and galactorrhea in secondary amenorrhea has been published by Seppälä et al. (1977).

4.9.1.4 Metrorrhagia with Elevated PRL: Response to Bromocriptine

A 24-year-old multiparous woman (Fig.4.13) was seen at the gynecology clinic because of persistent metrorrhagia and profuse galactorrhea following an abortion 3 months earlier. Therapy with exogenous sex steroids and a curettage had failed to influence the patient's condition and at the time of referral to the endocrine clinic hysterectomy was contemplated because of intractable bleeding. Shortly after treatment with bromocriptine, 2.5 mg twice daily, was initiated, vaginal bleeding ceased and normal periods ensued. This PRL dependency, found in some cases of irregular or persistent uterine bleeding (Jaszmann and Sternthal 1974), is difficult to explain since this lactogen does not influence blood clotting mechanisms. A possibility lies in the blocking effect of elevated PRL concentrations on vascular reactivity as proposed by Horrobin et al. (1976).

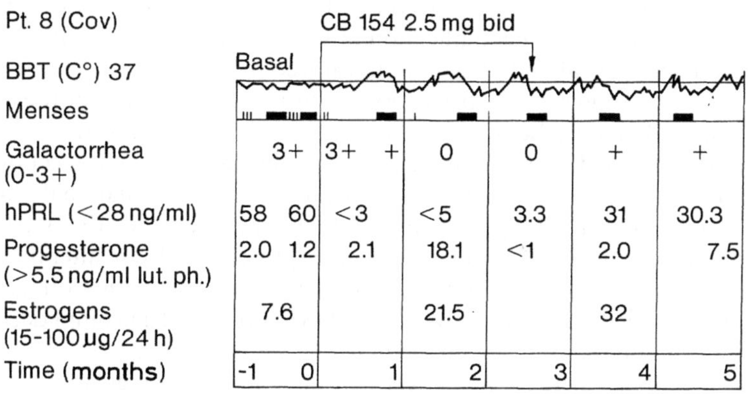

Fig. 4.13. Treatment with bromocriptine terminated uterine hemorrhage in a case of moderate hyperprolactinemia following a spontaneous abortion. (del Pozo et al. 1974)

4.9.1.5 Effect of Bromocriptine on the Anovulation of Polycystic Ovary Syndrome with Hyperprolactinemia

Thorner et al. (1974a) initially reported that the reduction of PRL secretion in galactorrhea patients with sclerocystic ovaries resulted in restoration of ovulatory cycles. In a preliminary study, Rocco et al. (1979) studied the effect of bromocriptine in hyperprolactinemic PCO. The criteria used for selection were: (a) enlarged ovaries on laparotomy, laparoscopy or pelvigraphy; (b) high pulsatile LH without ovulatory surge or low FSH; (c) elevated levels of testosterone and estrone, with normal or low estradiol. Out of ten patients studied, three had amenorrhea and seven had varying degrees of oligomenorrhea. Basal body temperature records were monophasic in all. Bromocriptine was administered in daily doses of 3.75–7.5 mg for several months. Biphasic basal temperatures were recorded in eight cases and ovulation was supported by plasma progesterone increases in all of them.

The mechanisms involved in this positive effect of dopaminergic stimulation on PCO anovulation are far from clear. Two hypotheses can be proposed: the PRL action on adrenal androgen synthesis and a possible effect of bromocriptine on central dopaminergic mechanisms. Thus, the events leading to the classic clinical and biochemical picture of PCO are depicted in Fig.4.14: exaggerated hypothalamic stimulation leads to inappropiately elevated circulating LH and subsequently to increased ovarian androgen synthesis in addition to an adrenal component. Target tissues transform androgens, mainly androstenedione into estrone, which in turn may disturb hypothalamic cyclicity. However, it cannot be ruled out that androgenic steroids may interfere with estrogen receptor synthesis to "masculinize" the hypothalamus and produce a male type of LH secreting pattern with wide pulsatile episodes and without cycling activity. On the other hand, pituitary exposure to augmented circulating estrogens may prime lactotrophe cells to oversecrete PRL. This lactogen would in turn enhance dehydroepiandrosterone synthesis by the adrenals, an effect demonstrated in idiopathic hyperprolactinemia.

This mechanism would be rendered inoperative under PRL suppression with bromocriptine. However, its validity can be criticized in view of the difficulty in discriminating between abnormal androgen synthesis due to inappropriate adrenal stimulation by PRL or due to the basic biochemical and enzymatic abnormalities inherent in PCO. The implication of sex steroids and its receptor interplay in this syndrome has already been discussed (see p. 112).

The dopaminergic theory offers an attractive alternative for the normalization of hypothalamic cyclic activity after bromocriptine administration. Recent studies have suggested that dopamine agonists exert an inhibitory effect on gonadotropin and PRL release in humans (Lachelin et al. 1977b) and that this effect can be correlated with endogenous estrogen levels (Judd et al. 1978). Although chronic dopaminergic stimulation with bromocriptine has failed to alter the cyclic pattern of LH (del Pozo et al. 1975a; Schulz et al. 1978), the basic situation may be different in PCO. Indeed, recent work by Yen (1980) has revealed a prompt decline in LH levels in response to a 4-h dopamine infusion, leading to the working hypothesis that a concomitant increase in LHRH neuronal activity brings about a decrease in the inhibitory influence of dopamine (Judd et al. 1979). This effect has also been observed after bromocriptine treatment (del Pozo and Falaschi 1980). Thus, the modulating effect of bromocriptine on LH in PCO could involve a dopaminergic mechanism unrelated to the well-known inhibition of PRL secretion.

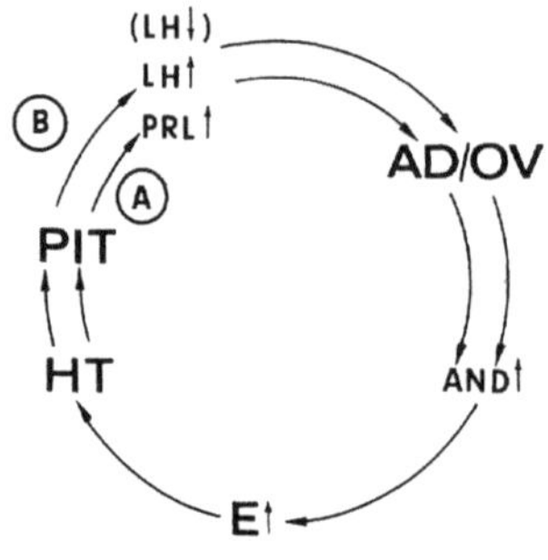

Fig. 4.14. Possible action of bromocriptine on PCO: *A* via PRL and adrenal androgens; *B* via dopaminergic LH reduction; *AD/OV*, adrenals ovary; *AND*, androgens; *HT*, hypothalamus; *PIT*, pituitary; *E*, estrogen. (del Pozo and Falaschi 1980)

4.9.1.6 Miscellaneous Effects of Bromocriptine on Reproduction

The effect of bromocriptine on normoprolactinemic anovulation has also been tested. A positive effect has been reported in some instances (Tolis and Naftolin 1976; Nencioni et al. 1978; Milligan et al. 1978) but controlled studies in larger series of euprolactinemic patients have failed to show clear differences between placebo and drug treatment (Crosignani et al. 1978a; Wright et al. 1979). Nevertheless, Seppälä et al. (1976a) found a number of bromocriptine responders in normoprolactinemic anovulation. Van der Steeg and Coelingh Bennink (1977) reported ovulations after treatment with bromocriptine in 74% of women with postpill anovulation. The authors attributed the favorable effect to the dopaminergic action of the drug facilitating the release of LHRH from the hypothalamus. This mechanism would be independent of the PRL-lowering effect and also would explain the positive results reported by other authors (Hirvonen et al. 1976; Wolf et al. 1979; Moggi et al. 1979; Coelingh Bennink 1979) in "converting" euprolactinemic antiestrogen resistant patients to sensitivity. Thus, the concomitant treatment with bromocriptine and antiestrogens led to ovulatory cycles in women previously resistant to the latter agent alone. Although the basic mechanism of this effect remains uncertain, the bromocriptine-induced elevation of plasma estrogen reported by Seppälä et al. (1976c) in normoprolactinemic amenorrhea may have sensitized the hypothalamus to the action of the antiestrogen. Lenton et al. (1977) recorded 63.4% pregnancies in apparently ovulating women unable to conceive after treatment with clomiphene or a progestogenic agent. The cumulative conception rate after bromocriptine was significantly higher than in a similar group of patients who were given no treatment. The authors speculated that the favorable effect of bromocriptine might be mediated by other mechanisms than suppression of PRL secretion alone. Indeed, Kort et al. (1980) demonstrated by ultrasound scanning that bromocriptine treatment would induce follicular enlargement paralleling plasma estradiol increases in normoprolactinemic anovulatory patients.

Sexual function is often impaired in patients on maintenance hemodialysis. This disturbance can undoubtedly have many causes involving both, psychological and hormonal factors. Basal plasma PRL has been found to be higher than normal in this condition (Nagel et al. 1973; Hagen et al. 1976; Sciarra et al. 1977; Czernichow et al. 1976) which is also associated with reduced libido and sexual activity (Levy and Wynbrandt 1975; Bommer et al. 1976; Holdsworth et al. 1978). In a controlled trial conducted by Bommer et al. (1979) the effect of bromocriptine on sexual activity was studied in male hemodialysis patients. Plasma PRL concentrations were consistently reduced while sexual function assessed by appropriate questionnaires was markedly improved.

Treatment with bromocriptine has been found to relieve mastodynia accompanying fibrocystic disease of the breast (Zippel et al. 1977; Mansel et al. 1978). Martin-Comin et al. (1976) were also able to substantiate a positive effect on breast lesions by radiologic examinations. The mechanism of this favorable action is not clear. Histochemical and morphological studies in breast tissue biopsies have failed to show significant changes after prolonged treatment with bromocriptine (Zippel et al. 1977). In another study, Palmer and Monteiro (1977) demonstrated a reduction in premenstrual breast engorgement and mastalgia in normally ovulating women. It can be speculated that the fall in circulating PRL led to a reduction in ductal pressure and subsequently to an improvement of clinical symptoms, mainly mastodynia.

4.9.2 Impact of Bromocriptine Treatment on Pregnancy and Fetal Development

Previously infertile women treated with bromocriptine may become pregnant, whereby medication can continue for some period after conception. A follow up report involving 448 cases has been previously published (Griffith et al. 1978). Until now, information has been collected from 1232 patients (Turkalj et al., unpublished). The mean daily drug intake was 5.7 mg (range 1 to 40 mg). Table 4.4 depicts the duration of treatment during pregnancy. It may be noted that a relatively large number of cases were treated for different periods during gestation, up to 8 weeks and longer. In 145 (11.4%) women spontaneous abortion occurred, and pregnancy was interrupted in 28 women for different reasons. The incidence of spontaneous termination is whithin the range recorded in the nontreated population (Griffith et al. 1978) and compares favorably with that quoted for pregnancies achieved after restoration of fertility with exogenous gonadotropins or clomiphene (Ahlgren et al. 1976; Kase 1973; Spadoni et al. 1974; Thompson and Hansen 1970).

As shown in Table 4.5 there is a great variability in the time interval between discontinuation of bromocriptine intake and the occurrence of spontaneous abortion, suggesting a lack of dependency between treatment and course of gestation. In addition, treatment with daily doses between 5 and 32 mg throughout pregnancy in eight cases failed to modify its course and produced no alterations in fetal development (del Pozo et al. 1977a; Bigazzi et al. 1979).

The multiple pregnancy rate in the collective of 1088 births studied was 2.3%, which is lower than the rate obtained with ovulation-inducing agents (Ahlgren et al. 1976; Kase 1973). Major and minor malformations occured in the bromocriptine series in 3.3% of cases. Smithells (1976) cites a frequency of 2.7% spontaneous malformations for a collecitve of over 600 000 control mothers. After reviewing the world literature covering 20 million births, Wilson (1973) cites Kennedy, who

Table 4.4. Exposure to Parlodel during pregnancy. (Turkalj et al., unpublished)

Duration of treatment[a] (weeks)	Number of cases n	%
0–1	46	4
1–2	80	7.0
2–3	303	26.7
3–4	268	23.6
4–5	177	15.6
5–6	108	9.5
6–7	53	4.7
7–8	27	2.4
>8	74	6.5
Total	1136	100.0

[a] Day 0 supposed day of conception *or* day 1 of last menstruation plus 13 days.

Table 4.5. Spontaneous abortions: exposure to Parlodel. (Turkalj et al., unpublished)

Duration of therapy (weeks)	Daily dose of Parlodel (mg)					Total	%
	<5.0	5.0	5.0-10.0	10.1-15.0	No data		
0–1	0	2	2	0	0	4	2.5
1–2	0	5	2	0	0	7	4.5
2–3	10	19	12	2	0	43	27.4
3–4	9	15	10	0	0	34	21.6
4–5	6	14	4	0	0	24	15.3
5–6	1	5	2	0	0	8	5.1
6–7	0	3	2	2	0	7	4.5
7–8	3	2	0	0	0	5	3.2
8–9	0	0	0	0	1	1	0.6
>9	2	2	1	0	0	5	3.2
No data	2	10	6	0	1	19	12.1
Total	33(21.0%)	77(49.0%)	41(26.1%)	4(2.6%)	2(1.3%)	157	100%

concluded that the frequency of individuals with some degree of congenital defect at birth almost certainly exeeds 2%, and indeed more intensive studies in about half a million infants yielded a mean rate for defective individuals of 4.5%. A report from Sweden gives a frequency of 3.3% for major malformations (Ekelund et al. 1970) and in the London area, a rate of 3.7% has been reported (Davie et al. 1972). A higher incidence of fetal malformations in newborn infants (major malformations, 4.1% and minor malformations, 6.5%) has been reported by Drew et al. (1977). In the United States the Collaborative Perinatal Project studied 50 282 pregnancies and recorded 6.5% malformations (Heinonen et al. 1977). It can be concluded that the incidence for Parlodel lies within normal limits. If ovulation is induced with clomiphene or with natural gonadotropin, increased frequency of malformations can occur. Thus, Ahlgren et al. (1976) reported eight major abnormalities in 148 births related to clomiphene therapy, i.e., a rate of 5.4%. Spadoni et al. (1984) found a 5.5% incidence of anomalies in 62 pregnancies following ovulation induction with human menopausal gonadotropin. On the other hand, some workers have reported no increased frequency of anomalies with these forms of therapy (Hack et al. 1970; Kase 1973).

4.9.3 Treatment of Male Hyperprolactinemia with Bromocriptine

Carter et al. (1978), treating 11 out of 22 hyperprolactinemic male patients with this compound, observed a major decline in plasma PRL and an improvement in potency. Plasma testosterone increased from a mean of 135 ng% to 396 ng%. It is interesting to note that in two subjects who remained impotent after testosterone replacement therapy potency returned when hyperprolactinemia was reduced with bromocriptine. In another series of 29 patients, Franks et al. (1978) also noted a return of potency in subjects treated with bromocriptine, and similar results have been reported by Fossati et al. (1976b) and Spark et al. (1980). It is striking that, in males, pituitary lesions (see next chapter) seem far more advanced and mean plasma PRL levels substantially higher than in hyperprolactinemic women. This is probably due to the absence of the alarming sign of cessation of cyclic activity in women which makes lesions clinically apparent at an earlier stage of development. As already mentioned, bromocriptine treatment has induced sexual development in a male subject with pubertal retardation and elevated plasma PRL (Koenig et al. 1977).

More difficult to assess is the relationship between oligospermia and PRL in normoprolactinemic individuals without other signs of endocrine abnormality. In a study of 18 oligospermic patients screened by the authors all had normal plasma PRL concentrations, but in another study (Roulier et al. 1976) elevated levels were found in 4 out of 51 (3%) cases. Also Falaschi et al. (1979) found that oligospermic men exhibited higher basal plasma PRL than age-matched normal controls. More recently Hovatta et al (1979) treated 40 oligospermic, low-androgen men with bromocriptine or placebo. While PRL levels were normalized during bromocriptine treatment and plasma testosterone and dihydrotestosterone levels increased slightly in both groups, no significant difference in sperm counts between the two treatments was seen. Further reports appeared recently to confirm this lack of action of dopaminergic stimulation on defective spermatogenesis and on sexual impotence (Ambrosi et al. 1977; Masala et al. 1979; Merino et al. 1980; Legros et al. 1980) despite isolated communications showing favorable results (Pierini and Nusimovich 1978).

4.10 Effect of Other Ergot Derivatives on Prolactin Secretion

A number of ergot compounds in current clinical practice have also been investigated for antigalactic properties. A screening of the PRL-suppressing action of ergot derivatives has been published previously (Flückiger and del Pozo 1980). The choice of a reference preparation, however, has been rendered difficult by the differences found in dose-effect relationships (e.g., uterotonic or vascular constrictive) specificity of action and route of administration. Dihydroergotamine failed to reduce plasma PRL when injected intravenously in a dose of 0.2 mg, whereas 0.25 mg of methylergonovine significantly depresses PRL (del Pozo et al. 1977a), but failed to influence postpartum lactation (del Pozo et al. 1975b). Methysergide, a drug characterized as a serotonin antagonist, has been shown to inhibit PRL secretion in single oral doses of 1–2 mg (del Pozo and Lancranjan 1978). Likewise, antigalactic properties have been found in hydergine, a compound of three ergot alkaloids (Flückiger and del Pozo 1980). Curiously, treatment with methysergide has failed to restore ovulation in the galactorrhea-amenorrhea syndrome (Crosignani et al. 1977), and the effect of hydergine is quite variable according to the author's experience.

Lergotrile is an ergoline of the clavine type. After its PRL-inhibiting properties were demonstrated in the experimental animal, studies in man showed doses of 2–6 mg to inhibit effectively serum PRL levels and also to delay the onset of perphenazine-induced hyperprolactinemia (Lemberger et al. 1974). This compound has also been found effective in suppressing nonpuerperal hyperprolactinemia (Cleary et al. 1975).

Lisuride and metergoline, two ergot derivatives characterized as serotonin antagonists, have been the subject of recent investigation. Chiodini et al. (1976) have reported a significant PRL fall in acromegalic individuals treated with metergoline.

On the basis of comparative studies with dopamine antagonists, they concluded that the suppressive effect of this compound on PRL was due to a dopaminergic mechanism of action rather than to serotonin blockade. Crosignani et al. (1978a) have reported a favorable action of metergoline in hyperprolactinemic anovulation, and pregnancy has been reported after treatment with this drug (Crosignani et al. 1977 and 1980).

Long-lasting dopaminergic activity was reported by Liuzzi et al. (1978) using lisuride-hydrogen maleate, a compound previously found to diminish PRL release after TRH stimulation (Horowsky et al. 1978). Furthermore, lisuride prevented the PRL increase that follows cimetidine administration (Bohnet et al. 1978). This drug has been characterized as a dopamine agonist after extensive endocrine investigation (Delitala et al. 1979). Its efficacy in galactorrhea-amenorrhea has been tested by de Cecco et al. (1978) and Bohnet et al. (1979). These authors have reported restoration of ovulatory cycles in such patients and pregnancies were recorded in some instances. The compound is administered usually twice daily in doses ranging from 50 to 200 μg. At the higher dosage strength, signs of dopaminergic overstimulation such as drowsiness and nausea are common (Delitala et al. 1979). More recently, Lemberger and Crabtree (1979), Kleinberg et al. (1980) and Callaghan et al. (1981) have reported prolonged PRL suppression after single or repeated administration of 100 to 400 μg pergolide mesylate, another ergot compound.

4.11 Effect of Nonergot Dopamine Agonists on Prolactin Secretion

Piribedil, a piperazine derivative, has been administered to postpartum women and to hyperprolactinemic patients in an acute trial (Müller et al. 1977). Although PRL secretion was reduced by this agent, its potency was somewhat lower in comparison with bromocriptine. More recently this compound has been claimed to prevent postpartum lactation. It is noteworthy that this effect was recorded despite persistence of plasma PRL in the lactational range (Alagna et al. 1979).

Re-uptake of endogenous catecholamines can be inhibited by nomifensine, an amino-tetrahydroisoquinoline (Hunt et al., 1974). This drug has been found to significantly inhibit basal PRL secretion without appreciable effect on the TRH-stimulated release of the hormone (Masala et al. 1980b).

The report by Foukas (1973) on the effectiveness of pyridoxine (Vitamine B_6) in suppressing puerperal lactation was followed shortly thereafter by another study claiming its efficacy in restoring reproductive function in hyperprolactinemic amenorrhea (McIntosh 1976). It was suggested that pyridoxine, a cofactor required for the conversion of L-dopa to dopamine, may increase endogenous dopaminergic activity with a resultant decrease in PRL secretion, and Delitala et al. (1976) observed a significant fall in plasma PRL after intravenous administration of 300 mg pyridoxal sulfate silfate to normal subjects. However, del Pozo et al. (1975c) del Pozo and Brun del Re (1979), Canales et al. (1976a), and McDonald et al. (1976) were unable to show an inhibitory effect of this vitamin on puerperal lactation or on postpartum plasma PRL.

Furthermore, Tolis et al. (1977), Spiegel et al. (1978), and Goodenow and Malarkey (1979) failed to restore ovulatory cycles in hyperprolactinemic patients whereas Lehtovirta et al. (1978) could see plasma PRL levels normalized in only 6 out of 22 women taking pyridoxine. Finally, Peters et al. (1978) reported that the TRH-induced PRL elevation recorded in normal volunteers was not significantly modified by the administration of vitamin B_6. Thus, there is a considerable body of evidence to support the claim that this substance has no use in the management of the galactorrhea-amenorrhea syndrome or in the suppression of puerperal lactation.

4.12 References

Adler RA, Noel GL, Wartofsky L, Frantz AG (1975) Failure of oral water loading and intravenous hypotonic saline to suppress plasma prolactin in man. J Clin Endocrinol Metab 41:383–389

Ahlgren M, Källen B, Rannevik G (1976) Outcome of pregnancy after clomiphene therapy. Acta Obstet Gynecol Scand 55:371–375

Alagna S, Lodico G, Devilla L, Masala A, Delitala G, Rovasio PP, Stoppelli I (1979) Piribedil inhibition of prolactin release induced by mechanical breast emptying in puerperal women. Clin Endocrinol (Oxf) 11:301–305

Ambrosi B, Bara R, Travaglini P, Weber G, Beck Peccoz P, Rondena M, Elli R, Faglia G (1977) Study of the effects of bromocriptine on sexual impotence. Clin Endocrinol (Oxf) 7:417–421

Aono T, Shioji T, Kohno M, Ueda G, Kurachi K (1976 a) Pregnancy following 2-bromo-α-ergocryptine (CB-154) – induced ovulation in an acromegalic patient with galactorrhea and amenorrhea. Fertil Steril 27:341–344

Aono T, Miyake A, Shioji T, Kinugasa T, Onishi T, Kurachi K (1976 b) Impaired LH release following exogenous estrogen administration in patients with amenorrhea-galactorrhea syndrome. J Clin Endocrinol Metab 42:696–702

Aono T, Shioji T, Miyai K, Onishi T, Kurachi K (1977 a) Effect of triiodothyronine treatment on prolactin secretion in patients with amenorrhea-galactorrhea. J Clin Endocrinol Metab 44:8–14

Aono T, Shioji T, Onishi T, Kurachi H, Miyai K, Kurachi K (1977 b) Pregnancy following thyroid hormone treatment in a patient with amenorrhea-galactorrhea due to primary hypothyroidism. Fertil Steril 28:878–880

Aono T, Yasuda M, Shioji T, Kondo K, Kurachi K (1978 a) Absence of inhibition by hyperprolactinemia on ovarian response to exogenous gonadotrophin. Acta Endocrinol (Copenh) 89:142–148

Aono T, Shioji T, Kinugasa T, Onishi T, Kurachi K (1978 b) Clinical and endocrinological analyses of patients with galactorrhea and menstrual disorders due to sulpiride or metoclopramide. J Clin Endocrinol Metab 47:675–680

Aono T, Shioji T, Aki T, Hirota K, Nomura A, Kurachi K (1979 a) Augmentation of puerperal lactation by oral administration of sulpiride. J Clin Endocrinol Metab 48:478–482

Aono T, Miyake A, Shioji T, Yasuda M, Koike K, Kurachi K (1979 b) Restoration of oestrogen positive feedback effect on LH release by bromocriptine in hyperprolactinemic patients with galactorrhea-amenorrhea. Acta Endocrinol (Copenh) 91:591–600

Apostolakis M, Kapetanakis S, Lazos G, Madena-Pyrgaki A (1972) Plasma prolactin activity in patients with galactorrhea after treatment with psychotropic drugs. In: Wolstenhome GEW, Knight J (eds). Lactogenic hormones. Livingstone, Edinburgh, pp 349–354

Arana G, Boyd AE, Reichlin S, Lippett D (1977) Prolactin levels in mild depression. Psychosom Med 39:193–197

Archer DF, Josimovich JB (1975) Response of serum prolactin to exogenous stimulation. Fertil Steril 26:627–633

Archer DF, Josimovich JB (1976) Ovarian response to exogenous gonadotropins in women with elevated serum prolactin. Obstet Gynecol 48:155–157

Archer DF, Nankin HR, Gabos PF, Maroon J, Nosetz S, Wadhwa SR, Josimovich JB (1974) Serum prolactin in patients with inappropriate lactation. Am J Obstet Gynecol 119:466–472

Archer DF, Sprong JW, Nankin HR, Josimovich JB (1976) Pituitary gonadotropic response in women with idiopathic hyperprolactinemia. Fertil Steril 27:1158–1164

Archer DF, Maroon JC, DuBois PJ (1978) Galactorrhea, amenorrhea, hyperprolactinemia, and empty sella. Obstet Gynecol [Suppl] 52:23s–27s

Argonz J, del Castillo EB (1953) Syndrome characterized by estrogenic insufficiency, galactorrhea and decreased urinary gonadotropin. J Clin Endocrinol 13:79–87

Ayalon D, Peyser MR, Toaff R, Cordova T, Harell A, Franchimont P, Lindner HR (1974) Effect of L-dopa on galactopoiesis and gonadotropin levels in the inappropriate lactation syndrome. Obstet Gynecol 44:159–170

Badano AR, Arcangeli OA, Mirkin A, Miechi H, Rodriguez A, Turner D, Aparicio N, Figueroa Casas PR (1977) La bromocriptina en el tratamiento del sindrome de amenorrea e hiperprolactinemia. Obstet Ginecol Lat Am 35:325–332

Bassi F, Giusti G, Borsi L, Cattaneo S, Giannotti P, Forti G, Pazzagli M, Vigiani C, Serio M (1977) Plasma androgens in women with hyperprolactinemic amenorrhoea. Clin Endocrinol (Oxf) 6:5–10

Bauer AGC, Wilson JHP, Lamberts SWJ (1980) The kidney is the main site of prolactin elimination in patients with liver disease. J Clin Endocrinol Metab 51:70–73

Bellmann O (1977) Prolaktin in der Schwangerschaft und im Wochenbett – zu Sekretion, Regulation und Funktion. Gynaekologe 10:66–83

Benakis A, Stefan Y (1977) Argument in favor of the direct action of psychotropic drugs on hypophysis prolactin and gonadotrope cells. In: Crosignani PG, Robyn C (eds). Prolactin and human reproduction. Academic Press, London, pp 125–134

Benoit E, Pearson-Murphy BE, Robert F, Marcovitz S, Hardy J, Tsoukas G, Gardiner RJ (1980) Hyperthyroidism due to a pituitary TSH secreting tumour with amenorrhea-galactorrhea. Clin Endocrinol (Oxf) 12:11–19

Bergh T, Nillius SJ, Wide L (1977) Hyperprolactinaemia in amenorrhoea – incidence and clinical significance. Acta Endocrinol (Copenh) 86:683–694

Berle P (1973) Comparative studies on the metabolic effects of some parameters of carbohydrate and lipid metabolism after intravenous administration of human placental lactogen, human prolactin and growth hormone. Acta Endocrinol [Suppl] (Copenh) 173:104

Berlinger EG, Ruder HJ, Wilber JF (1977) Cushing's syndrome associated with galactorrhea, amenorrhea, and hypothyroidism: a primary hypothalamic disorder. J Clin Endocrinol Metab 45:1205–1210

Besser GM, Parke L, Edwards CRW, Forsyth IA, McNeilly AS (1972) Galactorrhea: successful treatment with reduction of plasma prolactin levels by brom-ergocryptine. Br Med J 3:669–672

Beumont PJV, Gelder MG, Friesen HG, Harris GW, MacKinnon PCB, Mandelbrote BM, Wiles DH (1974 a) Effects of phenothiazines on endocrine function. I. Patients with inappropriate lactation and amenorrhea. Br J Psychiatry 124:413–419

Beumont PJV, Corker CS, Friesen HG, Kolakowska T, Mandelbrote BM, Marshall J, Murray MAF, Wiles DH (1974 b) The effects of phenotiazines on endocrine function. II. Effects in men and post-menopausal women. Br J Psychiatry 124:420–430

Beumont PJV, Friesen HG, Gelder MG, Kolakowska T (1974c) Plasma prolactin and luteinizing hormone levels in anorexia nervosa. Psychol Med 4:219–221

Beumont P, Bruwer J, Pimstone B, Vinik A, Utian W (1975) Bromoergocryptine in the treatment of phenothiazine-induced galactorrhea. Br J Psychiatry 126:285–288

Bigazzi M, Gheri R, Casciano S, Zurli A, Refetoff S (1974) Prolactin, pituitary trophins and peripheral hormones: measurements in a case of male galactorrhea. In: Giornata endocrinologica Pisana. Pacini, Pisa, pp 9–18

Bigazzi M, Ronga R, Lancranjan I, Ferraro S, Branconi F, Buzzoni P, Martorana G, Scarselli GF, del Pozo E (1979) A pregnancy in an acromegalic woman during bromocriptine treatment: effects on growth hormone and prolactin in the maternal, fetal, and amniotic compartments. J Clin Endocrinol Metab 48:9–12

Bischoff J, Rebhan E-M, Prestele H, Becker H (1980) Serum-Prolaktin und Anamnesevergleich bei Mammazysten und zystischer Mastopathie. Geburtshilfe Frauenheilkd 40:65–71

Bohnet HG, Dahlen HG, Schneider HPG (1974) Hyperprolactinemia and pulsatile LH fluctuation. Acta Endocrinol [Suppl] (Copenh) 184:109

Bohnet HG, Dahlen HG, Wuttke W, Schneider HPG (1976) Hyperprolactinemic anovulatory syndrome. J Clin Endocrinol Metab 42:132–143

Bohnet HG, Mühlenstedt D, Hanker JPF, Schneider HPG (1977) Prolactin oversuppression. Arch Gynaekol 223:173–178

Bohnet HG, Greiwe M, Hanker JP, Aragona C, Schneider HP (1978) Effects of cimetidine on prolactin, LH, and sex steroid secretion in male and female volunteers. Acta Endocrinol (Copenh) 88:428–434

Bohnet HG, Hanker JP, Horowski R, Wickings EJ, Schneider HPG (1979) Suppression of prolactin secretion by lisuride throughout the menstrual cycle and in hyperprolactinaemic menstrual disorders. Acta Endocrinol (Copenh) 92:8–19

Bommer J, Tschöpe W, Ritz E, Andrassy K (1976) Sexual behaviour of hemodialysed patients. Clin Nephrol 6:315–318
Bommer J, Ritz E, del Pozo E, Bommer G (1979) Improved sexual function in male haemodialysis patients on bromocriptine. Lancet 2:496–497
Boroditsky RS, Faiman C (1973) Galactorrhea-amenorrhea due to primary hypothyroidism. Am J Obstet Gynecol 116:661–665
Boucher D, Hermabessiere J, Doly M (1977) Prolactin secretion in infertile men before and after treatment with bromocriptine. Ann Biol Anim Biochim Biophys 17:483–498
Bowers CY, Friesen HG, Hwang P, Guyda HJ, Folkers K (1971) Prolactin and thyrotropin release in man by synthetic pyroglutamyl-prolinamide. Biochem Biophys Res Commun 45:1033–1041
Boyar RM, Hellman L (1974) Syndrome of benign nodular adrenal hyperplasia associated with feminization and hyperprolactinemia. Ann Intern Med 80:389–394
Boyar RM, Kapen S, Finkelstein JW, Perlow M, Sassin JF, Fukushima DK, Weitzman ED, Hellman L (1974) Hypothalamic-pituitary function in diverse hyperprolactinemic states. J Clin Invest 53:1588–1598
Boyd AE, Spare S, Bower B, Reichlin S (1978) Neurogenic galactorrhea-amenorrhea. J Clin Endocrinol Metab 47:1374–1377
Brambilla F, Guastalla A, Guerrini A, Rovere C, Legnani G, Sarno M, Riggi F (1976) Prolactin secretion in chronic schizophrenia. Acta Psychiatr Scand 54:275–286
Brauman H (1973) Stimulation of prolactin secretion in various clinical situations. In: Pasteels JL, Robyn C (eds). Human prolactin. Excerpta Medica, Amsterdam, pp 224–225
Brisson GR, Volle MA, DeCarufel D, Desharnais M, Tanaka M (1980) Exercise-induced dissociation of the blood prolactin response in young women according to their sports habits. Horm Metab Res 12:201–205
Brown DR, McMillin JM (1977) Post traumatic anterior hypopituitarism-revised: the role of TRH provocative release of prolactin in the confirmation of anterior pituitary damage. Pediatrics 59:948–950
Buckman MT, Peake GT (1973) Estrogen potentiation of phenotiazine-induced prolactin secretion in man. J Clin Endocrinol Metab 37:114–117
Buckman MT, Kaminsky N, Conway M, Peake GT (1973) Water load – a new test for prolactin suppression. Clin Res 21:250
Buckman MT, Peake GT, Srivastava S (1980) Periovulatory enhancement of spontaneous prolactin secretion in normal women. Metabolism 29:753–757
Callaghan JT, Cleary RE, Crabtree R, Lemberger L (1981) Clinical response of patients with galactorrhea to pergolide, a potent, long acting dopaminergic ergot derivative. Life Sci 28: 95–102
Camanni F, Genazzani AR, Massara F, de Leo V, Molinatti GM, Müller EE (1980) Prolactin responsiveness to nomifensine in patients with hyperprolactinemia of tumorous or uncertain etiology. J Clin Endocrinol Metab 51:650
Canales ES, Forsbach G, Soria J, Zarate A (1976a) Infertility due to hyperprolactinemia and its treatment with ergocryptine. Fertil Steril 27:1335–1336
Canales ES, Soria J, Zárate A, Mason M, Molina M (1976b) The influence of pyridoxine on prolactin secretion and milk production in women. Br J Obstet Gynaecol 83:387–388
Carlson HE, Ippoliti AF (1977) Cimetidine, an H2-antihistamine, stimulates prolactin secretion in man. J Clin Endocrinol Metab 45:367–370
Carlson HE, Jacobs LS, Daughaday WH (1973) Growth hormone, thyrotropin and prolactin responses to thyrotropin-releasing hormone following diethylstilbestrol pretreatment. J Clin Endocrinol Metab 37:488–490
Carlson HE, Brickman AS, Bottazzo GF (1977) Prolactin deficiency in pseudohypoparathyroidism. N Engl J Med 296:140–144
Caro JF, Woolf PD (1980) Pituitary-ovarian axis responsivity to prolonged gonadotropin-releasing hormone infusion in normal and hyperprolactinemic women. J Clin Endocrinol Metab 50:999–1004
Carr D, Mathie IK, Manners AR, Colman C (1979) Hyperprolactinemia in a patient with the McCune-Albright syndrome. Br J Obstet Gynaecol 86:330–331
Carter JN, Tyson JE, Warne GL, McNeilly AS, Faiman C, Friesen HG (1977) Adrenocortical function in hyperprolactinemic women. J Clin Endocrinol Metab 45:973–980
Carter JN, Tyson JE, Tolis G, Van-Vliet S, Faiman C, Friesen HG (1978) Prolactin-secreting tumors and hypogonadism in 22 men. N Engl J Med 299:847–852

Castro JH, Caro JR, Kim HJ, Glennon JA (1980) Effects of parathyroid hormone infusion and primary hyperparathyroidism on serum prolactin in man. J Clin Endocrinol Metab 51:397–398

Chalmers RJ, Bennie EH, Johnson RH, Masterton G (1978) Growth hormone, prolactin, and corticosteroid responses to insulin hypoglycaemia in alcoholics. Br Med J 1:745–748

Chiari J, Braun K, Späth J (1855) Klinik der Geburtshilfe und Gynäkologie. Erlangen

Chiodini PG, Liuzzi A, Müller EE, Botalla L, Cremascoli G, Oppizzi G, Verde G, Silvestrini F (1976) Inhibitory effect of an ergoline derivative, methergoline, on growth hormone and prolactin levels in acromegalic patients. J Clin Endocrinol Metab 43:356–363

Cleary RE, Crabtree R, Lemberger L (1975) The effect of lergotrile on galactorrhea and gonadotropin secretion. J Clin Endocrinol Metab 40:830–833

Coelingh Bennink HJT (1979) Intermittent bromocriptine treatment for the induction of ovulation in hyperprolactinemic patients. Fertil Steril 31:267–272

Corenblum B, Pairaudeau N, Shewchuk AG (1976) Prolactin hypersecretion and short luteal phase defects. Obstet Gynecol 47:486–488

Crosignani PG, Reschini E, Peracchi M, D'Alberton A, Lombroso GC (1977) Pregnancy following metergoline treatment in a patient with hyperprolactinaemia. Br J Obstet Gynaecol 84:386–388

Crosignani PG, Reschini E, Lombroso GC, Arosio M, Peracchi M (1978 a) Comparison of placebo and bromocriptine in the treatment of patients with normoprolactinemic amenorrhea. Br J Obstet Gynaecol 85:773–775

Crosignani PG, Peracchi M, Lombroso GC, Reschini E, Mattei A, Caccamo A, D'Alberton A (1978 b) Antiserotonin treatment of hyperprolactinemic amenorrhea: long-term follow-up with metergoline, methysergide and cyproheptadine. Am J Obstet Gynecol 132:307–311

Crosignani PG, Ferrari C, Malinverni A, Barbieri C, Mattei AM, Caldara R, Rocchetti M (1980) Effect of central nervous system dopaminergic activation on prolactin secretion in man: evidence for a common central defect in hyperprolactinemic patients with and without radiological signs of pituitary tumors. J Clin Endocrinol Metab 51:1068–1073

Czernichow P, Dauzet MC, Broyer M, Rappaport R (1976) Abnormal TSH, PRL and GH response to TSH releasing factor in chronic renal failure. J Clin Endocrinol Metab 43:630–637

D'Agata R, Gulizia S, Vicari E, Aliffi A, Polosa P (1979) Effect of androgens on prolactin (PRL) release in humans. Acta Endocrinol (Copenh) 90:409–413

Dagget PR, Nabarro JDN (1979) Measurements of the integrated concentration of prolactin in normal subjects and patients with hyperprolactinemia. J Endocrinol 83:91–94

Darragh A, O'Kelly DA, Browne ADH, del Pozo E (1976) Restoration of fertility in prolactin induced premature menopause. Ir J Med Sci 145:231–236

Davie R, Butler N, Goldstein H (1972) From birth to seven. The second report of the national child development study. Longman, London

De Cecco L, Foglia G, Ragni N, Possato P, Venturini PL (1978) The effect of lisuride hydrogen maleate in the hyperprolactinaemia-amenorrhoea syndrome: clinical and hormonal responses. Clin Endocrinol (Oxf) 9:491–498

Delitala G, Masala A, Alagna S, Devilla L (1976) Effect of pyridoxine on human hypophyseal trophic hormone release: a possible stimulation of hypothalamic dopaminergic pathway. J Clin Endocrinol Metab 42:603–605

Delitala G, Wass JAH, Stubbs WA, Jones A, Williams S, Besser GM (1979) The effect of lisuride hydrogen maleate, an ergot derivative on anterior pituitary hormone secretion in man. Clin Endocrinol (Oxf) 11:1–9

Delitala G, Devilla L, Lotti G (1980) Domperidone, an extracerebral inhibitor of dopamine receptors, stimulates thyrotropin and prolactin release in man. J Clin Endocrinol Metab 50:1127–1130

Delvoye P, Taubert H-D, Jurgensen O, L'Hermite M, Delogne J, Robyn C (1974) Serial measurements of gonadotrophins and progesterone in the luteal phase of the menstrual cycle during hyperprolactinaemia-induced by sulpiride. C R Acad Sci [D] (Paris) 279:1463–1466

De Rivera JL, Lal S, Ettigi P, Hontela S, Muller HF, Friesen HG (1976) Effect of acute and chronic neuroleptic therapy on serum prolactin levels in men and woman of different age groups. Clin Endocrinol (Oxf) 5:273–282

Diedrich K, Leidenberger F, Lehman F, Bettendorf G (1978) Klinisch experimentelle Studie zur Therapie ovarieller Funktionsstörungen mit 2-Br-α-Ergocryptin (Pravidel®). Geburthilfe Frauenheilkd 38:716–725

Döhler KD, Wuttke W (1974) Total blockade of phasic pituitary prolactin release in rats: effect on serum LH and progesterone during the estrous cycle and pregnancy. Endocrinology 94:1595–1601

Drew J, Parkinson P, Walstab JE, Beischer NA (1977) Incidences and types of malformations in newborn infants. Med J Aust 2:945–949
Duello TM, Halmi NS (1977) Pituitary adenoma producing thyrotrophin and prolactin. An immunocytochemical and electron microscopic study. Virchows Arch [Pathol Anat] 376:255–265
Duignan NM (1976) Polycystic ovarian disease. Br J Obstet Gynaecol 83:593–602
Ehara Y, Siler T, van den Berg G, Sinha YN, Yen SSC (1973) Circulating prolactin levels during the menstrual cycle: episodic release and diurnal variation. Am J Obstet Gynecol 117:962–970
Ekelund K, Kullander S, Källen B (1970) Major and minor malformations in newborns and infants up to one year of age. Acta Paediatr Scand 59:297–302
Evans WS, Rogol AD, MacLeod RM, Thorner MO (1980) Dopaminergic mechanisms and luteinizing hormone secretion. I. Acute administration of the dopamine agonist bromocriptine does not inhibit luteinizing hormone release in hyperprolactinemic women. J Clin Endocrinol Metab 50: 103–107
Faglia G, Beck-Peccoz P, Travaglini P, Ambrosi B, Rondena M, Paracchi A, Spada A, Weber G, Bara R, Bouzin A (1977) Functional studies in hyperprolactinemic states. In: Crosignani PG, Robyn C (eds). Prolactin and human reproduction. Academic Press, London, pp 225–238
Falaschi P, Frajese G, Rocco A, Toscano V, Sciarra D (1977) Polycystic ovary syndrome and hyperprolactinemia. J Steroid Biochem 8:xiii
Falaschi P, Frajese G, Sciarra F, Rocco A, Conti C (1978) Influence of hyperprolactinaemia due to metoclopramide on gonadal function in men. Clin Endocrinol (Oxf) 8:427–433
Falaschi P, Frajese G, Rocco A, Besser GM, Rees LH (1979) Prolactin and idiopathic oligospermia. Lancet 1:667
Falaschi P, del Pozo E, Rocco A, Toscano V, Petrangeli E, Pompei P, Frajese G (1980) Prolactin release in polycystic ovary. Obstet Gynecol 55:579–582
Fave GFD, delle Tamburrano G, deMagistris L, Natoli C, Santoro ML, Carratu R, Torsoli A (1977) Gynaecomastia with cimetidine. Lancet 1:1319
Feek CM, Sawers JSA, Brown NS, Seth J, Irvine WJ, Toft AD (1980) Influence of thyroid status on dopaminergic inhibition of thyrotropin and prolactin secretion: evidence for an additional feedback mechanism in the control of thyroid hormone secretion. J Clin Endocrinol Metab 51:585–589
Ferrari C, Crosignani PG, Caldara R, Picciotti MC, Malinverni A, Barattini B, Rampini P, Telloli P (1979) Failure of nomifensine administration to discriminate between tumorous and nontumorous hyperprolactinemia. J Clin Endocrinol Metab 50:23–26
Flückiger E, del Pozo E (1978) Influence on the endocrine system. In: Berde B, Schild HO (eds). Ergot alkaloids and related compounds. Springer, Berlin Heidelberg New York (Handbook of experimental pharmacology, vol 49, pp 615–690)
Flückiger E, del Pozo E (1980) Ergot derivatives and pituitary hormones. In: Müller EE (ed). Neuroactive drugs in endocrinology. Elsevier/North-Holland, Amsterdam, pp 169–190
Foley TP, Jacobs LS, Hoffman W, Daughaday WH, Blizzard RM (1972) Human prolactin and thyrotropin concentration in the serum of normal and hypopituitary children before and after the administration of synthetic thyrotropin releasing hormone. J Clin Invest 51:2143–2150
Fonzo D, Angeli A, Sivieri R, Andriolo S, Frajria R, Ceresa F (1977) Hyperprolactinemia in girls with idiopathic precocious puberty under prolonged treatment with cyproterone acetate. J Clin Endocrinol Metab 45:164–168
Forbes AP, Henneman PH, Griswold GC, Albright F (1954) Syndrome characterized by galactorrhea, amenorrhea, and low urinary FSH. J Clin Endocrinol 14:265–271
Forsbach G, Soria J, Canales ES, Guzman V, Cabezas A, Zarate A (1977) Gonadotropic responsiveness to clomiphene, LRH, estradiol, and bromocriptine in galactorrheic women. Obstet Gynecol 50:139–144
Fossati P, L'Hermite M, Derrien G, Goldstein J, Vanhaelst L, Robyn C, Linquette M (1975) La prolactinémie chez les myxoedémateuses. Ann Endocrinol (Paris) 36:145–151
Fossati P, L'Hermite M, Buvat J, Cappoen JP, Grenier JL, Linquette M (1976a) Etude de la fonction prolactinique dans 14 cas de galactorrhée avec hypogonadisme et micro-déformation de la selle turcique. Ann Endocrinol (Paris) 37:157–169
Fossati P, Strauch G, Tourniaire J (1976b) Etude de l'activité de la bromocriptine dans les états d'hyperprolactinémie. Résultats d'un essai coopératif chez 135 patients. Nouv Presse Med 5: 1687–1691
Foukas MD (1973) An antilactogenic effect of pyridoxine. J Obstet Gynaecol Br Commonw 80:718–720
Franks S, Jacobs HS (1977) Characterization and management of hyperprolactinemic amenorrhea. In:

Crosignani PG, Robyn C (eds) Prolactin and human reproduction. Academic Press, London, pp 245–258

Franks S, Murray MAF, Jequier AM, Steele SJ, Nabarro JDN, Jacobs HS (1975) Incidence and significance of hyperprolactinemia in women with amenorrhea. Clin Endocrinol (Oxf) 4:597–607

Franks S, Jacobs HS, Nabarro JDN (1976) Prolactin concentrations in patients with acromegaly: clinical significance and response to surgery. Clin Endocrinol (Oxf) 5:63–69

Franks S, Jacobs HS, Martin N, Nabarro JDN (1978) Hyperprolactinemia and impotence. Clin Endocrinol (Oxf) 8:277–287

Frantz AG (1973) The regulation of prolactin secretion in humans. In: Ganong WF, Martini L (eds). Frontiers in neuroendocrinology. Oxford University Press, New York, pp 337–374

Frantz AG, Habif DV, Hyman GA, Suh HK (1972a) Remission of metastatic breast cancer after reduction of circulating prolactin in patients treated with L-dopa. Clin Res 20:864

Frantz AG, Kleinberg DL, Noel GL (1972b) Studies on prolactin in man. Recent Prog Horm Res 28:527–590

Fredricsson B, Bjork G, Carlstrom K (1977) Short luteal phase and prolactin. Lancet 1:1210

Friedman S, Goldfien A (1969) Amenorrhea and galactorrhea following oral contraceptive therapy. JAMA 210:1888–1891

Friesen HG, Tolis G (1977) The use of bromocriptine in the galactorrhea-amenorrhea syndromes: the Canadian cooperative study. Clin Endocrinol [Suppl 1] (Oxf) 6:91s–97s

Friesen H, Hwang P, Guyda H, Tolis G, Tyson J, Myers R (1972) A radioimmunoassay for human prolactin. In: Boyns AR, Griffiths K (eds) Prolactin and carcinogenesis. Alpha Omega Alpha, Cardiff, pp 64–80

Friesen HG, Tolis G, Shiu R, Hwang P (1973) Studies on human prolactin: chemistry, radioreceptor assay and clinical significance. In: Pasteels JL, Robyn C (eds). Human prolactin. Excerpta Medica, Amsterdam, pp 11–23

Frohman LA, Sievertsen G, Fine S, Lim VS, Kathpalia S (1979) Comparison of the altered dopaminergic mechanisms affecting prolactin secretion in the ovulatory disturbances of prolactin-secreting tumors and chronic renal failure. In: Zichella L, Pancheri P (eds) Psychoneuroendocrinology in reproduction. Elsevier/North-Holland, Biomedical Press, Amsterdam, pp 349–357

Fromantin M, Gautier D, Querency E, Kamalodine T (1975) Hyperthyroidism, sarcoidosis, amenorrhea-galactorrhea syndrome appearing simultaneously in the same patients. Ann Med Interne 126:853–856

Frommel R (1882) Über puerperale Atrophie des Uterus. Z Geburtshilfe Gynaekol 7:305–313

Futterweit W, Krieger DT (1979) Pituitary tumors associated with hyperprolactinemia and polycystic ovarian disease. Fertil Steril 31:608–613

Gala RR, van de Walle C, Hoffman WH, Lawson DM, Pieper DR, Smith SW, Subramanian MG (1977) Lack of circannual cycle of daytime serum prolactin in man and monkey. Acta Endocrinol (Copenh) 86:257–262

Gambrell RD, Greenblatt RB, Mahesh VB (1971) Post-pill and pill-related amenorrhea-galactorrhea. Am J Obstet Gynecol 110:838–848

Gates RB, Friesen H, Samaan NA: Inappropriate lactation and amenorrhoea: pathological and diagnostic considerations. Acta Endocrinol (Copenh) 72:101–114

Gaulden EC, Littlefield DC, Putoff OE, Seivert AL (1964) Menstrual abnormalities associated with heroin addiction. Am J Obstet Cynecol 90:155–160

Gautvik KM, Weintraub BD, Graeber CT, Maloof F, Zuckerman JE, Tashjian AH (1973) Serum prolactin and TSH: effects of nursing and pyroGlu-His-ProNH administration in postpartum women. J Clin Endocrinol Metab 37:135–139

Gershengorn MC, Marcus-Samuels BE, Geras E (1979) Estrogens increase the number of thyrotropin-releasing hormone receptors on mammotropic cells in culture. Endocrinology 105:171–176

Giusti G, Bassi F, Forti G, Giannotti P, Calabresi E, Pazzagli M, Fiorelli G, Mannelli M, Misciglia N, Serio M (1978) Effects of prolactin on androgen secretion by the human adrenal cortex. In: Robyn C, Harter M (eds) Progress in prolactin physiology and pathology. Elsevier/North-Holland Biomedical Press, Amsterdam, pp 293–303

Glass MR, Shaw RW, Butt WR, Logan Edwards R, London DR (1975) An abnormality of oestrogen feedback in amenorrhea-galactorrhea. Br Med J 3:274–275

Glass MR, Shaw RW, Butt WR, Logan Edwards R, London DR (1979) Oestrogen-gonadotropin feedback abnormalities in hyperprolactinaemic amenorrhoea demonstrated by the response to clomiphene administration. Br J Obstet Gynecol 86:64–68

Gold MS, Donabedian RK, Dillard M, Slobetz FW, Riordan CE, Kleber HD (1977) Antipsychotic effect of opiate agonists. Lancet 2:398–399

Gomez F, Reyes FI, Faiman C (1977) Nonpuerperal galactorrhea and hyperprolactinemia. Clinical findings, endocrine features and therapeutic responses in 56 cases. Am J Med 62:648–660

Goodenow TJ, Malarkey WB (1979) Ineffectiveness of pyridoxine in evaluation and treatment of the hyperprolactinemic amenorrhea-galactorrhea syndrome. Am J Obstet Gynecol 133:161–164

Gräf K-J, Schmidt-Gollwitzer M, Koch UJ, Lorenz F, Hammerstein J (1978) Hyperprolactinemia induced by cyproterone acetate in human subjects. Acta Endocrinol (Copenh) 87:s96

Griffith RW, Turkalj I, Braun P (1978) Outcome of pregnancy in mothers given bromocriptine. Br J Clin Pharmacol 5:227–231

Groom GV, Hughes LE, Mansel RE, Pashby NL, Wisbey JR (1980) The effect of withdrawal of bromocriptine on serum prolactin in patients with benign breast disease. J Physiol (Lond) 306:39P

Gropp C, Havemann K, Scheuer A (1980) Ectopic hormones in lung cancer patients at diagnosis and during therapy. Cancer 46:347–354

Guitelman A, Aparicio NJ, Mancini A, Debeljuk L (1978) Release of prolactin during pregnancy: effect of sulpiride. Fertil Steril 30:42–44

Hack M, Brish M, Serr DM, Insler V, Lunenfeld B (1970) Outcome of pregnancy after induced ovulation. Follow-up of pregnancies and children born after gonadotropin therapy. JAMA 211:791–797

Hagen C, Ølgaard K, McNeilly AS, Fischer S (1976) Prolactin and the pituitary-gonadal axis in male uraemic patients on regular dialysis. Acta Endocrinol (Copenh) 82:29–38

Hagen C, Brandt MR, Kehlet H (1980) Prolactin, LH, FSH, GH and cortisol response to surgery and the effect of epidural analgesia. Acta Endocrinol (Copenh) 94:151–154

Halbreich U, Grünhaus L, Ben-David M (1979) Twenty-four-hour rhythm of prolactin in depressive patients. Arch Gen Psychiatry 36:1183–1186

Harrower ADB, Yap PL, Nairn IM, Walton HJ, Strong JA, Craig A (1977) Growth hormone, insulin, and prolactin secretion in anorexia nervosa and obesity during bromocriptine treatment. Br Med J 2:156–159

Harter M, Krebs B, Balarac N, Kozlowski JM, Chichmanian RM, Strulo S, Canivet B (1978) Interrelations between prolactin and carbohydrates. In: Robyn C, Harter M (eds) Progress in prolactin physiology and pathology. Elsevier/North-Holland, Biomedical Press, Amsterdam, pp 331–349

Haus E, Lakatua DJ, Halberg F, Halberg E, Cornelissen G, Sackett LL, Berg JG, Kawasaki T, Ueno M, Uezono K, Matsuoka M, Omae T (1980) Chronobiological studies of plasma prolactin in women in Kyushu, Japan, and Minnesota, USA. J Clin Endocrinol Metab 51:632–640

Healy DL, Burger GH (1977) Increased prolactin and thyrotrophin secretion following oral metoclopramide: Dose-response relationships. Clin Endocrinol (Oxf) 7:195–201

Healy DL, Pepperell RJ, Stockdale J, Bremner WJ, Burger HG (1977) Pituitary autonomy in hyperprolactinemic secondary amenorrhea: results of hypothalamic-pituitary testing. J Clin Endocrinol Metab 44:809–819

Healy DL, Burger HG, Korman MG, Stern A (1979) Long-term cimetidine does not alter serum prolactin. Horm Res 11:53–55

Heinonen OP, Slone D, Shapiro S (1977) Birth defects and drugs in pregnancy. Publishing Sciences Group, Littleton

Heras F de las, Negro-Vilar A (1979) Effect of aromatizable androgens and estradiol on prolactin secretion in prepuberal male rats. Arch Androl 2:135–139

Hirvonen E, Ranta T, Seppälä M (1976) Prolactin suppression stimulates clomiphene responsiveness. Int J Fertil 21:255–257

Holdsworth SR, de Kretser DM, Atkins RC (1978) A comparison of hemodialysis and transplantation in reversing the uremic disturbance of male reproductive function. Clin Nephrol 10:146–150

Horn K, Erhardt F, Fahlbusch R, Pickardt R, Werder DV, Scriba PC (1976) Recurrent goitre hyperthyroidism galactorrhea and amenorrhea due to a thyrotropin and prolactin producing pituitary tumor. J Clin Endocrinol Metab 43:137–143

Horowski R, Wendt H, Gräf KJ (1978) Prolactin-lowering effect of low doses of lisuride in man. Acta Endocrinol (Copenh) 87:234–240

Horrobin DF, Mtabaji JP, Manku MS (1976) Physiological cortisol levels block the inhibition of vascular reactivity produced by prolactin. Endocrinology 99:406–410

Hovatta O, Koskimies AI, Ranta T, Stenman UH, Seppälä M (1979) Bromocriptine treatment of oligospermia: a double blind study. Clin Endocrinol (Oxf) 11:377–382

Hughes P, Gillespie A, Dewhurst CJ (1972) Amenorrhea and galactorrhea. Obstet. Gynecol 40:147–151

Hunt P, Kannengiesser M-H, Raynaud J-P (1974) Nomifensine: a new potent inhibitor of dopamine uptake into synaptosomes from rat brain corpus striatum. J Pharm Pharmacol 26:370–371

Hyyppä MT, Långvik V-A, Rinne UK (1978) Plasma pituitary hormones in patients with Parkinson's disease treated with bromocriptine. J Neural Transm 42:151–157

Isaac R, Merceron RE, Caillens G, Raymond JP, Ardaillou R (1978) Effect of parathyroid hormone on plasma prolactin in man. J Clin Endocrinol Metab 47:18–22

Isaacs AJ, Leslie RDG, Gomez J, Bayliss R (1980) The effect of weight gain on gonado trophins and prolactin in anorexia nervosa. Acat Endocrinol (Copenh) 94:145–150

Jacobs LS, Snyder PJ, Wilber JF, Utiger RD, Daughaday WH (1971) Increased serum prolactin after administration of synthetic thyrotropin releasing hormone (TRH) in man. J Clin Endocrinol Metab 33:996–998

Jacobs LS, Snyder PJ, Utiger RD, Daughaday WH (1973) Prolactin response to thyrotropin releasing hormone in normal subjects. J Clin Endocrinol Metab 36:1069–1073

Jacobs LS, Franks S, Murray MAF, Hull MGR, Steele SJ, Nabarro JDN (1976) Clinical and endocrine features of hyperprolactinaemic amenorrhea. Clin Endocrinol (Oxf) 5, 439–454

Jarrel J, Franks S, McInnes R, Gemayel K, Guyda H, Arronet GH, Naftolin F (1980) Breast examination does not elevate serum prolactin. Fertil Steril 33:49–51

Jaszmann L, Sternthal V (1974) Advances in the treatment of the galactorrhea-amenorrhea syndrome: induction of ovulation and inhibition of lactation with 2-Br-alpha-ergocryptine. Geburtshilfe Frauenheilkd 34:59–64

Jeffcoate WJ, Matthews RW, Edwards CRW, Field LH, Besser GM (1980) The effect of cyproterone acetate on serum testosterone, LH, FSH, and prolactin in male sexual offenders. Clin Endocrinol (Oxf) 13:189–195

Jenkins JS, Gilbert CJ, Ang V (1976) Hypothalamic-pituitary function in patients with craniopharyngiomas. J Clin Endocrinol Metab 43:394–399

Johnston DG, Alberti KGMM, Nattrass M, Burrin JM, Blesa-Malpica G, Hall K, Hall R (1980) Hyperinsulinaemia in hyperprolactinaemic women. Clin Endocrinol (Oxf) 13:361–368

Jones JR, Gentile GP (1975) Incidence of galactorrhea in ovulatory and anovulatory females. Obstet Gynecol 45:13–14

Judd JS, Rakoff JS, Yen SSC (1978) Inhibition of gonadotropin and prolactin release by dopamine: Effect of endogenous estradiol levels. J Clin Endocrinol Metab 47:494–498

Judd SL, Lazarus L, Smythe G (1976) Prolactin secretion by metoclopramide in man. J Clin Endocrinol Metab 43:313–317

Judd SJ, Rigg LA, Yen SSC (1979) The effects of ovariectomy and estrogen treatment on the dopamine inhibition of gonadotropin and prolactin release. J Clin Endocrinol Metab 49:182–184

Jürgensen O, Wirth C, von Rauch J, Dericks-Tan SE, von Wild K, Taubert HD (1976) Bromocryptin (CB 154) in der Behandlung der sekundären Amenorrhoe mit und ohne Hyperprolactinämie. Nota Bene Med 6:38–46

Jung Y, Russfeld AB (1972) Prolactin cells in the hypophysis of cirrhotic patients. Arch Pathol 94:265–269

Kalliala K, Laatikainen T, Vihko R (1970) Neural steroid sulphates in human ovarian blood. J Clin Endocrinol Metab 30:533–535

Kandeel FR, Butt WR, London DR (1977) Regulation of serum prolactin levels by oestrone and oestrone sulphate in normally menstruating women. Acta Endocrinol [Suppl] (Copenh) 212:42

Kandeel FR, Butt WR, Rudd BT, Lynch SS, London DR, Logan Edwards R (1979) Oestrogen modulation of gonadotrophin and prolactin release in women with anovulation and their responses to clomiphene. Clin Endocrinol (Oxf) 10:619–635

Kapcala LP, Molitch ME, Post KD, Biller BJ, Prager RJ, Jackson IMD, Reichlin S (1980) Galactorrhea, oligo/amenorrhea, and hyperprolactinemia in patients with craniopharyngiomas. J Clin Endocrinol Metab 51:798–800

Kapen S, Boyar R, Freeman R, Frantz A, Hellman L, Weitzman ED (1975) Twenty-four hour secretory patterns of gonadotropins and prolactin in a case of Chiari-Frommel syndrome. J Clin Endocrinol Metab 40:234–239

Kase N (1973) Induction of ovulation with clomiphene citrate. Clin Obstet Gynecol 16:192–200

Kemmann E, Jones JR (1979) Hyperprolactinemia and primary amenorrhea. Obstet Gynecol 54:692–694

Kemmann E, Gemzell CA, Beinert WC, Jones JR (1977) Plasma prolactin changes during the administration of human menopausal gonadotropins in nonovulatory women. Am J Obstet Gynecol 129:145–149

Keye WR, Ho Yuen B, Knopf RF, Jaffe RB (1976) Amenorrhea, hyperprolactinemia and pituitary enlargement secondary to primary hypothyroidism. Successful treatment with thyroid replacement. Obstet Gynecol 48:697–702

Kleinberg DL, Noel GD, Frantz AG (1977) Galactorrhea: a study of 235 cases, including 48 with pituitary tumors. N Engl J Med 296:589–600

Kleinberg DL, Lieberman A, Todd J, Greising J, Neophytides A, Kupersmith M (1980) Pergolide mesylate: a potent day-long inhibitor of prolactin in rhesus monkeys and patients with parkinson's disease. J Clin Endocrinol Metab 51:152–154

Klibanski A, Neer RM, Beitins IZ, Ridgway EC, Zervas NT, McArthur JW (1980) Decreased bone density in hyperprolactinemic women. N Engl J Med 303:1511–1514

Klindt J, Davis SL, Ohlson DL (1979) Plasma concentration of thyrotropin-releasing hormone, thyrotropin, prolactin, and growth hormone during five-day osmotic pump infusion of thyrotropin-releasing hormone. Endocrinology 104:45–49

Kollin E, Gergely I, Szücs J, Hollo I (1981) Decreased bone density in hyperprolactinemic women. N Engl J Med 304:1362

Koenig MP, Zuppinger K, Liechti B (1977) Hyperprolactinemia as a cause of delayed puberty: successful treatment with bromocriptine. J Clin Endocrinol 45:825–827

Kolakowska T, Wiles DH, McNeilly AS, Gelder MG (1975) Correlation between plasma levels of prolactin and chlorpromazine in psychiatric patients. Psychol Med 5:214–226

Kolodny RC, Jacobs LS, Daughaday WH (1972) Mammary stimulation causes prolactin secretion in non-lactating women. Nature 238:284–286

Koninckx P (1978) Stress hyperprolactinemia in clinical practice. Lancet 1:273

Kopelman PG, Pilkington TRE, White N, Jeffcoate SL (1979) Impaired hypothalamic control of prolactin secretion in massive obesity. Lancet 1:747–750

Kort HI, Obbins JC, Decherney BV (1980) Ovarian function and morphology following treatment with bromoergocryptine. Fertil Steril [Suppl] 33:236

Krieger DT, Howanitz PJ, Frantz AG (1976) Absence of nocturnal elevation of plasma prolactin concentrations in Cushing's disease. J Clin Endocrinol Metab 42:260–272

Labrie F, Beaulieu M, Caron MG, Raymond V (1978) The adenohypophyseal dopamine receptor: Specificity and modulation of its activity by estradiol. In: Robyn C, Harter M (eds) Progress in prolactin physiology and pathology. Elsevier/North-Holland, Biomedical Press, Amsterdam, pp 121–136

Lachelin GCL, Abu-Fadil S, Yen SSC (1977a) Functional aberration of hyperprolactinemic-amenorrhea. J Clin Endocrinol Metab 44:1163–1174

Lachelin GCL, Leblanc H, Yen SSC (1977b) The inhibitory effect of dopamine agonists on LH release in women. J Clin Endocrinol Metab 44:728–732

Landgraf L, Landgraf MMC, Weissmann A, Hörl R, von Werder K, Scriba PC (1977) Prolactin: A diobetogenic hormone. Diabetologia 13:99–104

Langer G, Sachar EJ, Halpern FS, Gruen PH, Solomon M (1977) The prolactin response to neuroleptic drugs. A test of dopaminergic blockade: neuroendocrine studies in normal men. J Clin Endocrinol Metab 45:996–1002

Large DM, Anderson DC, Laing I (1980) Twenty-four hour profiles of serum prolactin during male puberty with and without gynecomastia. Clin Endocrinol (Oxf) 12:293–302

Laurent MF, Peillon F, Dreyfus G (1973) Primary hypothyroidism of the adult and non-tumoral amenorrhea-galactorrhea syndrome. Sem Hop Paris 49:2840–2844

Lavric MV (1969) Galactorrhea and amenorrhea with polycystic ovaries. Am J Obstet Gynecol 104:814–817

Lavric MV (1975) Non-puerperal breast secretion. Its relation to the use of oral contraceptives. Obstet Gynecol 46:12–14

Lederer J, Marchandise B (1971) Galactorrhée sans aménorrhée dans un cas d'insuffisance thyroidienne primaire. Ann Endocrinol (Paris 32:210–213

Lee PA, Kelly MR, Wallin JD (1976) Increased prolactin levels during reserpine treatment of hypertensive patients. JAMA 235:2316–2317

Legros JJ, Chiodera P, Mormont C, Servais J (1980) A psychoneuroendocrinological study of sexual impotence in patients with abnormal reaction to a glucose tolerance test. Adv Biol Psychiatry 5:117–124

Lehtovirta P, Ranta T, Seppälä M (1978) Pyridoxine treatment of galactorrhea-amenorrhea syndromes. Acta Endocrinol (Copenh) 87:682–686

Leighton PC, McNeilly AS, Chard T (1976) Shortterm variation in blood levels of prolactin in women. J Endocrinol 68:177–178

Lemberger L, Crabtree RE (1979) Pharmacological effects in man of a potent, long-acting dopamine receptor agonist. Science 205:1151–1153

Lemberger L, Crabtree RE, Clemens J, Dyke RW, Woodburn RT (1974) The inhibitory effect of an ergoline derivative (Lergotrile, compound 83536) on prolactin secretion in man. J Clin Endocrinol Metab 39:579–584

Lenton EA, Sobowale OS, Cooke ID (1977) Prolactin concentrations in ovulatory but infertile women: treatment with bromocriptine. Br Med J 2:1179–1181

Levy CL, Carlson HE (1978) Decreased prolactin reserve in hemochromatosis. J Clin Endocrinol Metab 47:444–446

Levy NB, Wynbrandt GD (1975) The quality of life on maintenance haemodialysis. Lancet 2: 1328–1330

Leyendecker G, Nocke W, Schmidt-Gollwitzer M, Entzian W, del Pozo E (1977) Klinik der hyperprolaktinämischen Amenorrhoe. Gynaekologe 10:93–108

L'Hermite M, Delvoye P, Nokin J, Vekemans M, Robyn C (1972) Human prolactin secretion, as studied by radioimmunoassay: some aspects of its regulation. In: Boyns AR, Griffiths K (eds) Prolactin and carcinogenesis. Alpha Omega Alpha, Cardiff, pp 81–97

L'Hermite M, Denayer P, Golstein J, Virasoro E, Vanhaelst L, Copinschi G, Robyn C (1978 a) Acute endocrine profile of sulpiride in the human. Clin Endocrinol (Oxf) 9:195–204

L'Hermite M, Delogne-Desnoeck J, Michaux-Duchene A, Robyn C (1978 b) Alteration of feedback mechanism of estrogen on gonadotropin by sulpiride-induced hyperprolactinemia. J Clin Endocrinol Metab 47:1132–1136

L'Hermite M, Michaux-Duchêne A, Robyn C (1979) Tiapride-induced chronic hyperprolactinaemia: interference with the human menstrual cycle. Acta Endocrinol (Copenh) 92:214–227

Lim VS, Kathpalia SC, Frohman LA (1979) Hyperprolactinemia and impaired pituitary response to suppression and stimulation in chronis renal failure: reversal after transplantation. J Clin Endocrinol Metab 48:101–107

Lister FC, Underwood LE, Marshall RN, Friesen HG, van Wyk JJ (1974) Evidence for a direct effect of thyrotropin releasing hormone (TRH) on prolactin release in humans. J Clin Endocrinol Metab 39:1148–1150

Liuzzi A, Chiodini PG, Oppizzi G, Botalla L, Verde G, de Stefano L, Colussi G, Gräf KJ, Horowski R (1978) Lisuride hydrogen maleate: evidence for a long lasting dopaminergic activity in humans. J Clin Endocrinol 46:196–202

Lloyd SJ, Josimovich JB, Archer DF (1975) Amenorrhea and galactorrhea: results of therapy with 2-Brom-α-ergocryptine (CB 154). Am J Obstet Gynecol 122:85–89

Lunenfeld B, Farin D, Serr DM, Mashiach S, Oelsner G, Blankstein J (1980) Catamnestic studies in 700 women following HMG/HCG treatment. Acta Endocrinol [Suppl 234] (Copenh) 94:170–172

Lutterbeck PM, Pryor JS, Varga L, Wenner R (1971) Treatment of non-puerperal galactorrhoea with an ergot alkaloid. Br Med J 3:228–229

Maas JM (1967) Amenorrhea-galactorrhea syndrome. Before, during, and after pregnancy. Fertil Steril 18:857–865

Macaron C, Wilber JF, Green O, Freinkel N (1978) Studies of growth hormone (GH), thyrotropin (TSH) and prolactin (PRL) secretion in anorexia nervosa. Psychoneuroendocrinology 3:181–185

MacDonald HN, Collins YD, Tobin MJW, Wijayaratne DN (1976) The failure of pyridoxine in suppression of puerperal lactation. Br J Obstet Gynecol 83:54–55

Mahesh VB, Pria SD, Greenblatt RB (1969) Abnormal lactation with Cushing's syndrome – a case report. J Clin Endocrinol 29:978–981

Majumdar SK, Thomson AD, Shaw GK (1978) Cimetidine and serum prolactin. Br Med J 1:409

Malarkey WB (1975) Nonpuerperal lactation and normal prolactin regulation. J Clin Endocrinol Metab 40:198–204

Malarkey WB, Kataria YP (1974) Sarcoidosis and hyperprolactinemia. Ann Intern Med 81:116

Malarkey WB, Jacobs LS, Daughaday WH (1971) Levodopa suppression of prolactin in nonpuerperal galactorrhea. N Engl J Med 285:1160–1163

Mancini AM, Guitelman A, Vargas CS, Debeljuk L, Aparicio NJ (1976) Effect of sulpiride on serum prolactin levels in humans. J Clin Endocrinol Metab 42:181–184

Mansel RE, Preece PE, Hughes LE (1978) A doubleblind trial of the prolactin inhibitor bromocriptine in painful benign breast disease. Br J Surg 65:724–727

March CM, Dvajan V, Mishell DR (1979) Ovulation induction in amenorrheic women. Obstet Gynecol 53:8–11

Martin-Comin J, Pujol-Amat P, Cararach V, Davi E, Robyn C (1976) Treatment of fibrocystic disease if the breast with a prolactin inhibitor: 2-Br-Alpha-Ergocryptine (CB 154). Obstet Gynecol 48:703–706

Martin-Perez J, del Pozo E, Clarenbach P, Brownell J, Derrer F (1979) Lack of dependency between androgen production and prolactin in man. Acta Endocrinol [Suppl] (Copenh) 225:174

Masala A, Delitala G, Alagna S, Delitala L, Rovasio PP (1979) Effects of long-term treatment with metergoline in patients with idiopathic oligospermia. Clin Endocrinol (Oxf) 11:349–352

Masala A, Alagna S, Faedda R, Satta A, Rovasio P (1980 a) Prolactin secretion in man following acute and long-term cimetidine administration. Acta Endocrinol (Copenh) 93:392–395

Masala A, Alagna S, Devilla L, Delitala G, Rovasio PP (1980 b) Inhibition of prolactin secretion by nomifensine in man. Clin Endocrinol (Oxf) 12:237–241

Massara F, Camanni F, Belforte L, Campagnoli C, Novi RF, Molinatti GM (1980) Effect of sulpiride on plasma prolactin levels in women with puerperal or pathological hyperprolactinemia. Acta Endocrinol (Copenh) 95:145–150

McCallum RW, Sowers JR, Hershman JM, Sturdevant RAL (1976) Metoclopramide stimulates prolactin secretion in man. J Clin Endocrinol Metab 42:1148–1152

McIntosh IH (1976) Treatment of women with the galactorrhea-amenorrhea syndrome with pyridoxine (vitamine B_6). J Clin Endocrinol Metab 42:1192–1195

McNatty KP, Sawers RS, McNeilly AS (1974) A possible role for prolactin in the control of steroid secretion by the human Graafian follicle. Nature 250:654–655

McNeilly AS, Chard R (1974) Circulating levels of prolactin during the menstrual cycle. Clin Enocrinol (Oxf) 3:105–112

McNeilly AS, Hagen C (1974) Prolactin, TSH, LH and FSH responses to a combined LHRH-TRH test at different stages of the menstrual cycle. Clin Endocrinol (Oxf) 3:427–435

McNeilly AS, Thoner MO, Volans G, Besser GM (1974 a) Metoclopramide and prolactin. Br Med J 2:729

McNeilly AS, Sturdy J, Evans DG, Chard T (1974 b) Short-term variation in blood levels of prolactin, luteinizing hormone and follicle-stimulating hormone in normal men throughout the day. J Endocrinol 61:301–302

Meltzer HY, Fang VS (1976) The effect of neuroleptics on serum prolactin in schizophrenic patients. Arch Gen Psychiatry 33:279–288

Meltzer HY, Sachar EJ, Frantz AG (1974) Serum prolactin levels in newly admitted psychiatric patients. Psychopharmacol Bull 10:14–15

Mendlewicz J, Linkowaski P, Braunman H (1977) Growth hormone and prolactin response to levodopa in affective illness. Lancet 1:652–653

Merino G, Canales ES, Vadillo ML, Forsbach G, Solis J, Zarate A (1980) Abnormal prolactin levels in serum and seminal plasma in infertile men. Arch Androl 4:353–355

Milligan MP, Moore DM, Singh MM, Buckingham M, Elstein M (1978) Bromocriptine in the treatment of secondary amenorrhea and ovarian dysfunction in hyper- and normo-prolactinaemic patients. Curr Med Res Opin 5:734–738

Moggi G, Giampietro O, Chisci R, Brunori I, Simonini N (1979) Pregnancy induction after bromocriptine-cyclofenil treatment in some normoprolactinemic anovulatory women. Fertil Steril 32:289–293

Mohini P, Raj BJ, Chapekar TN (1980) Effect of ovine luteinizing hormone and prolactin on progesterone secretion by goat granulosa cells in culture. J Endocrinol 84:311–313

Morgan MY, Jakobovits AW, Gore MBR, Wills MR, Sherlock S (1978) Serum prolactin in liver disease and its relationship to gynecomastia. Gut 19:170–174

Moriondo P, Travaglini P, Nissim M, Faglia G (1980) Evaluation of two inhibitory tests (nomifensine and L-dopa + carbidopa) for the diagnosis of hyperprolactinaemic states. Clin Endocrinol (Oxf) 13:525–533

Mühlenstedt D, Meissner M, Schneider HPG (1979) Prolaktin und Lutealphasendefekt. Geburtshilfe Frauenheilkd 39:580–587

Müller EE, Genazzani A, Murru S, Fioretti P (1977) Effect of piribedil on plasma prolactin levels in women with puerperal orpathological hyperprolactinaemia. Acta Endocrinol (Copenh) 86:33–41

Müller EE, Genazzani AR, Murru S (1978) Nomifensine: Diagnostic test in hyperprolactinemic states. J Clin Endocrinol Metab 47:1352–1357

Müller EE, Camanni F, Cocchi D, Genazzani AR, Massara F, Casanueva F, De Leo, La Rosa R (1980) Dopamine agonists and antagonist drugs in the diagnosis of hyperprolactinemic states. In: Müller EE (ed) Neuroactive drugs in endocrinology. Elsevier/North Holland, Amsterdam, pp 245–262

Nagel TC, Freinkel N, Bell RH, Friesen H, Wilber JF, Metzger BE (1973) Gynecomastia, prolactin, and other peptide hormones in patients undergoing chronic hemodialysis. J Clin Endocrinol Metab 36:428–432

Nagulesparen M, Ang V, Jenkins JS (1978) Bromocriptine treatment of males with pituitary tumours, hyperprolactinaemia, and hypogonadism. Clin Endocrinol (Oxf) 9:73–79

Nencioni T, Miragoli A, Dorato F, Polvani F (1978) Bromocriptine induced pregnancy in two cases of euprolactinemic hypothalamic amenorrhea. J Endocrinol Invest 1:65–67

Nillius SJ, Wide L (1971) Effects of progesterone on the serum levels of FSH and LH in postmenopausal women treated with oestrogen. Acta Endocrinol (Copenh) 67:362–370

Noel GL, Suh HK, Stone JG, Frantz AG (1972) Human prolactin and growth hormone release during surgery and other conditions of stress. J Clin Endocrinol 35:840–851

Noel GL, Dimond RC, Wartofsky L, Earll JM, Frantz AG (1974a) Studies of prolactin and TSH secretion by continuous infusion of small amounts of TRH. J Clin Endocrinol Metab 39:6–17

Noel GL, Suh HK, Frantz AG (1974b) Prolactin release during nursing and breast stimulation in postpartum and nonpostpartum subjects. J Clin Endocrinol Metab 38:413–423

Odell WD, Swerdloff RS (1968) Progestogen-induced luteinizing and follicle-stimulating hormone surge in post menopausal women: a stimulated ovulatory peak. Proc Natl Acad Sci USA 61:529–536

Palmer BV, Monteiro JCMP (1977) Bromocriptine for severe mastalgia. Br Med J 1:1083

Panerai AE, Salerno F, Manneschi M, Cocchi D, Müller EE (1977) Growth hormone and prolactin responses to thyrotropin-releasing hormone in patients with severe liver disease. J Clin Endocrinol Metab 45:134–140

Parker DC, Rossman LG, Vanderlaan EF (1973) Sleep related, nycthemeral and briefly episodic variation in human plasma prolactin concentrations. J Clin Endocrinol Metab 36:1119–1124

Peillon F, Philippon J, Brandi AM, Fohanno D, Laplanc D, Dubois MP, Decourt J (1979) Prolactin-secreting pituitary adenoma in a man with gigantism: a case report. Acta Endocrinol (Copenh) 92:627–639

Pelosi MA, Sama JC, Caterini H, Kaminetzky HA (1974) Galactorrhea amenorrhea syndrome associated with heroin addiction. Am J Obstet Gynecol 118:966–970

Pentland B, Sawyers JS (1980) Galactorrhea after withdrawal of bromocriptine. Br Med J 281:716

Pepperell RJ, Evans JH, Brown JB, Smith MA, Healy D, Burger HG (1977a) Serum prolactin levels and the value of bromocriptine in the treatment of anovulatory infertility. Br J Obstet Gynecol 84:58–66

Pepperell RJ, Evans JH, Brown JB, Bright MI, Smilth M, Burger HG, Healy D (1977a) A study of the effects of bromocriptine on serum prolactin, follicle stimulating hormone and luteinizing hormone and on ovarian responsiveness to exogenous gonadotropins in anovulatory women. Br J Obstet Gynaecol 84:456–463

Perez C, Novoa J, Alcañiz J, Salto L, Barcelo B (1980): Leydig cell tumor of the testis with gynaecomastia and elevated oestrogen, progesterone and prolactin levels: case report. Clin Endocrinol 13:409–412

Peters F, Zimmermann G, Breckwoldt M (1978) Effect of pyridoxine on plasma levels of HGH, PRL, and TSH in normal women. Acta Endocrinol (Copenh) 89:217–220

Pierini AA, Nusimovich B (1978) Bromocriptina y oligospermia. Sem Med 152:426–428

Polatti F, Bolis PF, Ravagni-Probizer MF, Baruffini A, Cafalleri A (1978) Treatment of hyperprolactinemic amenorrhea by intermittent administration of bromocriptine (CB 154). Am J Obstet Gynecol 131:792–796

Polleri A, Carolei A, Rolandi E, Masturzo P, Meco G, Agnoli A (1977) Changes in pituitary hormones serum levels in bromocryptine-treated patients. Neuropsychobiology 3:42–48

Pourmand M, Rodriguez-Arnao MD, Weightman DR, Hall R, Cook DB, Lewis M, Scanlon MF (1980) Domperidone: a novel agent for the investigation of anterior pituitary function and control in man. Clin Endocrinol (Oxf) 12:211–215

Powell EDU, Frantz AG, Rabkin MT, Field RA (1966) Growth hormone in relation to diabetic retinopathy. N Engl J Med 275:922–925

Pozo E del (1982) Hyperprolactinemia in male infertility: treatment with bromocriptine. In: Bain F, Schwarzstein L, Schill WB (eds) Treatment of male infertility. Springer, Berlin Heidelberg New York (in press)

Pozo E del, Brun del Re R (1979) Letter to the editor. N Engl J Med 301:107

Pozo E del, Falaschi P (1980) Role of prolactin in the polycystic ovary syndrome. Scott Med J 25: S89–S93

Pozo E del, Köbberling J (1981) Effect of a metenkephalin (FK 33-824) on the LH response to LHRH and on the LH pulsatile pattern in humans. In: Cortes Prieto J, Campos da Paz A (eds) Research on fertility and sterility. MTP Lancaster 205–211

Pozo E del, Lancranjan I (1978) Clinical use of drugs modifying the release of anterior pituitary hormones. In: Ganong WF, Martini L (eds) Frontiers of neuroendocrinology, vol 5. Raven, New York, pp 207–247

Pozo E del, Brun del Re R, Varga L, Friesen H (1972) The inhibition of prolactin secretion in man by CB-154 (2-Br-α-ergocryptine). J Clin Endocrinol Metab 35:768–771

Pozo E del, Varga L, Wyss H, Tolis G, Friesen H, Wenner R, Vetter L, Uettwiler A (1974) Clinical and hormonal response to bromocriptine (CB-154) in the galactorrhea syndromes. J Clin Endocrinol Metab 39:18–26

Pozo E del, Goldstein M, Friesen H, Brun del Re R, Eppenberger U (1975 a) Lack of action of prolactin suppression on the regulation of the human menstrual cycle. Am J Obstet Gynecol 123:719–723

Pozo E del, Brun del Re R, Hinselmann M (1975 b) Lack of effect of methyl-ergnonovine on postpartum lactation. Am J Obstet Gynecol 123:845–846

Pozo E del, Brun del Re R, Hinselmann M, Wyss H (1975 a) Wirkung von Methergin®, Oxytocin, Pyridoxin (Vitamin B_6) und Quinestrol auf die Prolaktin- und Milchsekretion im Puerperium. Arch Gynaekol 219:469–470

Pozo E del, Varga L, Schulz KD, Künzig HJ, Marbach P, Lopez del Campr G, Eppenberger U (1975 a) Pituitary and ovarian response patterns to stimulation in the postpartum and in galactorrhea-amenorrhea. Obstet Gynecol 46:539–543

Pozo E del, Wyss H, Lancranjan I, Obolensky W, Varga L (1976) Prolactin-induced luteal insufficiency and its treatment with bromocriptine: preliminary results. In: Crosignani PG (ed) Proceedings on "Ovulation in the Human", Freiburg. Academic Press, London, pp 297–299

Pozo E del, Darragh A, Lancranjan I, Ebeling D, Burmeister P, Bühler F, Marbach P, Braun P (1977 a) Effect of bromocriptine on the endocrine system and fetal development. Clin Endocrinol [Suppl] (Oxf) 6:47s–55s

Pozo E del, Hiba J, Lancranjan I, Kunzig HJ (1977 b) Prolactin measurements throughout the life cycle: endocrine correlations. In: Crosignani PG, Robyn C (eds) Prolactin and human reproduction. Academic Press, London, pp 61–69

Pozo E del, Wyss H, Tolis G, Alcañiz J, Campana A, Naftolin F (1979) Prolactin and deficient luteal function. Obstet Gynecol 53:282–286

Pozo E del, Bigazzi M, Calaf J (1980) Induced human gestational hypoprolactinemia: Lack of action on fetal adrenal androgen synthesis. J Clin Endocrinol Metab 51:936–938

Rabello MM, Snyder PJ, Utiger RD (1974) Effects on the pituitary-thyroid axis and prolactin-secretion of single and repetitive oral doses of TRH. J Clin. Endricrinol Metab 39:571–578

Rakoff JS, Yen SSC (1978) Progesterone induced acute release of prolactin in estrogen primed ovariectomized women. J Clin Endocrinol Metab 47:918–921

Ramirez G, O'Neill WM, Bloomer HA, Jubiz W (1977) Abnormalities in the regulation of prolactin in patients with chronic renal failure. J Clin Endocrinol Metab 45:658–661

Rankin IS, Goldfarb AF, Rakoff AE (1969) Galactorrhea-amenorrhea syndromes: Postpartum galactorrhea-amenorrhea in the absence of intracraneal neoplasm. Obstet Gynecol 33:1–10

Ranta T, Lehtovirta P, Stenman U-H, Seppälä M (1979) Serum prolactin and progesterone concentrations in ovulatory infertility. J Endocrinol Invest 2:71–73

Refetoff S, Fang VS, Rapoport B, Friesen HG (1974) Interrelationships in the regulation of TSH and prolactin secretion in man: effects of L-dopa, TRH and thyroid hormone in various combinations. J Clin Endocrinol Metab 38:450–457

Reymond M, Lemarchand-Beraud T (1976) Effects of oestrogen on prolactin and thyrotropin responses to TRH in women during the menstrual cycle and under oral contraceptive treatment. Clin Endocrinol (Oxf) 5:429–437

Ridgeway EC, Kourides IA, Kliman B, Bigos T, Maloof F (1975) Thyrotropin and prolactin pituitary reserve in the "empty sella syndrome". J Clin Endocrinol Metab 41:968–973

Riese W (1928) Milchsekretion und Zwischenhirn. Klin Wochenschr 7: 1954–1955

Rifkin I, Nachtigall LE, Beckman EM (1972) Amenorrhea following use of oral contraceptives. Am J Obstet Gynecol 113:420–432

Rjosk HK, von Werder K, Fahlbusch R (1976) Hyperprolaktinämische Amenorrhoe. Geburtshilfe Frauenheilkd 36:575–587

Robyn C, Delvoye P, Nokin J, Vekemans M, Badawi M, Perez-Lopez FR, L'Hermite M (1973) Prolactin and human reproduction. In: Pasteels JL, Robyn C (eds) Human prolactin. Excerpta Medica, Amsterdam, pp 167–188

Robyn C, Vekemans M, Delvoye P, Joostens-Defleur V, Caufriez A, L'Hermite M (1976) Prolactin and fertility control in women. In: Pecile A, Müller EE (eds) Growth hormone and related peptides. Excerpta Medica, Amsterdam, pp 396–406

Robyn C, Delvoye P, Vekemans M, Caufriez A, Michaux-Duchêne A, L'Hermite M (1977) Physiological aspects of human prolactin secretion. Prog Reprod Biol 2:204–222

Rocco A, Falaschi P, Pompei P, del Pozo E, Frajese G (1979) Chronic anovulation in polycystic ovary syndrome: Role of hyperprolactinemia and its suppression with bromocriptine. In: Zichella L, Pancheri P (eds) Psychoneuroendocrinology in reproduction. Elsevier/North-Holland, Amsterdam, pp 387–394

Rolland R, Schellekens LA, Lequin RM (1974) Successful treatment of galactorrhea and amenorrhea and subsequent restoration of ovarian function by a new ergot alcaloid 2-brom-alpha-ergocryptine. Clin Endocrinol (Oxf) 3:155–166

Rollet J, Guinet P, Girod C, Trouillas J, Cure M (1975) A case of oncocytic adenoma of the pituitary gland with galactorrhea and hyperprolactinemia. Ann Endocrinol (Paris) 38:221–222

Ross F, Nusynowitz ML (1968) A syndrome of primary hypothyroidism, amenorrhea and galactorrhea. J Clin Endocrinol Metab 28:591–595

Rotrosen J, Angrist B, Clark C, Gershon S, Halpern FS, Sachar EJ (1978) Suppression of prolactin by dopamine agonists in schizophrenics and controls. Am J Psychiatry 135:949–951

Roulier R, Mattei A, Reuter A, Franchimont P (1976) Taux de prolactine dans les stérilités et hypogonadismes masculines. Nouv Presse Med 5:1911

Sachar EJ (1976) The use of neuroendocrine techniques in psychopharmacological research. In: Sachar EJ (ed) Hormones, behaviour and psychopathology. Raven, New York, pp 161–176

Sanchez-Franco F, Cacicedo GL, Martin-Zurro A, Escobar del Ray F (1973) Influence of sex and phase of the menstrual cycle on TSH response to TRH, J Clin Endocrinol Metab 37:736–740

Sandler R, Gardner HM (1973) Successfully treated nonpuerperal galactorrhea-amenorrhea: a comment on the fallacy of eponymic classification. Am J Obstet Gynecol 115:861–863

Sarris S, Swyer GIM, MacGarrigle HHG, Lawrence DM, Little V, Lachelin GCL (1978) Prolactin and luteal insufficiency. Clin Endocrinol (Oxf) 9:543–547

Saunders DM, Hunter JC, Haase HR, Wilson GR (1979) Treatment of luteal phase inadequacy with bromocriptine. Obstet Gynecol 53:287–289

Sawin CT, Hershman JM, Boyd AE III., Longcope C, Bacharach P (1978) The relationship of changes in serum estradiol and progesterone during the menstrual cycle to the thyrotropin and prolactin responses to thyrotropin-releasing hormone. J Clin Endocrinol Metab 47:1296–1302

Scanlon MF, Gomez-Pan A, Mora B, Cook DB, Dewar JH, Hildyard A, Weightman DR, Evered DC, Hall R (1977) Effects of nomifensine, an inhibitor of endogenous catecholamine re-uptake, in acromegaly, in hyperprolactinemia, and against stimulated prolactin release in man. Br J Clin Pharmacol 4:191S–197S

Schaison G, Mowszowicz I, Cesselin E, Lagoguey M (1977) Actions antigonadotropes de la prolactine. Etude de dix cas féminins d'hyperprolactinémie. Nouv Presse Med 6:425–428

Schroeder LL, Johnson JC, Malarkey WB (1976) Cerebrospinal fluid prolactin: a reflection of abnormal prolactin secretion in patients with pituitary tumors. J Clin Endocrinol Metab 43:1255–1260

Schulz K-D, del Pozo E, Lose KH, Künzig HJ, Geiger W (1975) Successful treatment of mastodynia with the prolactin inhibitor bromocryptine (CB 154). Arch Gynaekol 220:83–87

Schulz K-D, Geiger W, del Pozo E, Kunzig HJ, (1978) Pattern of sexual steroids, prolactin, and gonadotropic hormones during prolactin inhibition in normally cycling women. Am J Obstet Gynecol 132:561–566

Sciarra F, Sparano F, Giaquinto G, Odoardi A, Mercuri MA, Concolino G (1977) Prolactinemia during hemodialysis: lack of correlation with sodium, potassium and renin-aldosterone system. Horm Res 8:129–138

Seki K, Seki M (1974) Successful ovulation and pregnancy achieved by CB 154 (2-Br-alpha-ergocryptine) in a woman with Chiari-Frommel syndrome. J Clin Endocrinol Metab 38:508–509

Seppälä M, Hirvonen E (1975) Raised serum prolactin levels associated with hirsutism and amenorrhea. Br Med J 4:144–145

Seppälä M, Hirvonen E, Ranta T (1976 a) Bromocriptine treatment of secondary amenorrhea. Lancet 1:1154–1156
Seppälä M, Hirvonen E, Ranta R (1976 b) Hyperprolactinemia and luteal insufficiency. Lancet 1:229–230
Seppälä M, Unnérus H-A, Hirvonen E, Ranta T (1976 c) Bromocriptine increases plasma estradiol-17β concentration in amenorrhea patients with normal serum prolactin. J Clin Endocrinol Metab 43:474–477
Seppälä M, Lehtovirta P, Ranta T (1977) Discordant patterns of hyperprolactinemia and galactorrhoea in secondary amenorrhoea. Acta Endocrinol (Copenh) 86:457–463
Shearman RP, Fraser IS (1977) Impact of new diagnostic methods on the differential diagnosis and treatment of secondary amenorrhea. Lancet 1:1195–1197
Shearman RP, Smith ID (1972) Statistical analysis of relationship between oral contraceptives, secondary amenorrhea and galactorrhea. J Obstet Gynecol 79:654–656
Shearman RP, Turtle JR (1970) Secondary amenorrhea with inappropriate lactation. Am J Obstet Gynecol 106:818–827
Sievertsen GD, Lim VS, Nakawatase C, Frohman LA (1980) Metabolic clearance and secretion rates of human prolactin in normal subjects and in patients with chronic renal failure. J Clin Endocrinol Metab 50:846–852
Smithells RW (1976) Environmental teratogens of man. Br Med Bull 32:27–33
Smythe GA, Compton PJ, Lazarus L (1975) The stimulation of human prolactin secretion by 3-Iodo-L-tyrosine. J Clin Endocrinol Metab 40:714–716
Snyder PJ, Jacobs LS, Rabello MM, Sterling FH, Shore RN, Utiger RD, Daughaday WH (1974) Diagnostic value of thyrotropin releasing hormone in pituitary and hypothalamic diseases. Assessment of thyrotropin and prolactin secretion in 100 patients. Ann Intern Med 81:751–757
Sowers JR, MacCallum RW, Hershman JM, Carlson HE, Sturdevant RAL, Meyer N (1976) Comparison of metoclopramide with other dynamic tests of prolactin secretion. J Clin Endocrinol Metab 43:679–681
Spadoni LR, Cox DW, Smith DC (1974) Use of human menopausal gonadotropin for the induction of ovulation. Am J Obstet Gynecol 120:988–993
Spark RF, White RA, Connolly PB (1980) Impotence is not always psychogenetic. Newer insights into hypothalamic-pituitary-gonadal dysfunction. JAMA 243:750–755
Spiegel AM, Rosen SW, Weintraub BD, Marynick SP (1978) Effect of intravenous pyridoxine on plasma prolactin in hyperprolactinemie subjects. J Clin Endocrinol Metab 46:686–688
Spitz IM, Landau H, Almaliach U, Rosen E, Brautbar N, Russell A (1977) Diminished prolactin reserve: a case report. J Clin Endocrinol Metab 45:412–418
Starup J (1972) Amenorrhea following oral contraception. Acta Obstet Gynecol Scand 51:341–345
Steiner J, Cassar J, Mashiter K, Dawes I, Russell Fraser T, Breckenridge A (1976) Effects of methyldopa on prolactin and growth hormone. Br Med J 1:1186–1188
Steininger H, Sulm O (1978) Chromophobe Hypophysenadenome mit Hyperprolaktinämie: Postoperative Behandlung mit Bromocriptin. In: Geyer GV (ed) Symposium on 1st April 1978, vienna. Sandoz, Vienna, pp 97–106
Strauch G, Balcke JC, Mahoudeau JA, Bricaire H (1977 a) Hormonal changes induced by bromocriptine (CB 154) at the early stage of treatment. J Clin Endocrinol 44:588–590
Strauch G, Mahoudeau J, Valcke J-C, Thoquenne D, Pique L, Bricaire H (1977 b) Studies on reversal and mechanisms of prolactin-induced anovulation. In: Crosignani PG, Robyn C (eds) Prolactin and human reproduction. Academic Press, London New York, pp 161–178
Tashjian AH, Barowsky NJ, Jensen DK (1971) Thyrotropin-releasing hormone: direct evidence for stimulation of prolactin production by pituitary cells in culture. Biochem Biophys Res Commun 43:516–523
Thompson CR, Hansen LM (1970) Pergonal (menotropins): a summary of clinical experience in the induction of ovulation and pregnancy. Fertil Steril 21:844–843
Thorner MO, Besser GM (1977) Hyperprolactinemia and gonadal function: results of bromocriptine treatment. In: Crosignani PG, Robyn C (eds) Prolactin and human reproduction. Academic Press, London, pp 285–301
Thorner MO, McNeilly AS, Hagan C, Besser GM (1974 a) Long-term treatment of galactorrhea and hypogonadism with bromocriptine. Br Med J 2:419–422
Thorner MO, Besser GM, Hagan C, McNeilly AS (1974 b) Introduction of a new stimulation test for prolactin: the sulpiride test: a comparison with other dynamic function tests. J Endocrinol 61:1xxxii

Thorner MO, Besser GM, Jones A, Dacie J, Jones AE (1975) Bromocriptine treatment of female infertility: report of 13 pregnancies. Br Med J 4:694–697

Tolis G, Naftolin F (1976) Induction of menstruation with bromocryptine in patients with euprolactinemic amenorrhea. Am J Obstet Gynecol 126:426–429

Tolis G, Goldstein M, Friesen HG (1973) Functional evaluation of prolactin secretion in patients with hypothalamic-pituitary disorders. J Clin Invest 52:783–788

Tolis G, Somma M, van Campenhout J, Friesen H (1974) Prolactin secretion in sixty-five patients with galactorrhea. Am J Obstet Gynecol 118:91–101

Tolis G, Hickey J, Guyda H (1975) Effects of morphine on serum growth hormone, cortisol, prolactin and thyroid-stimulating hormone in man. J Clin Endocrinol Metab 41:797–800

Tolis G, Laliberté R, Guyda H, Naftolin F (1977) Ineffectiveness of pyridoxine (B_6) to alter secretion of growth hormone and prolactin and absence of therapeutic effects on galactorrhea-amenorrhea syndromes. J Clin Endocrinol 44:1197–1199

Tolis G, Hoyte K, McKenzie JM, Mason B, Robb P (1978 a) Clinical biochemical, and radiologic reversibility of hyperprolactinemic galactorrhea-amenorrhea and abnormal sella by thyroxine in a patient with primary hypothyroidism. Am J Obstet Gynecol 131:850–852

Tolis G, Bertrand G, Carpenter S, McKenzie JM (1978 b) Acromegaly and galactorrhea-amenorrhea with two pituitary adenomas secreting growth hormone or prolactin. Ann Intern Med 89:345–348

Tourniaire J, Pallo D, Pousset S, Bizollon C, Bachelot I (1974) Diminution de la tolérance glucidique et hyperinsulinisme dans l'adénome à prolactine. Nouv Presse Med 3:1705–1707

Travaglini P, Elli R, Ambrosi B, Ballabio M, Moriondo P, Faglia G (1979) Serum LH increase after estradiol and progesterone administration in hyperprolactinemic women. J Endocrinol Invest 2:407–411

Turkington RW (1971) Ectopic production of prolactin. N Engl J Med 285:1455–1458

Turkington RW, Underwood LE, Van Wyk JI (1971) Elevated serum prolactin levels after pituitary-stalk section in man. N Eng J Med 285:707–710

Turkington RW (1972 a) Phenotiazine stimulation test for prolactin reserve: the syndrome of isolated deficiency. J Clin Endocrinol 34:246–249

Turkington RW (1972 b) Inhibition of prolactin secretion and successful therapy of the Forbes-Albright syndrome with L-dopa. J Clin Endocrinol 34:306–311

Turkington RW (1972 c) Serum prolactin levels in patients with gynecomastia. J Clin Endocrinol 34:62–66

Turkington RW (1972 d) Prolactin secretion in patients treated with various drugs: phenothiazines, triciclic antidepressants, reserpine and methyldopa. Arch Intern Med 130:349–354

Turkington RW (1972 e) Secretion of PRL by patients with pituitary and hypothalamic tumors. J Clin Endocrinol Metab 34:159–164

Turkington RW, MacIndoe J (1972) Hyperprolactinemia in sarcoidosis. Ann Intern Med 76:545–549

Turpin G, Heshmati HM, Roger M, Legrand S, Guillemant S, de Gennes JL (1979) La fonction corticosurrénale dans les hyperprolactinémies. Ann Endocrinol (Paris) 40:553–554

Tyson JE, Andreasson B, Huth J, Smith B, Zacur H (1975) Neuroendocrine dysfunction in galactorrhea-amenorrhea after oral contraceptive use. J Obstet Gynecol 46:1–11

Van der Gugten AA, Sala M, Kwa HG (1970) Effect of oestrone pellet implantation on plasma levels of prolactin. Acta Endocrinol (Copenh) 64:265–272

Van der Steeg HJ, Coelingh Bennink HJT (1977) Bromocriptine for induction of ovulation in normoprolactinemic post-pill anovulation. Lancet 1:502–504

Van Look PFA, McNeilly AS, Hunter WM, Baird DT (1977) The role of prolactin in secondary amenorrhea. In: Crosignani PG, Robyn C (eds) Prolactin and human reproduction. Academic Press, London New York San Francisco, pp 217–224

Varga L, Wenner R, del Pozo E (1973) Treatment of galactorrhea-amenorrhea syndrome with Br-ergocryptine (CB 154): Restoration of ovulatory function and fertility. Am J Obstet Gynecol 117:75–79

Vermeulen A, Ando S (1978) Prolactin and adrenal androgen secretion. Clin Endocrinol (Oxf) 8:295–303

Vermeulen A, Suy E, Rubens R (1977) Effect of prolactin on plasma DHEA(S) levels. J Clin Endocrinol Metab 44:1222–1225

Wajchenberg BL, Lerario AC, Lieberman B, Marino R, Leonardo Lancho MC (1979) Blunted response to prolactin stimulation tests and evaluation of L-dopa depression in patients with hyperprolactinemic-galactorrhea. J Endocrinol Invest 2:427–431

Wartofsky L, Dimond RC, Noel GL, Frantz AG, Earll RM (1976) Effect of acute increases in serum triiodothyronine on TSH and prolactin responses to TRH, and estimates of pituitary stores of TSH and prolactin in normal subjects and in patients with primary hypothyroidism. J Clin Endocrinol Metab 42:443–458

Weinstein D, Ben-David M, Polishuk WZ (1976) Serum prolactin and the suppression of lactation. Br J Obstet Gynecol 83:679–682

Weir JH (1971) Postthoracotomy galactorrhea successfully treated with comiphene citrate. Am J Obstet Gynecol 111:106–107

Weissel M, Stummvoll HK, Kolbe H, Höfer R (1979) Basal and TRH-stimulated thyroid and pituitary hormones in various degrees of renal insufficiency. Acta Endocrinol (Copenh) 90:23–32

Wenner R (1975) Les antiprolactines. Actual Gynecol 6:91–93

Wernze H, Burghardt W (1979) Hyperprolactinämie bei Lebererkrankungen. Med Klin 74:1615–1623

Wernze H, Schmitz E (1977) Plasma prolactin and prolactin release in liver cirrhosis. Acta Hepatogastroenterol (Stuttg) 24:97–101

Wider JA, Marshall JR, Ross GT (1969) Familial galactorrhea in 3 sisters with oligo-ovulation. JAMA 209:669–671

Wilson JG (1973) Environment and birth defects. Academic Press, New York

Wilson RG, Singhal VK, Percy-Robb I, Forrest APM, Cole EN, Boyns AR, Griffiths K (1972) Response of plasma prolactin and growth hormone to insulin hypoglycaemia. Lancet 2:1283–1285

Wolf AS, Musch K, del Pozo E (1979) Conversion of clomiphene response by bromocriptine therapy in normoprolactinemic amenorrhea. Acta Endocrinol [Suppl] (Copenh) 225:186

Wright CS, Steele SJ, Jacobs HS (1979) Value of bromocriptine in unexplained primary infertility: a double-blind controlled trial. Br Med J 1:1037–1039

Yamaji T (1974) Modulation of prolactin release by altered levels of thyroid hormones. Metabolism 23:745–752

Yen SSC (1978) Chronic anovulation. In: Yen SSC, Jaffi RB (eds) Reproductive endocrinology. Sounders, Philadelphia London Toronto, pp 341–372

Yen SSC (1980) The polycystic ovary syndrome. Clin Endocrinol (Oxf) 12:177–208

Yen SSC, Ehara Y, Siler RM (1974) Augmentation of prolactin secretion by estrogen in hypogonadal women. J Clin Invest 53:652–655

Ylikorkala O, Kivinen S, Reinilä M (1979) Serial prolactin and thyrotropin responses to thyrotropin-releasing hormone throughout normal human pregnancy. J Clin Endocrinol Metab 48:288–292

Young RL, Bradley EM, Goldzieher JW, Myers PW, Lecoq FR (1967) Spectrum of nonpuerperal galactorrhea: report of two cases evolving through the various syndromes. J Clin Endocrinol 27:461–466

Zanoboni A, Zanoboni-Muciaccia W (1975) Gynecomastia in alcoholic cirrhosis. Lancet 4:876

Zarate A, Villalobos M, Valenzuela S (1968) A case of Chiari-Frommel syndrome associated with polycystic ovaries. Am J Obstet Gynecol 101:1131–1132

Zarate A, Canales ES, Jacobs LS, Maneiro PJ, Soria J, Daughaday WH (1973 a) Restoration of ovarian function in patients with the amenorrhea-galactorrhea syndrome after long-term therapy with L-dopa. Fertil Steril 24:340–344

Zarate A, Jacobs LS, Canales ES, Schally AV, de la Cruz A, Soria J, Daughaday WH (1973 b) Functional evaluation of pituitary reserve in patients with the amenorrhea-galactorrhea syndrome utilizing luteinizing hormone-releasing hormone (LH-RH), L-Dopa and chlorpromazine. J Clin Endocrinol Metab 37:855–859

Zarate A, Canales ES, Forsbach G, Fernandez-Lazala R (1978) Bromocriptine. Clinical experience in the induction of pregnancy in galactorrhea-amenorrhea. Obstet Gynecol 52:442–444

Zippel HH, Schulz K-D, del Pozo E (1977) Morphological studies in human proliferating mammary tissues under treatment with the prolactin inhibitor bromocryptine (CB 154). Acta Endocrinol [Suppl 208] (Copenh) 84:42

Zondek B, Bromberg YM, Rozin S (1951) An anterior pituitary hyperhormonotrophie syndrome. J Obstet Gynecol 58:525–537

Zuckerman M, Carmel S (1973) The inhibition of lactation by clomiphene. J Obstet Gynecol Br Commonw 80:822–823

5 Pituitary Prolactinomas

5.1 Morphology of the Prolactin Secreting Cell

5.1.1 Normal Lactotroph

Conventional staining of the anterior pituitary with hematoxylin-eosin reveals only three cell types which differ in color of their granules, i.e., the acidophilic, the basophilic and the chromophobe cell. However, the anterior pituitary secretes more than three hormones, each of which is secreted from specific cells which are dispersed among the other cells of the anterior pituitary lobe. Thus it is obvious that hematoxylin-eosin staining does not allow differentiation between anterior pituitary cells according to their endocrine function. Furthermore, it has been shown that chromophobe and acidophilic cells can secrete the same hormones (Landolt 1975, 1978; Racadot et al. 1975; Saeger 1977). Thus, acidophilic and chromophobe adenomas can secrete GH, PRL or both, and therefore do not reflect the observed clinical picture (Landolt 1975, 1978; Olivier et al. 1975; Peillon et al. 1978). Because of these serious drawbacks of the staining procedure, the distinction between acidophilic, basophilic or mucoid, and chromophobe cells has become less useful. Identification of anterior pituitary cells is now mainly based on new staining techniques, immunohistochemistry, and electron microscopy (Pasteels 1972; Olivier et al. 1975; Landolt 1975; Racadot et al. 1975). The morphological findings can now be correlated with dynamic studies in vivo since the development of radioimmunoassay has made it possible to measure all known anterior pituitary hormones in body fluids.

PRL was the last anterior pituitary hormone in man to be identified as a product of a separate cell, although more recently endogenous opioid peptides, the endorphins, have been discovered. However, these peptides belong to the lipotropin-ACTH family and are secreted by ACTH-producing corticotrophic cells, identified earlier (Guillemin et al. 1977; Weber et al. 1978). However, cells in the anterior pituitary lobe implicated in pregnancy and lactation (Erdheim and Stumme 1909; Bailey 1932; Herlant 1943) were described before the discovery of human PRL as a separate hormone. In fact Romeis (1940) gave a detailed description of the pregnancy cells, proposing a new nomenclature of the adenohypophyseal cells in his monograph on the human pituitary. But only after the discovery of human PRL as a separate hormone has detailed information about the morphology of the PRL-secreting cell become available (Racadot et al. 1971; Pasteels 1972).

In the following pages specific features of the PRL-secreting cell, the normal lactotroph, and the characteristic features of these cells in PRL-secreting adenomas – so-called prolactinomas – shall be reviewed.

PRL-secreting cells can be stained by acid dyes such as orange G, erythrosin and eosin (Pasteels 1972). The affinity of the lactotroph to acid dyes is similar to the somatotrophic cells and is explained by the free amino groups of the hormones present in the secretory granules. This identification is based on the property of the secretory product and not on cytologic features such as shape and size of cytoplasmic organelles (Pasteels 1972). If the cells are very active they appear degranulated and the secretory material cannot be identified. They appear as "chromophobes". In agreement with this finding, treatment with bromocriptine leads to retention of the secretory granules which in light microscopy may be documented as a change from a chromophobe to a chromophile cell (Landolt et al. 1978). However, the introduction of more sophisticated staining techniques such as Herlant's tetrachrome staining (Herlant 1960; Herlant and Pasteels 1967) and the PAS-orange G staining has allowed identification of PRL-containing granules with greater sensitivity and specificity. Thus with Herlant's tetrachrome a differential staining of PRL- and GH-secreting cells is possible. The lactotrophs appear bright red, whereas the GH-producing cells appear orange-yellow. With this technique prolactin cells have been revealed in the pituitaries of all mammals studied during lactation (Pasteels 1972). In humans PRL-secreting cells were first documented by this technique in tissue cultures (Pasteels 1963) and in pituitary tissue of pregnant women (Pasteels 1963; Herlant and Pasteels 1967), as well as in PRL-secreting adenomas (Herlant et al. 1965, 1966). More conclusive evidence about the content of the granules is derived from immunofluorescent techniques, which allow localization of very small amounts of PRL in the cell (Pasteels et al. 1972a, b). This sensitivity and selectivity is due to the high specificity of the prolactin antiserum which does not crossreact with any of the other anterior pituitary hormones including GH (Hwang et al. 1971; Sinha et al. 1973; von Werder et al. 1978b). The immunofluorescent technique can be combined with the sophisticated staining where counter-stains can bring out further morphological details. With this technique it has been shown that PRL-secreting cells are found considerably more frequently in the human pituitary than had been anticipated from the content of human PRL in the pituitary measured by radioimmunoassay (Zimmerman et al. 1974). The PRL content of an adult human pituitary varies from 120 to 300 μg per gland (Lewis et al. 1971; Hwang et al. 1973; Lewis and Singh 1973, Zimmerman et al. 1974). The content of GH is 50–100 times higher than PRL ranging from 5 to 15 mg per gland (Guyda et al. 1973; Zimmerman et al. 1974). In contrast PRL levels in the serum are ten times higher than GH levels. The fact that in the adenohypophysis the amount of GH and PRL cells identified by immunofluorescence is not significantly different underlines the fact that PRL is a high turnover hormone. Whereas GH is stored in the great number of uniform secretory granules in the pituitary, PRL is released as soon as it is synthesized. Contrary to the classical theory "one cell – one hormone", evidence for the possible simultaneous production of PRL and GH in normal human pituitary cells has been presented (Zimmerman et al. 1974), though this could not be shown in GH- and PRL-producing adenomas (Guyda et al. 1973; Zimmerman et al. 1974; Robert and Hardy 1975).

The most detailed information about the morphology of prolactin secretion can be obtained from electron microscopic studies. Since the first ultrastructural studies in animal pituitaries (Rinehart and Farquhar 1953; Farquhar and Rinehart 1954a, b) several authors have contributed to the elucidation of the ultrastructural morphology of the PRL-secreting cell (Foncin 1966; Racadot et al. 1971; Pasteels 1972; Olivier et al. 1975).

Prolactin cells can be identified by very distinctive morphological features, such as the lamellar and often concentric pattern of the rich rough endoplasmic reticulum with enlarged Golgi complexes and significant polymorphism of the secretory granules with a diameter ranging from 100 to 600, sometimes up to 1200 nm (Racadot et al. 1971; Pasteels 1972; Landolt 1975, 1978; Bloodworth et al. 1979).

The secretory granules of the lactotroph are the largest to be found in the pituitary (Bloodworth et al. 1979). The nuclei are oval, often in an eccentric position, and sometimes indented and the nucleoli are very prominent. Though the most prominent and distinctive features of the lactotrophs are the polymorphic granules which vary in size and shape within the same cell, it must be stated that these granules vary not only according to the species but also according to their functional condition (Pasteels 1972). Perhaps because of the morphological heterogeneity of the prolactin cells the amount of lactotrophic cells in the human pituitary detected by electron microscopy is considerably less compared to the immunofluorescence techniques, which reveals that up to 40% of the total cell mass of the anterior lobe is composed of PRL-producing cells (Zimmerman et al. 1974). This is in contrast to other authors who have only rarely encountered PRL-producing cells in the normal human pituitary (Landolt 1975). Only after stimulation by estrogens the characteristic polymorphic granulation of the typical lactotroph is revealed by electron microscopy (Pasteels 1972). These cells are comparable to the numerous lactotrophic cells found in the human pituitary during pregnancy and in male patients treated with estrogens for prostatic cancer (Hachmeister et al. 1972).

Thus not only the ultrastructure but also the amount of PRL-secreting pituitary cells varies according to the physiologic circumstances. In the fetus one can identify PRL-secreting cells after the 16th week of pregnancy from which time they increase until term (Pasteels et al. 1972a). This correlates with the pituitary PRL concentration and the PRL levels in the umbilical vein. Prolactin can be identified by radioimmunoassay in the fetal pituitary as early as in the tenth week of gestation (Aubert et al. 1975). At term, hypertrophy of the lactotrophs is encountered (Pasteels et al. 1972a).

In the adult male only very careful examination by nonimmunological means reveals PRL-secreting cells similar to those found in the nonpregnant female (Herlant and Pasteels 1967; Goluboff and Ezrin 1969; Landolt 1975).

The distribution of the PRL-secreting cells in the adenohypophysis is not uniform (Pasteels 1972). Like the somatotrophic cells the lactotrophs are predominantly found in the lateral and anteromedial part of the pituitary lobe (Paiz and Hennigar 1970).

Whereas there is some controversy about the amount of lactotrophs in nonpregnant human pituitaries the observed enlargement of the anterior pituitary during pregnancy is undoubtedly due to hypertrophy and hyperplasia of the lactotrophic cells (Goluboff and Ezrin 1969; Pasteels et al. 1972a). The morphological changes of the pituitary include increased vascularization of the entire gland leading to further enlargement during pregnancy (Linquette et al. 1970). The enlargement of the pituitary gland during pregnancy was first described in 1909 by Erdheim and Stumme, who studied the pituitaries of 118 pregnant women, drawing attention to the numerous pregnancy cells (Schwangerschaftszellen) in the anterior lobe. According to Erdheim and Stumme (1909) multiparous women have in general more pregnancy cells and larger pituitaries than primiparas. In the pituitaries of some women with several pregnancies the growth of the pregnancy cells was suggestive of adenoma formation. Genuine intrasellar microadenomas were observed in 14 of Erdheim and Stumme's 118 cases. At that time the functional meaning of these pregnancy cells or adenomas which have been later confirmed by several investigators (Romeis 1940; Herlant and Pasteels 1967) was not clear. The hypertrophy and hyperplasia of the lactotrophs begins at the end of the first trimester (Herlant and Pasteels 1967; Goluboff and Ezrin 1969). With the development of the PRL radioimmunoassay the functional correlation between the pituitary growth during pregnancy and rising PRL levels in the circulation was clarified. With electron microscopy techniques it could be shown that these pregnancy cells were identical with the cells observed in females and males under estrogen treatment in pharmacologic doses. Thus, the conclusion that these pregnancy cells are a consequence of increased estrogen secretion during pregnancy was obvious. This is also underlined by the reversibility of the pituitary enlargement after termination of pregnancy correlating again with normalization of PRL levels (L'Hermite et al. 1972) (see also Chap. 3).

5.1.2 Prolactinomas

Persistent amenorrhea after pregnancy accompanied by persistent galactorrhea with and without evidence of pituitary tumor has been described before the discovery of human prolactin (Chiari et al. 1855; Frommel 1882; Forbes et al. 1954; Herlant et al. 1965) It is not clear if in these cases the enhanced estrogen secretion of pregnancy has induced an adenomatous growth of the prolactin cells which has become autonomous or if these patients already had an autonomous adenomatous lactotrophic lesion which has grown under the influence of high estrogen concentration. That existing prolactinomas may increase in size during pregnancy has been well documented (Emperaire et al. 1972; Falconer and Stafford-Bell 1975; Child et al. 1975a; Bergh et al. 1978; Gemzell and Wang 1979; Lamberts et al. 1979b).

There is little doubt that the endogenous estrogens play an important role in the development of prolactinomas (Peillon et al. 1970) although there is a variable response of plasma PRL to an estrogen challenge among these patients (White et al. 1981). Accordingly microprolactinomas are detected more frequently in females but remain a rare disease in males (Peillon et al. 1978; Derome er al. 1979; von Werder et al. 1980a). However, there is no evidence so far that estrogenic compounds in antiovulatory medication induce pituitary adenoma formation (Annegers et al. 1978) as has been suggested by Sherman et al. (1978). However, in animals estrogens have been shown to induce pituitary tumors (El-Etreby and Günzel 1973).

Prolactinomas have been detected more frequently in recent years in males and females alike (Peillon et al. 1978; Reichlin 1979). According to their size and radiologic appearance prolactinomas are divided into micro- and macroprolactinomas (Hardy 1969, 1973, 1980; Vezina and Sutton 1974; Fahlbusch et al. 1980; see Sect. 2.4.1).

Prolactin levels correlate well with the size of the adenoma (Malarkey and Johnson 1976; Nader et al. 1976; Fahlbusch et al. 1978, 1980; Kelly et al. 1978; von Werder et al. 1978a, 1980a). Thus patients with microprolactinomas with a diameter of less than 10 mm have, in general, lower PRL levels than those with large prolactinomas. In cases with no evidence for a pituitary adenoma using the most sophisticated radiologic technique the hyperprolactinemia is even more moderate than in those with microprolactinomas (Gomez et al. 1977; Kleinberg et al. 1977; von Werder et al. 1980a). In these patients hyperprolactinemia is regarded as functional if other causes are excluded, though a very small microadenoma, which does not lead to any changes in the osseous structure, cannot be ruled out (Tolis 1977; McKenna et al. 1978; Silvestrini et al. 1978; Editorial 1980).

As has been pointed out above microadenomas and so-called functional hyperprolactinemia appear almost exclusively in females (Peillon et al. 1978; von Werder et al. 1980a), whereas larger prolactin cell adenomas seem to be evenly distributed among females and males (Derome et al. 1979; von Werder et al. 1980a, Fig. 5.9).

The macroprolactinomas in females, very often associated with galactorrhea and amenorrhea, were believed to be the source of a lactogenic hormone before prolactin had been identified as a separate peptide in the human (Forbes et al. 1954). In the male, where galactorrhea is less apparent (Besser et al. 1977; Thorner 1977; von Werder et al. 1978a) macroprolactinomas were believed to be functionless tumors (Child et al. 1975b; Landolt 1975; Franks et al. 1977; Osamura et al. 1978), though galactorrhea in the male has been found before to be associated with hypertrophy of the pituitary in individual cases (Haenel 1928).

In all tumors, i.e., micro- and macroprolactinomas associated with hyperprolactinemia, PRL can be identified by immunofluorescence regardless of the clinical picture (Zimmerman et al. 1974; Osamura et al. 1978).

Ultrastructural studies reveal similar polymorphic granules in most instances varying from 100 to 500 nm and rarely up to 1200 nm in normal lactotrophic cells and in those of microadenomas and macroadenomas (Fig.5.1). According to Saeger (1975), Kovacs et al. (1975), Landolt (1975), Robert and Hardy (1975), Kovacs (1977), Bloodworth et al. (1979) these tumors can be separated into sparsely and densely granulated prolactin cell adenomas. The sparsely granulated type occurs more frequently than the densely granulated type (Kovacs 1977) both being associated with hyperprolactinemia. Both cell types can be found in the same tumor. There is no difference in the ultrastructure of adenomas which cause amenorrhea and galactorrhea and those which remain clinically silent (Landolt 1978).

In contrast to nontumorous prolactin cells where small immature secretory granules fuse into larger granules which represent the secretory product ready for exocytosis the sparsely granulated PRL-producing adenoma cells (Fig.5.1) contain smaller granules. Sometimes secretory granules are found between the two plasma membranes just opposite the site facing the capillary, which has been called misplaced exocytosis (Horvath and Kovacs 1974). The rough endoplasmatic reticulum is very prominent and the Golgi complex is very large (Fig.5.1) and may occupy up to one-third of the cytoplasm (Bloodworth et al. 1979). In addition nucleolar hypertrophy is found (Peillon et al. 1978). This can be taken as morphological evidence for an even higher turnover in the sparsely granulated adenoma compared to a normal lactotrophic cell (McComb and Kovacs 1978).

The ultrastructural picture of the large macroprolactinomas shows more mitoses compared to the smaller adenomas (Peillon et al. 1978). This seems to be the only

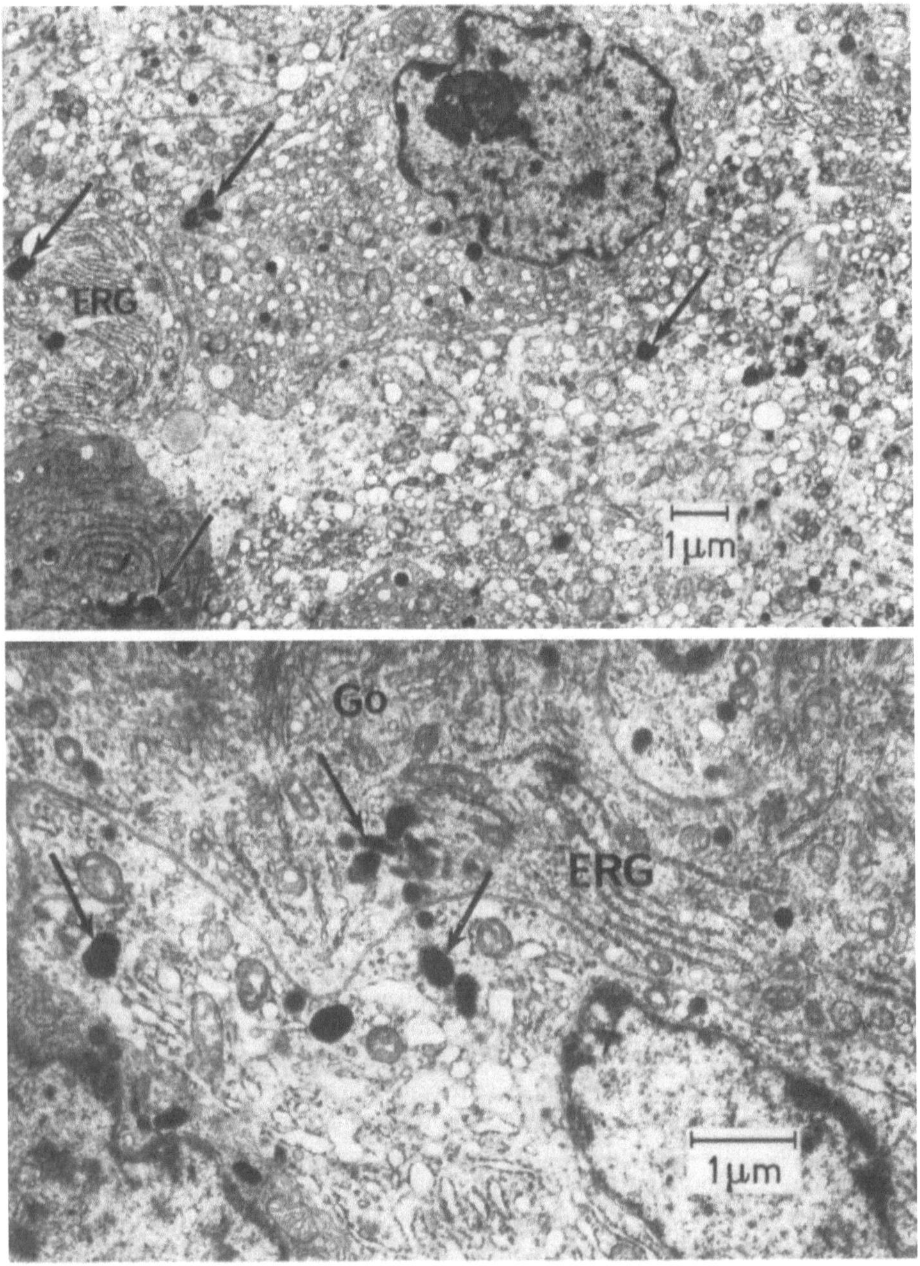

Fig. 5.1. Prolactinoma; *Top:* The cells are poorly granulated (*arrows*). A well-developed lamellar ergastoplasma (ERG) is visible. The nucleus is indented with a prominent nucleolus. *Bottom:* The cells contain pleomorphic and large granules (*arrows*) and a lamellar rough ERG. The Golgi apparatus (GO) is typically prominent (× 15 000). (A. M. Brandi, Laboratory of Prof. Racadot, 105 Bd Hôpital, F 75 Paris 13[e])

clear-cut morphological difference between these two types of tumors. More frequent cell division in one adenoma may lead to a large, possibly invasively growing adenoma, whereas in the other situation the limited growth potential leads to a small adenoma, i.e., a microadenoma. However, the longer history of hyperprolactinemia of the patients with large adenomas may also play a role in the development of these tumors (Peillon et al. 1978).

Enlarged Golgi cisterns and well-developed RER have also been found in cases which had been called chromophobe adenomas and had shown progression with headaches and visual problems under estrogen treatment (Peillon et al. 1970; Racadot et al. 1971).

In these adenomas the parenchyma is very often disrupted by hemorrhagic areas. This again may lead to pituitary apoplexy which occurs sometimes in such patients (Peillon et al. 1978). The situation may be compared to the pituitary apoplexy which can be observed in patients harboring prolactinomas during pregnancy. The initial conclusion that these chromophobe adenomas which grow under the influence of estrogens produce PRL (Peillon et al. 1970; Racadot et al. 1971) has turned out to be valid (Friesen et al. 1973; Peillon et al. 1975b; Franks et al. 1977; Osamura et al. 1978).

The estrogen-induced growth of a prolactinoma in the absence of galactorrhea has been called the Peillon-Racadot syndrome (Landolt 1975) in contrast to the Forbes-Albright syndrome, in which the prolactinoma is accompanied by galactorrhea and amenorrhea (Forbes et al. 1954). Since conventional staining techniques, immunofluorescence, ultrastructure and dynamics of PRL secretion in the Peillon-Racadot syndrome and the Forbes-Albright syndrome are not significantly different, it is justifiable to call all prolactin-producing adenomas of the pituitary, prolactinomas. In addition the differentiation of the galactorrhea syndrome into Forbes-Albright syndrome associated with pituitary enlargement (Forbes et al. 1954), Ahumada-del Castillo syndrome associated with a normal sella turcica (Ahumada and del Castillo 1932; Argonz and del Castillo 1953), and the Chiari-Frommel syndrome (Chiari et al. 1855; Frommel 1882) with postpartum persistent galactorrhea is no longer useful because this terminology does not take into consideration PRL secretion and pituitary morphology. Thus galactorrhea does not necessarily mean elevated PRL levels and a normal sella turcica does not exclude a PRL-producing microadenoma (Kleinberg et al. 1977; McKenna et al. 1978). Even the combination of hyperprolactinemia and a pituitary tumor does not mean that the patient suffers from a PRL-producing adenoma, since endocrine inactive adenomas with suprasellar extension may lead to inhibition of PIF transport and therefore disinhibition of the normal pituitary function resulting in enhanced PRL secretion (Turkington et al. 1971; MacLeod 1976; Illig et al. 1978; Frantz 1978; von Werder and Rjosk 1979).

The study of the paraadenomatous tissue of the anterior pituitary lobe in patients harboring prolactin cell adenomas reveals in most cases numerous prolactin cells identified by immunohistology or electron microscopy (Saeger 1977; Kovacs et al. 1978). The finding of prolactin cells in nontumorous tissue may suggest a hypothalamic origin of the pituitary neoplasia. On the other hand suprasellar extension and compression of the pituitary stalk or local disturbance of portal circulation leading to secondary disinhibition of the nontumorous lactotrophs may be responsible. The fact that excessive PRL release from the prolactinoma leads to inhibition of prolactin secretion in the nontumorous tissue via an ultrashort feedback (Jacobs et al. 1976; Herbert et al. 1979; Lamberts and MacLeod 1979) is not substantiated by these morphological findings. However, since the prolactin cells in the nontumorous tissue are not much increased compared to normal anterior pituitaries, the lack of normalization of PRL secretion after supposedly complete and selective adenomectomy (von Werder et al. 1978a; Fahlbusch et al. 1978, 1980; Derome et al. 1979) remains unexplained.

5.2 Pathophysiology of Prolactinomas

5.2.1 Prolactin Secretion

5.2.1.1 Basal Prolactin Levels

The most useful indicator for abnormal PRL secretion is an elevated basal PRL level (Turkington 1972a; Jacobs and Daughaday 1973; Friesen et al. 1973; Kleinberg et al. 1977; von Werder et al. 1977). PRL secretion in man follows a 24-h secretory pattern with an increase in PRL secretion at night, (Nokin et al. 1972; Sassin et al. 1972, 1973; Parker et al. 1973). In addition to the sleep-associated PRL secretion short pulses of PRL secretion without fixed periodicity can be observed (Boyar et al. 1976). In general a fall of about 70% in PRL levels from 8 A.M. to 12 noon will be encountered (Sassin et al. 1972: Parker et al. 1973). Prolactin is also a stress hormone. Thus during periods of psychological and physical stress prolactin levels rise (Noel et al. 1972; Eversmann et al. 1978). In addition many drugs can cause hypersecretion of PRL leading to clinically relevant hyperprolactinemia (for review see MacLeod 1976; Thorner 1977; Frantz 1978; von Werder and Rjosk 1979; Clemens and Shaar 1980, and Chap. 4). This must be kept in mind if a single elevated PRL level has to be interpreted or patients with moderate hyperprolactinemia are followed up on an outpatient basis during different times of the day.

The reasons for the development of hyperprolactinemia are manyfold (for review see MacLeod 1976; Thorner 1977; Frantz 1978; von Werder and Rjosk 1979). Thus, the patient who has elevated PRL levels in the presence of enlargement of the sella turcica does not necessarily suffer from a prolactinoma (Turkington 1972b; Saeger 1975; von Werder et al. 1977). However, PRL levels above 250 ng/ml or 5000 μU A 71/222/ml are almost always indicative of a PRL-secreting adenoma (Friesen et al. 1973; del Pozo et al. 1974; Nader et al. 1976; von Werder et al. 1977; Gomez et al. 1977; Kleinberg et al. 1977; Derome et al. 1979), although such adenomas may occur with PRL levels below this limit. The clinical experience that microadenomas are more frequently found in females (von Werder et al. 1977, 1980a; Peillon et al. 1978; Derome et al. 1979) and the fact that the larger macroprolactinomas are usually found in males (Peillon et al. 1975b, 1977; Carter et al. 1978; Franks et al. 1978; Landolt 1978; Derome et al. 1979) is also reflected in the usually higher basal PRL levels which are encountered in male patients (Friesen et al. 1973; Peillon et al. 1977; Carter et al. 1978; von Werder et al. 1980a). In the male PRL levels above 20000 ng/ml have been reported (Friesen et al. 1973; Carter et al. 1978; von Werder et al. 1978a). Hyperprolactinemia caused by ectopic PRL secretion is extremely rare (Turkington 1971; Rosen and Weintraub 1977), but can be traced by measuring PRL in venous blood obtained from different sites (McKenna et al. 1978).

Immunoreactive basal PRL is found in different molecular forms (see Chap. 1). Though in some patients an extremely high proportion of big prolactin has been observed (Rogol and Rosen, 1974; Schneider et al. 1975), in most patients with prolactinomas 75% of total prolactin immunoreactivity always elutes as little or monomeric prolactin (von Werder and Clemm 1974). Partial pituitary adenomectomy leading to a decrease of prolactin levels to half of the initial value does not lead to a significant change in the distribution of prolactin immunoreactivity (von Werder and Clemm 1974).

5.2.1.2 Hypothalamic Hypophyseotropic Hormones

Though high PRL levels are always suggestive of a prolactinoma even in the absence of radiologic changes of the sella turcica, the presence of a PRL-secreting adenoma in a patient with moderate hyperprolactinemia cannot be ruled out. In order to develop functional tests for the differentiation between prolactinoma-caused hyperprolactinemia and "functional" hyperprolactinemia several pharmacologic tools influencing PRL secretion in normal man (see also chapter 4) have been investigated in hyperprolactinemic patients (Friesen et al. 1973; Thorner 1977; Kleinberg et al. 1977; Editorial 1980).

Thyrotropin releasing hormone (TRH) stimulates prolactin secretion in normal males and females in a dose-dependent manner (Jacobs and Daughaday 1973). In patients with proven prolactinomas TRH leads to variable PRL responses (Lamberts et al. 1976; Samaan et al. 1977). In most cases with macroprolactinomas the rise of PRL is reduced (von Werder 1975; Kleinberg et al. 1977). But also in females with mild hyperprolactinemia the PRL response after TRH is blunted in most instances (Fig. 5.2). Only in patients with amenorrhea and galactorrhea and basal PRL levels within the normal range is a regular rise in PRL after TRH observed (del Pozo et al. 1974; Kleinberg et al. 1977). These findings suggest that a blunted PRL response to TRH is not necessarily indicative for autonomy of PRL secretion though patients directly postpartum with proven nontumorous hyperprolactinemia will in general respond with an adequate rise of PRL (Tyson et al. 1972). The TRH nonresponsiveness of PRL secretion in hyperprolactinemic states may be explained by an ultrashort feedback operating between circulating PRL and the lactotroph as has been suggested by Jacobs et al. (1976) and has been documented in animal studies (Herbert et al. 1979; Lamberts and MacLeod 1979). This would be in agreement with the more often positive PRL response to TRH in post partum and prolactinoma patients who have only moderate hyperprolactinemia compared to those with excessively elevated PRL levels (Kleinberg et al. 1977). The fact that TRH stimulates PRL release from a prolactinoma in vitro (Knazek and Skyler 1976) is also agreement with the hypothesis that circulating PRL, which in the in vitro experiments is not present, inhibits pituitary prolactin secretion from the normal as well as from the adenomatous lactotroph. These findings demonstrate that evaluation of TRH-stimulated PRL secretion adds no more information to the basal PRL level for the discrimination between "functional" and adenomatous hyperprolactinemia.

Inappropriate responses of GH to TRH (Faglia et al. 1973; Samaan et al. 1974; von Werder 1975) as well as to gonadotropin releasing hormone (Rubin et al. 1973) have been reported. This has been interpreted as a derangement of the receptor for growth hormone releasing hormone of the adenomatous somatotroph (von Werder

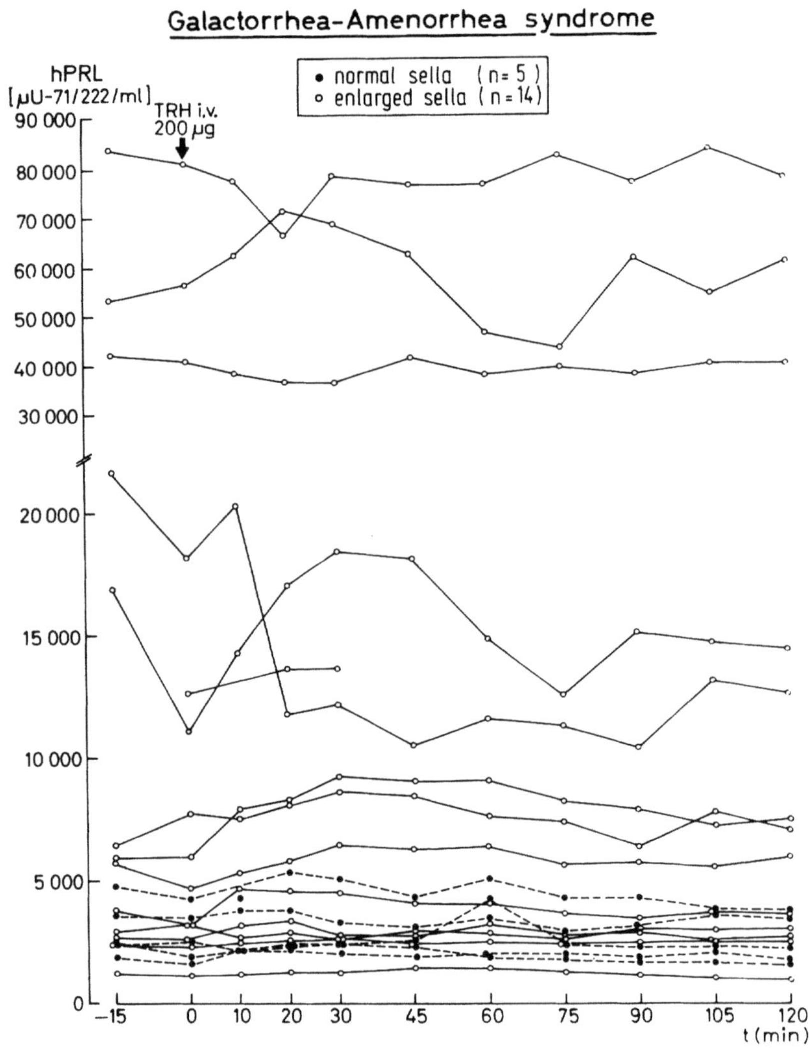

Fig. 5.2. Prolactin levels after TRH administration in hyperprolactinemic females (normal level <500 μU/ml). In general patients with proven pituitary tumor (○——○) and those with "functional" hyperprolactinemia (●----●) show no rise of prolactin after 200 μg TRH. (von Werder 1975)

and Fahlbusch 1978). In normal man GnRH does not stimulate PRL secretion and the majority of patients with prolactinomas show also no rise in elevated PRL after GnRH stimulation (unpublished observation). However, Catania et al. (1976) reported a response of PRL to GnRH in three out of eight normoprolactinemic acromegalics.

Somatostatin, which fails to inhibit TRH-induced PRL release (Siler et al. 1974) has been reported to lower PRL levels in acromegalics (Yen et al. 1974), although Wiegelmann et al. (1975) did not find any effect of somatostatin on basal PRL levels in normal subjects and acromegalics. The same seems to be true for prolactinoma patients in whom somatostatin in general exerts no inhibitory effect on PRL secretion (Fig. 5.3). However, Silvestrini et al. (1978) have reported an inhibitory effect of somatostatin, also on PRL levels, in some prolactinoma patients.

5.2.1.3 Insulin Hypoglycemia

Insulin hypoglycemia stimulates PRL release (Frantz et al. 1972; see Chap. 4). In patients with "functional" hyperprolactinemia and in patients with radiologically proven prolactinomas the response of PRL to insulin hyoglycemia is usually blunted or completely absent (L'Hermite et al. 1978).

5.2.1.4 Dopamine Antagonists

The dopamine antagonist chlorpromazine (Kleinberg et al. 1971) which stimulates PRL secretion in normal males and females rarely leads to a rise in PRL in hyperprolactinemic subjects (Zarate et al. 1975; Kleinberg et al. 1977). Chlorpromazine-stimulated PRL secretion may therefore be helpful in evaluating patients with various hypothalamic disorders (Tolis et al. 1973) but is of no use in the differential diagnosis of hyperprolactinemia (see Chap. 4).

Sulpiride has been shown to stimulate PRL secretion in men (Robyn et al. 1973). Silvestrini et al. (1978) reported a variable response in hyperprolactinemic subjects which is in contrast to the almost invariable negative response to chlorpromazine and to the most potent dopamine antagonistic PRL-releasing substance metoclopramide. The stimulation of PRL secretion in hyperprolactinemic patients seems to depend mainly on the height of the PRL level and not on the cause of PRL hypersecretion regardless of the stimulatory agent used (see Chap. 4). Therefore no stimulatory test alone for PRL secretion with any of the agents known to stimulate PRL in healthy subjects has turned out to be useful for the differential diagnosis of

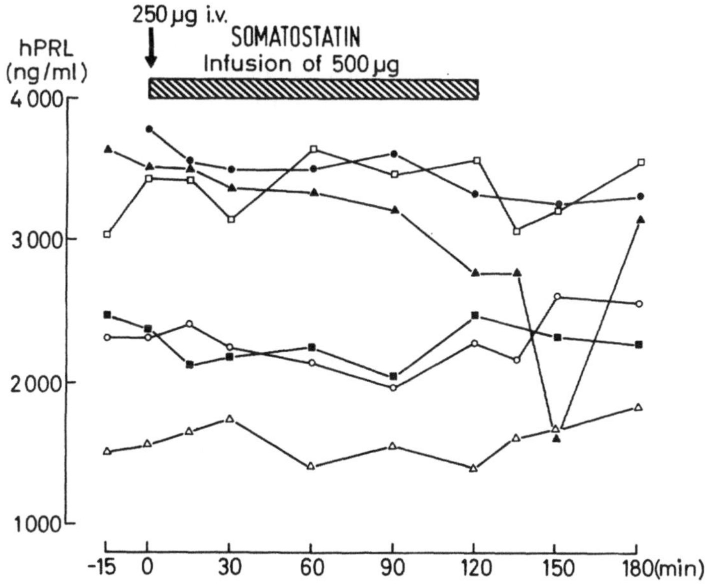

Fig. 5.3. Prolactin levels in six male patients with prolactinomas during administration of somatostatin (von Werder, unpublished)

"functional" and tumorous hyperprolactinemia (Frantz 1978; von Werder and Rjosk 1979; Editorial 1980). Recently Cowden et al. (1979) proposed a combination of measuring the circadian PRL profile together with TRH and metoclopramide stimulation. With this test combination they could make the preoperative diagnosis of prolactinoma in seven patients with normal tomography of the pituitary fossa. However, the number of patients investigated so far is small and the results of other investigators have been contradictory (Kleinberg et al. 1977; Frantz 1978; Pineda et al. 1979; Faglia et al. 1980b).

Recently Scanlon et al. (1981) demonstrated an enhanced TSH response after the administration of the dopamine antagonist domperidone in hyperprolactinemic females. This indicates altered dopaminergic regulation with increased dopaminergic inhibition of TSH secretion. The latter may be caused by PRL-induced stimulation of hypothalamic dopamine turnover, which has no effect in return on the autonomously functioning prolactinoma but only on TSH, which can be unmasked by domperidone. However, experience with this test using a large group of patients is still not available.

5.2.1.5 Dopamine Agonists

Dopamine agonists lead to a suppression of the basal PRL level regardless of the origin of the PRL excess. Thus L-dopa inhibits PRL secretion in hyperprolactinemic patients with normal sella turcica and those with a radiologically proven prolactinoma (Turkington 1972a; von Werder 1975; Frantz and Kleinberg 1978). However, the magnitude of the response to acute L-dopa administration seems to be more variable in hyperprolactinemic patients (Friesen et al. 1973) though most patients with a proven adenoma show a significant fall in PRL levels after L-dopa administration (von Werder 1975). After administration of bromocriptine and lisuride a prompt reduction in elevated PRL levels is always observed (Besser et al. 1972; Friesen et al. 1973; del Pozo et al. 1973; von Werder 1975; Horowski et al. 1978; Leebaw et al. 1978; De Cecco et al. 1978; Frantz and Kleinberg 1978). Because of this uniform suppression L-dopa and other dopamine agonists are not useful in the differential diagnosis of the cause of hyperprolactinemia.

Recently Müller et al. (1978) have suggested that nomifensine, an inhibitor of dopamine reuptake, is able to suppress PRL levels in "functional" hyperprolactinemia without an effect on tumorous PRL elevation (Müller et al. 1978; Genazzani et al. 1980). However, this has not been supported by others (Faglia et al. 1980b). This subject has been extensively discussed in Chap. 4. The combined administration of carbidopa and L-dopa has also been proposed to make this differential diagnosis (Fine and Frohman 1978). Thus it has been shown that after peripheral dopa decarboxylase inhibition with carbidopa, L-dopa administration is no longer able to inhibit PRL secretion in most patients with tumorous hyperprolactinemia, whereas full PRL suppression is retained in normal controls (Fine and Frohman 1978). However, Faglia et al. (1980b) reported five patients with pituitary microadenomas who significantly reduced serum PRL after the combined administration of carbidopa and L-dopa.

The usefulness of the water load test has been discussed in Chap. 4.

5.2.2 Pituitary Function

Endocrine active pituitary adenomas can lead to impairment of pituitary function by two different mechanisms. Disturbance of pituitary function can result from the growth of the adenoma leading to compression and destruction of the normal paraadenomatous tissue. It is obvious that such destruction of the anterior pituitary will only be encountered in patients with large macroadenomas who also show destruction of the osseous structure of the sella turcica. However, pituitary function can also be impaired as a consequence of the metabolic influence of the excessively secreted hormone. Thus growth hormone secretion is impaired in patients with ACTH-secreting microadenomas and Cushing's disease. In these patients no destruction of the somatotroph has occurred as is documented by the normalization of GH secretion after removal of the cortisol excess. In the following section we will therefore differentiate between impairment of pituitary function resulting from the space-occupying lesion and from hyperprolactinemia itself.

5.2.2.1 Growth Hormone Secretion

The evaluation of GH secretion with various stimulation tests is the most sensitive parameter of the endocrinologic investigation of pituitary function (von Werder 1975; Solbach et al. 1978). Hyperprolactinemic patients with a radiologically normal sella turcica have normal GH responses to insulin-induced hypoglycemia and arginine stimulation. This documents the fact that PRL itself has no effect on GH secretion (von Werder et al. 1973; Thorner et al. 1974). Therefore patients with radiologically proven microprolactinomas which do not compress or destroy anterior pituitary tissue also have normal GH secretion (Boyar et al. 1976; L'Hermite et al. 1978; von Werder et al. 1978a). If the sella turcica is enlarged, suggesting the presence of a macroprolactinoma, impairment of growth hormone secretion or complete somatotrophic insufficiency is encountered (Besser et al. 1972; Friesen et al. 1972; Boyar et al. 1974; von Werder et al. 1978a; L'Hermite et al. 1978; Hagen et al. 1979).

In patients with GH hypersecretion, hyperprolactinemia is frequently observed (Guyda et al. 1973; Zimmerman et al. 1974; Franks et al. 1975a; Tolis et al. 1978; Silvestrini et al. 1978). PRL hypersecretion could stem from a pituitary adenoma producing both PRL and GH or from normal anterior pituitary tissue which is cut off from inhibitory hypothalamic influences (Turkington et al. 1971; MacLeod 1976). Prolactin levels in acromegalics rarely exceeds 250 ng/ml which could give clear cut biochemical evidence for the presence of a PRL-producing adenoma (Friesen et al. 1973; Silvestrini et al. 1978). However, the observation that GH and PRL in hyperprolactinemic acromegalics behave very similarly, i.e., can be stimulated by TRH and are suppressed by dopamine agonistic drugs (Liuzzi et al. 1974, 1976) has led some investigators to hypothesize that GH and PRL are synthesized and secreted from the same mammosomatotrophic cell. However, the morphological basis for this hypothesis is still lacking (Landolt 1978). Zimmerman et al. (1974) presented evidence for GH and PRL synthesis in the same cell of a normal pituitary. However, both hormones in one single cell of adenomatous tissue could not be located by immunofluorescence techniques. Guyda et al. (1973) have also shown separate somatotrophs and lactotrophs in one single adenoma of a patient with acromegalic gigantism and hyperprolactinemia. Since no morphological evidence for enhanced PRL secretion in paraadenomatous tissue has been demonstrated (Zimmerman et al. 1974; Saeger 1977) one has to assume that in most instances PRL hypersecretion stems from a separate PRL-secreting cell of the mixed somatotrophic and lactotrophic adenoma. Thus mixed adenomas kept in culture demonstrate in vitro secretion of both GH and PRL with the same prepondernace as can be anticipated from the amount of lactotrophs and somatotrophs identified by immunostaining (Mashiter et al. 1979).

Tolis et al. (1978) reported a patient with acromegaly and galactorrhea-amenorrhea in whom a prolactinoma had been removed by transsphenoidal surgery. This led to normalization of the basal PRL level, but the GH levels remained elevated. At a second transsphenoidal operation a separate somatotrophic adenoma could be removed, thus suggesting that in some patients with hyperprolactinemic acromegaly two separate adenomas may lead to the dual hormone excess.

5.2.2.2 Gonadotropin Secretion

After GH secretion the next most sensitive indicator for anterior pituitary failure is gonadal function (Solbach et al. 1978). However, impaired secretion of LH and FSH in patients with hyperprolactinemia does not necessarily mean that gonadotrophic cells are destroyed. Thus, anovulation or amenorrhea is also found in only mildly hyperprolactinemic females with no enlargement of the sella turcica. For a detailed discussion of the impact of PRL hypersecretion on LH and FSH see Chap. 4.

Consistent with a functional derangement of gonadotropin secretion in hyperprolactinemic states is the restoration of cyclic gonadotropin secretion after normalization of PRL levels (Besser et al. 1972; del Pozo et al. 1974; Thorner et al. 1974; Rolland et al. 1974; von Werder 1975; Rjosk et al. 1976; Asfour et al. 1977; Gomez et al. 1977).

The responses of gonadotropin to GnRH in hyperprolactinemic subjects also varies considerably. Thus normal, exaggerated but also blunted or absent responses have been encountered (Robyn et al. 1973; Boyar et al. 1974; del Pozo et al. 1974; Peillon et al. 1975a; Zarate et al. 1975; Rjosk et al. 1976; Lachelin et al. 1977; Asfour et al. 1977; Buvat et al. 1978; L'Hermite et al. 1978). Patients with large prolactinomas and destructions of the gonadotrophs have no or blunted responses after GnRH administration (Peillon et al. 1975a; L'Hermite et al. 1978; von Werder et al. 1978a). However, a blunted response to GnRH does not necessarily suggest destruction of the gonadotrophs. Thus Rjosk et al. (1976) found blunted responses of LH to GnRH in long-standing hyperprolactinemia which became normal after normalization of PRL levels. When hyperprolactinemia reccurred after bromocriptine withdrawal, LH responses to GnRH remained normal in these patients (Rjosk et al. 1979b).

These findings are suggestive of a lack of endogenous GnRH which may lead to secondary atrophy of the gonadotrophs in long-standing hyperprolactinemia. There is some experimental evidence which suggests that prolactin-induced enhanced dopamine turnover in the hypothalamus may impair GnRH secretion (Lachelin et al. 1977; Yen 1978; Quigley et al. 1979).

Whatever the explanation may be, the interpretation of a GnRH-stimulation test cannot be based on the same criteria in evaluating patients with prolactinomas as in patients with other pituitary adenomas (Müller and von Werder 1977).

5.2.2.3 Thyrotropin Secretion

Thyrotropin (TSH) belongs, together with ACTH, to the most resistant anterior pituitary hormones (von Werder 1975). Therefore most of the patients even with large macroprolactinomas retain the ability to secret TSH under basal conditions and after TRH stimulation and have normal thyroid function. Brendel (1978) found only three patients with secondary hypothyroidism out of 50 females harboring a prolactinoma. Thus, whereas the PRL response is usually blunted, the TSH response to TRH is normal in hyperprolactinemic patients.

In severely hypothyroid patients with elevated basal TSH levels and excessive TSH response after TRH stimulation hyperprolactinemia can be encountered (Herlant et al. 1966; Edwards et al. 1971; Kleinberg et al. 1977). However, most of the hypothyroid patients have normal basal PRL levels (L'Hermite et al. 1974; Fossati et al. 1975) though an exaggerated rise in PRL together with a pronounced rise in TSH after TRH is generally observed (von Werder 1975). In contrast patients with hyperthyroidism and suppressed TSH secretion have a subnormal rise of PRL after TRH administration (von Werder 1975).

Careful roentgenologic evaluation may reveal an enlargement of the sella turcica in over 80% of women with primary hypothyroidism. The sella volume seems to correlate with the magnitude of the TSH elevation (Yamada et al. 1976). In these patients elevated PRL levels are due to continous endogenous TRH stimulation and not to an autonomous prolactinoma documented by normalization of hyperprolactinemia after substitutional therapy with thyroxine (Edwards et al. 1971). Autonomous TSH-producing tumors leading to hyperthyroidism have been reported (Horn et al. 1976; Reschini et al. 1976). In some patients TSH can be hypersecreted together with PRL by mixed thyrotrophic lactotrophic adenomas, leading to the clinical picture of amenorrhea and galactorrhea and recurrent hyperthyroid goiter (Horn et al. 1976; Werner 1979).

5.2.2.4 ACTH Secretion

ACTH secretion is the anterior pituitary function most resistant to pituitary damage (von Werder 1975). Hyperprolactinemia does not seem to have a specific influence on pituitary ACTH secretion. Thus, normal cortisol levels and a normal rise of ACTH or cortisol during insulin hypoglycemia in hyperprolactinemic subjects has been found (von Werder 1975; L'Hermite et al. 1978). Von Werder et al. (1973) found a normal funtioning hypothalamo-pituitary adrenal axis in all six patients investigated with hyperprolactinemia and a pituitary adenoma. Thorner et al. (1974) found only one out of six females with a prolactinoma who had impaired ACTH secretion. Similar results were reported by Boyar et al. (1974) and Friesen et al. (1973). Brendel (1978) reviewing the records of 50 female patients with prolactinomas found secondary adrenal insufficiency in ten patients, all harboring large macroprolactinomas. All these patients suffered from complete anterior pituitary failure with lack of FSH and LH and GH secretion and reduced TSH response after appropriate stimulation (Brendel 1978).

In contrast to the lack of effect of elevated PRL levels on ACTH secretion, a direct effect of PRL on the adrenal gland has been documented (see Chap. 4).

In patients with cortisol hypersecretion due to pituitary ACTH excess lactation has been observed in the presence of normal PRL levels (Kleinberg et al. 1977).

Similar cases, however without PRL determination have been reported (Levin et al. 1959; Mahesh et al. 1969). However, in general, patients with Cushing's disease due to an ACTH-producing adenoma as well as patients with Nelson's syndrome have normal PRL levels (Krieger and Luria 1976). TRH-stimulated prolactin release is also not impaired (L'Hermite et al. 1978) whereas the nocturnal surge of PRL is absent in patients with Cushing's disease (Krieger et al. 1976).

Müller et al. (1978) reported one patient with pituitary ACTH-dependent Cushing's disease, acromegaly and PRL levels above 500 ng/ml. This patient with a triple hormone excess responded to L-dopa and bromocriptine with suppression of PRL, GH and ACTH, whereas somatostatin inhibited GH and ACTH secretion but did not affect PRL. This may suggest that in this case disturbance of hypothalamic dopaminergic control led to hyperplasiogenic adenoma formation of the cortico-trophic, lactotrophic and somatotrophic cells together.

5.2.2.5 Posterior Pituitary Function

Though PRL is a very important regulator of water and mineral-homeostasis in several species (for review see Dellman et al 1977), no effect on water and electrolyte balance has been observed in man (Baumann and Loriaux 1976). Some of the effects observed earlier may have been due to contamination of PRL preparations with antidiuretic hormone (Vorherr 1979).

Therefore measurable changes of ADH are not encountered in hyperprolacti-nemic patients. This pertains also to patients with large pituitary adenomas in whom diabetes insipidus is not observed (Buchborn and Irmscher 1969; von Werder 1975; Derome et al. 1979). Only after radical surgery, particularly after cryotherapy transient and permanent diabetes insipidus may be encountered (Buchborn and Irmscher 1969; von Werder 1975).

For other metabolic effects of hyperprolactinemia see Chap. 4

5.2.3 Local Tumor Symptoms

A space-occupying lesion in the pituitary initially induces compression of the residual adenohypophysis, which may lead to dysfunction of anterior pituitary hormone secretion, as has been discussed above. Therefore signs of anterior pituitary failure such as GH and gonadotropin deficiency are regarded as early signs of pituitary pathology (von Werder 1975; Solbach et al. 1978). However, in the adult, impairment of GH secretion, although the most sensitive indicator of pituitary damage (Solbach et al. 1978; Hagen et al. 1979), remains clinically silent. Therefore in earlier observations (Schwarz 1969) hypogonadism was the first obvious symptom in patients with endocrine inactive adenomas. Since in these patients PRL levels had not been measured, many of the "inactive" pituitary tumors might in fact have been secreting PRL (Child et al. 1975b; Franks et al. 1977).

Since hyperprolactinemia itself causes hypogonadism (Besser et al. 1972, 1977; Asfour et al. 1977; Thorner and Besser 1977; von Werder et al. 1977; Frantz 1978), hypogonadism in patients with prolactinomas does not necessarily mean that the pituitary gonadotrophs are mechanically impaired. Only complete endocrine evaluation of the pituitary can clarify whether or not damage to the residual anterior pituitary has occurred (Müller and von Werder 1977). On the other hand, even small adenomas can be detected by sophisticated radiographic techniques now available. In addition, in patients with large tumors extending into the suprasellar area, visual disturbances can be encountered,

5.2.3.1 Sella Turcica

The diagnosis of a pituitary macroadenoma with a diameter larger than 10 mm can usually be made by plain skull X-rays (Deck 1973; Vezina and Sutton 1974; Doyle and McLachlan 1977; Newton and Richmond 1980). For the detection of microadenomas with a diameter of less than 10 mm thin section tomography has to be performed. The latter should be performed in two planes using the hypocycloidal technique, which picks up minor changes which may be missed using plain X-rays alone. A radiographic picture of the sella turcica represents a cut through the osseous cylinder at the base of the skull. The roof of this cylinder is represented by the diaphragm, tissue which is undetectable by conventional X-ray techniques. The open ends of the cylinder are closed on both sides by the cavernous sinus, the carotid artery and connective tissue, which are again radiographically undetectable. Therefore neither plain skull films nor any kind of tomography of the sella turcica allow detection of parasellar and suprasellar extension of a primary intrasellar adenoma. To evaluate the anatomic situation in this area additional techniques such as angiography and pneumoencephalography are indicated. However, the advances in cranial computer tomography have simplified the diagnostic approach considerably (Kazner et al. 1978). Computerized tomography (CT) has replaced pneumoencephalography in the evaluation of suprasellar adenoma extension (Fahlbusch et al. 1976; Jelden et al. 1976; Katzner et al. 1978; Danoff et al. 1978), whereas cerebral angiography remains indicated only if a carotid aneurysm is suspected or the lateral, parasellar extension of the adenoma has to be delineated (Deck 1973; Schürmann et al. 1978).

The size of the normal sella turcica may vary considerably, the length ranging from 9.3 mm to 16.1 mm, the height from 6.6 mm to 12 mm (Loepp and Lorenz 1971). Because of this considerable variation it is sometimes very difficult to determine whether a pituitary fossa is normal or abnormal (Besser 1976; Du Boulay et al., 1976; Dubois 1979). In addition asymmetry of the septum of the sphenoid sinus may simulate an abnormality of the floor of the sella turcica (Bruneton et al. 1979; Dubois et al. 1979). According to Banna et al. (1978) it is impossible to draw a sharp line between the normal variations and minor pathologic changes of the sella turcica. For systematic reasons Hardy (1969, 1973) and Vezina and Sutton (1974) have introduced their classification of pituitary adenomas according to their radiologic presentation and operative findings. The enclosed adenoma is a tumor which remains within the anatomic bounderies of the osteoaponeural sheets of the sella turcica (Hardy 1973; Vezina and Sutton 1974). Grade I of the enclosed adenoma is characterized by normal size of the sella turcica with asymmetry of the sella floor (Fig. 5.4), revealed by careful 2 mm cut tomography (Vezina and Sutton 1974, Newton and Richmond 1980). In these cases the adenomas have a diameter of less than 10 mm and are referred to as microadenomas (Hardy 1969, 1973, 1980). Grade II of the enclosed adenoma is characterized by an enlargement of the sella turcica like a ballooning or a typical double floor (Fig. 5.5). However suprasellar extension of the enclosed grade II adenoma is possible and has to be evaluated by CT. The enclosed adenoma grade I is equivalent to the intrahypophyseal adenoma and the grade II enclosed adenoma without suprasellar

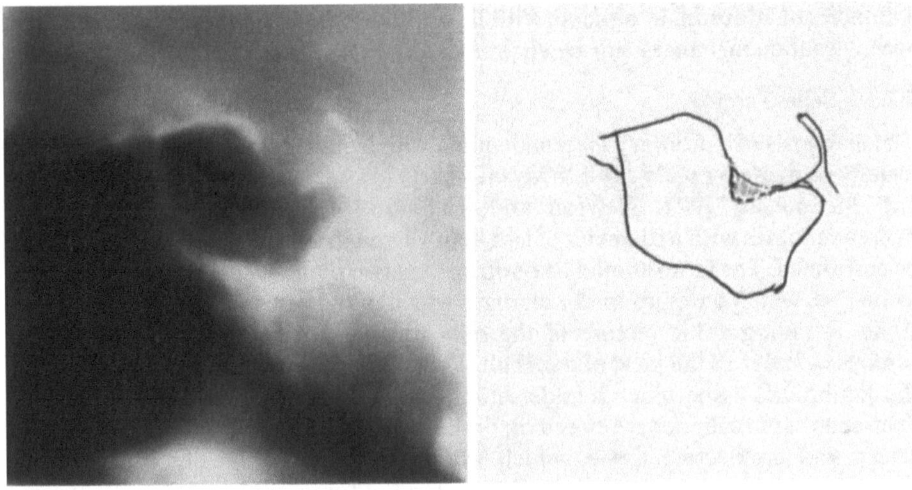

Fig. 5.4. Typical microprolactinoma as revealed by sella tomography in a female (23 years) with amenorrhea-galactorrhea. The PRL level was 6900 μU/ml. After transsphenoidal surgery (Dr. R. Fahlbusch, Dept. of Neurosurgery, Univ. of Munich) confirming the presence of a microadenoma, the PRL level fell to 430 μU/ml

extension is equivalent to the intrasellar adenoma according to the classification of Guiot (1978). The enclosed adenoma with suprasellar extension is the expansive adenoma in Guiot's classification. Enclosed adenomas have to be separated from the so-called invasive adenomas (Hardy 1969, 1973; Vezina and Sutton 1974; Guiot 1978). In these patients plain skull films and sella tomography show an erosion of the sella floor with possible herniation of the adenomatous tissue into the sphenoid sinus. If the erosion is localized the adenoma belongs to grade III, if the entire floor of the sella turcica is destroyed or the boundaries of the sella turcica have totally disappeared grade IV of the Hardy classification is assigned. All adenomas grade II to grade IV or intrasellar expansive or invasive adenomas (Guiot 1978), regardless of whether they grow invasively into sphenoid sinus or into the suprasellar area or remain enclosed, are called macroadenomas.

Microprolactinomas are usually located in the lateral wing of the gland (Hardy 1973). In these patients sellar tomography is helpful not only to reveal the presence of a microadenoma but also to localize the adenoma (Hardy 1973; Vezina and Sutton 1974). If sellar tomography does not reveal an abnormality of the sella turcica, an intrahypophyseal microadenoma may nevertheless be present (McKenna et al. 1978; Tyrrell et al. 1978).

This can only be proven by surgical intervention (McKenna et al. 1978), since no endocrine tests are available to document an autonomous PRL secreting microadenoma with absolute certainty (see Sect. 5.2.1).

However, sagittal and coronal CT with equipment now available may reveal even microadenomas within the sella turcica, which have not led to any radiologic changes of the osseous structure (Robertson 1978; Leonardi et al. 1980). Typical microadenomas documented by tomography of the sella turcica never lead to suprasellar extension. Therefore additional diagnostic procedures such as CT or pneumoencephalography in these patients are not necessary. In contrast to the

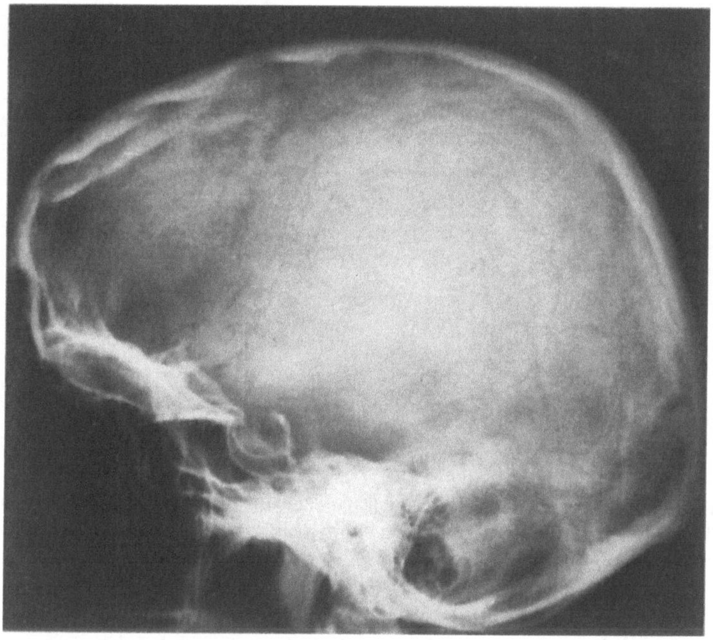

Fig. 5.5. Macroprolactinoma of the intrasellar type (or grade II). The double floor signals a large, asymmetrically growing adenoma. PRL levels were 3100 μU/ml. Normal CT and visual fields. This patient had an uneventful pregnancy after normalization of PRL levels with bromocriptine

microadenomas the macroadenomas are always visible on plain skull films (Deck 1973). However, in these cases sellar tomography is also recommended. The latter can demonstrate the location of defect of the sella floor and in cases of asymmetric enlargement the site of the expansive lesion (Fig. 5.6). In such cases sagittal as well as frontal cuts are particularly suitable for localisation and delineation of adenoma size (Deck 1973). If the sella turcica is completely destroyed the inferior extension along the clivus may be obscured by bone structure, again necessitating tomography. Any opacity in the sphenoid sinus should also be clarified by sellar tomography in order to differentiate intrasphenoidal extension of a pituitary adenoma from sinus carcinomas and mucoceles (Deck 1973).

Since in all macroadenomas, whether noninvasive or invasive, supra- and parasellar extension is possible, CT is always indicated (Kazner et al. 1978). Since this technique is noninvasive and as efficient as pneumoencephalography it should be the method of choice (Fahlbusch et al. 1976; Kazner et al. 1978).

Parasellar extension is quite frequently observed in macroprolactinomas (Fahlbusch et al. 1978; Derome et al. 1979) and may not be detectable by CT if the tumor herniates into the cavernous sinus. If parasellar adenoma growth is suspected, carotid angiography and phlebography is necessary before neurosurgical intervention is undertaken (Deck 1973; Fahlbusch et al. 1978; Schürmann et al. 1978). With presently available radiologic techniques the size of the prolactinoma can be estimated with great accuracy (Hardy 1973; Fahlbusch et al. 1978). Tumor size correlates well with basal PRL levels (Nader et al. 1976; Malarkey and Johnson 1976; Fahlbusch et al. 1978). In cases in which PRL levels are found to be

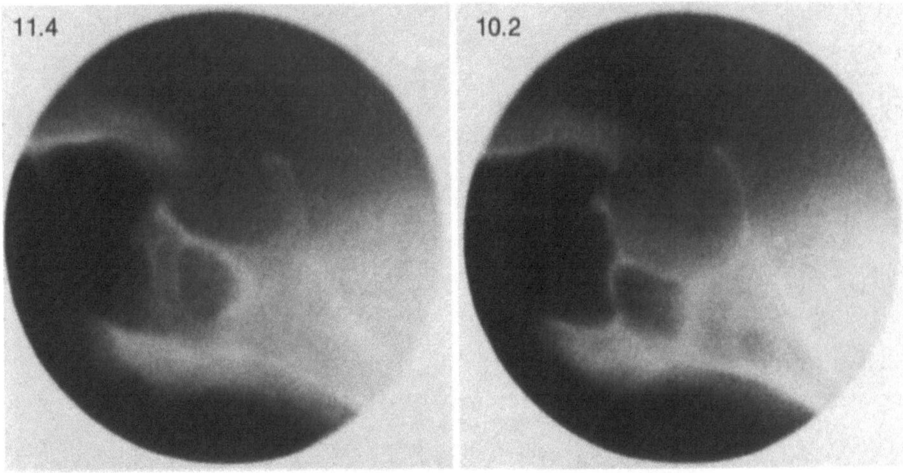

Fig. 5.6. Macroprolactinoma in a female patient with amenorrhea-galactorrhea and hyperprolactinemia (PRL 5590 μU/ml). The tomographs show clearly the asymmetry of the sella cylinder with normal configuration on the *left* (11.4) and massive enlargement on the *right* (10.2)

considerably below those anticipated from the radiologically delineated tumor size, cystic adenomas or bleeding within the adenoma with reduced lactotrophic cell mass have been found (Fahlbusch et al. 1978).

A partially empty sella is a finding frequently associated with intrahypophyseal microadenomas (Kleinberg et al. 1977; Domingue et al. 1978; Newton and Richmond 1980). The radiographic features of the empty sella syndrome are best revealed by pneumoencephalography (Kaufman et al. 1973). The term primary empty sella is used to describe intrasellar arachnoid invagination of the cysterna optico-chiasmatica due to congenital malformation of the diaphragm. Arachnoidal cysts and spontaneous necrosis of pituitary adenomas can also lead to an empty sella (Schaison and Metzger 1976; Jordan et al. 1977). Jordan et al. (1977) reported one, while Schaison and Metzger (1976) reported two of twelve patients with empty sella syndrome harboring prolactinomas. Kleinberg et al. (1977) found, out of 53 patients with enlarged sella turcica, five with empty sella syndrome. However these patients, though presenting clinically with galactorrhea, had normal PRL levels (Kleinberg et al. 1977). Bar et al. (1975) reported one female patient with galactorrhea who had primary empty sella with normal basal PRL levels, whereas the secretory profile during 24 h demonstrated an increased mean PRL concentration. Usually patients with empty sella syndrome have normal anterior pituitary function (Jordan et al. 1977). That PRL is detectable in cerebrospinal fluid in these patients (Assies et al. 1978) has not been confirmed (Schroeder et al. 1976; Jordan et al. 1976).

However, in 21 of 22 patients with suprasellar extension of pituitary adenomas, one or more adenohypophyseal hormones have been detected in CSF. This suggests direct secretion into the CSF from the adenoma, which has expanded into the basilar cysterna (Schroeder et al. 1976; Jordan et al. 1976; Assies et al. 1978).

The fact that an empty sella syndrome is rather frequently found in patients with micro- and macroprolactinomas emphasizes the need to document this syndrome. Whereas axial CT does not always detect this condition, coronal CT scanning allows its diagnosis (Kazner et al. 1978). However, pneumocephalography, which is still superior to CT in diagnosing this condition (Kaufman et al. 1973; Kazner et al. 1978), should be employed in questionable cases prior to surgery.

5.2.3.2 Chiasma Syndrome

Due to the particular location of the optic nerve and the ocular nerves at the base of the skull, suprasellar extension of pituitary adenomas can lead to devastating effects on the visual apparatus. However most patients are unaware of decreasing vision or visual field defects. Thus, out of 701 patients with various pituitary adenomas in whom visual field defects could be objectively demonstrated, 109 did not notice any visual changes (Hollenhorst and Younge 1973). This emphasizes the need to subject every patient with suspected macroadenoma to a visual field examination. Though the most typical visual field defect is bitemporal hemianopia followed by superior bitemporal defect (Hollenhorst and Younge 1973) other defects, such as central or temporal scotomas, can be encountered in patients with pituitary adenomas. Complete amaurosis of one eye with a temporal defect in the other eye were found in 81 out of 1 000 unselected pituitary adenoma patients (Hollenhorst and Younge 1973). In these patients fundoscopy revealed pallor of the optic disk or complete atrophy of the papilla. Papilledema is rarely observed in patients with pituitary adenomas. Only very large adenomas invading the lateral or the third ventricle may be accompanied by papilledema.

Parasellar extension of the adenoma can lead to extraocular muscle palsy. The cranial nerves III and VI are the most frequently involved. Again this is observed only in patients with very large adenomas or as a complication or surgical treatment (Fahlbusch et al. 1978, Guiot 1978), heavy particle therapy (Kjellberg and Kliman 1973) or pituitary apoplexy (Weisberg 1977).

In patients with microprolactinomas no visual field defects are found (Gomez et al. 1977; von Werder et al. 1978a; Derome et al. 1979; Fahlbusch et al. 1980). However, in all patients with an enlarged sella turcica suggesting any kind of macroprolactinoma, visual fields analyses must be beformed. Gomez et al. (1977) found unilateral or bilateral visual field defects in 7 out of 26 female patients with nonpuerperal galactorrhea and hyperprolactinemia, all with radiologic evidence of a macroprolactinoma. Out of 51 female patients with PRL levels ranging from 1100 to 88 000 μU/ml and radiologic evidence for prolactinoma, 25% had visual field defects demonstrated by routine perimetry (von Werder et al. 1978a). However, visual field defects were found in none of 108 patients with PRL levels ranging from 600 to 4 000 μU/ml without radiologic evidence of pituitary tumor. In contrast 70% of the male patients with prolactinomas and PRL concentrations ranging from 6 000 to 400 000 μU/ml had visual field defects. In all these patients suprasellar extension could be documented by CT scan (von Werder et al. 1978a). The higher frequency of visual disturbances in male patients with prolactinomas is explained by the usually larger adenomas found in males. Thus Derome et al. (1979) found visual field defects in 17% of 120 female patients with radiologically proven prolactinomas, whereas visual field defects were demonstrated in 53% of 30 male prolactinoma patients, two of whom showed also extraoccular muscle palsies.

Landolt (1978) reported only one out of 21 male prolactinoma patients as having normal vision, whereas 13 out of 31 females had normal visual fields. In a series of 20 female patients with hyperprolactinemia and radiologically documented enlargement of the sella turcica only one was found to have visual field defects (Kelly et al. 1978). Again Carter et al. (1978) found 9 out of 22 men with PRL-secreting adenomas accompanied by visual impairment.

In nonpregnant females harboring microprolactinomas visual field defects are nonexistent and usually less than 25% of all female patients with macroprolactinomas have visual impairment. However, pregnancy may increase the risk of visual field problems. Thus 25% of 73 females with untreated PRL-secreting pituitary adenomas developed visual disturbances, which had been found in only 5% receiving treatment before pregnancy (Magyar and Marshall 1978). This emphasizes the need to perform regular perimetry in female prolactinoma patients during pregnancy when frequent radiologic examinations are not feasible (Thorner et al. 1975a; Lamberts et al. 1977a; Bergh et al. 1978; Jewelewicz and Zimmerman 1978; Mornex et al. 1978; Rjosk et al. 1978, 1979b; von Werder et al. 1978a; see also Sect. 5.5: Pregnancy and Prolactinomas).

5.3 Clinical Signs and Diagnosis of Hyperprolactinemia Due to Prolactinomas

5.3.1 Females with Hyperprolactinemia

Whereas hyperprolactinemia in male patients is still a rare finding, elevated PRL levels in the female are quite common (Friesen et al. 1973; Thorner et al. 1974; Seppälä et al. 1975; Gomez et al. 1977; Reichlin 1979; Rjosk et al. 1979b). Two hypotheses are offered to explain the more frequent finding of hyperprolactinemia in females compared to males.

1) There is in fact no difference between females and males regarding the occurrence of hyperprolactinemia. In females hyperprolactinemia causes easily-recognizable symptoms such as oligomenorrhea or amenorrhea, sterility, and galactorrhea, which lead to endocrinologic investigation and detection of hyperprolactinemia. However, hyperprolactinemia in the male causes only unspecific symptoms such as gradual loss of libido and sexual potency, and may remain undetected (Besser et al. 1977; Thorner 1977; von Werder et al. 1977; Peillon et al. 1977; Franks et al. 1978; Nagulesparen et al. 1978).

2) The higher frequency of hyperprolactinemia and microadenoma formation in females is probably due to the higher endogenous estrogen levels, which are known to stimulate lactotrophic cell growth as well as PRL secretion (Peillon et al. 1970; El-Etreby and Günzel 1973; D'Agata and Scapagnini 1979). The latter hypothesis is supported by the fact that in amenorrhea out-patient clinics more than 15% of females are found to have hyperprolactinemia (Tolis et al. 1974; Franks et al. 1975b; Seppälä et al. 1975; Bohnet et al. 1976; Rjosk et al. 1976; Bergh et al. 1977; Moore et al. 1978), whereas careful screening of male patients in psychiatric impotence clinics (von Werder et al. 1980a) and infertility clinics (Rjosk and Schill 1979) has not led to detection of a significant number of male hyperprolactinemic patients.

5.3.1.1 Microprolactinomas

5.3.1.1.1 Incidence

The majority of female patients with hyperprolactinemia have a completely normal sella turcica documented by sella tomography. Thus, von Werder et al (1980a) found 133 women with normal sella turcica out of 258 hyperprolactinemic females. Sixty-eight of these patients had evidence of a microadenoma and in 57 females sella enlargement documenting a macroprolactinoma was found (Fig. 5.7). The mean age of the females with microprolactinomas at discovery of hyperprolactinemia was 33 years, ranging from 22 to 40 years. In a series presented by Bergh et al. (1977) 48% of the hyperprolactinemic women had evidence of a pituitary adenoma. Kleinberg et al. (1977) reported that 34% of his female patients with amenorrhea and galactorrhea had radiologic evidence of a pituitary tumor, though not all of them were found to have hyperprolactinemia. In another series 30% of hyperprolactinemic females with galactorrhea had radiologic evidence of a pituitary microadenoma, whereas only one had an overall enlarged sella turcica suggesting a macroprolactinoma (Gomez et al. 1977). However, Franks et al. (1975b) reported approximately 20% of hyperprolactinemia in 106 women with amenorrhea. Of the latter group more than half had radiologic evidence of a pituitary adenoma (Franks et al. 1975b). These findings document the fact that the frequency of radiologic signs suggestive of a prolactinoma vary considerably, depending on the speciality of the clinic where the patient is first seen. Thus out of 120 hyperprolactinemic females seen in a department of neurosurgery all had an abnormal sella turcica. Only 21%

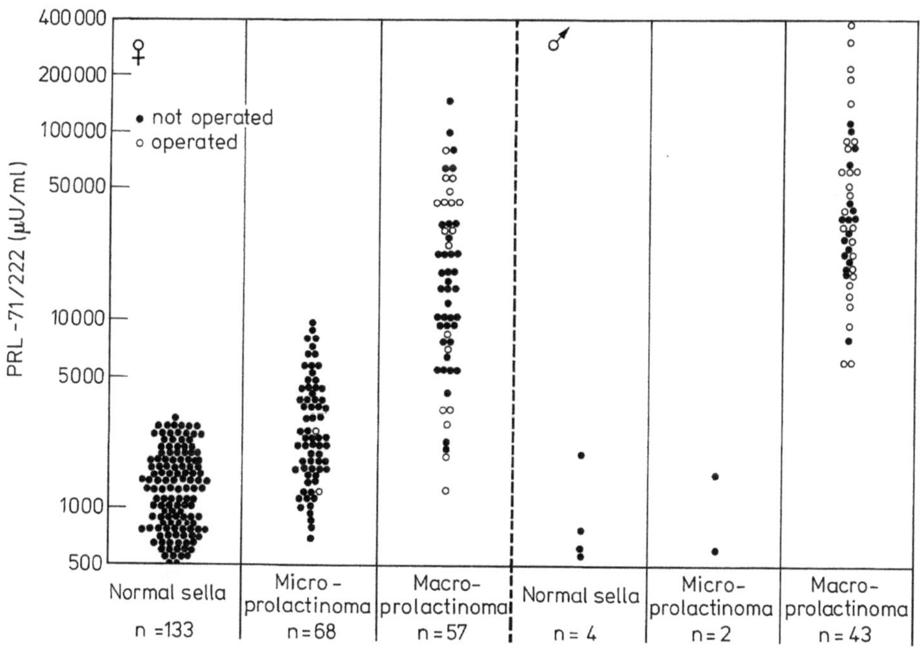

Fig. 5.7. Prolactin levels in 307 hyperprolactinemic patients without and with radiologic evidence of a pituitary tumor. (von Werder et al. 1980a)

had a pituitary microadenoma, whereas 75% had macroadenomas frequently of the invasive type with suprasellar extension (Derome et al. 1979). This is in contrast to the frequency of radiologic changes found in hyperprolactinemic females seen in departments of obstetrics and gynecology, where females with radiologic evidence of prolactinomas rarely represent more than 35% of all hyperprolactinemic patients (Jacobs et al. 1976; Rjosk et al. 1976, 1979b; Seppälä et al. 1977).

However the true incidence of microprolactinomas in cases with hyperprolactinemia is still unknown (Editorial 1980), since appropriate tests which can distinguish so-called functional hyperprolactinemia from pituitary microadenomas without changes of the osseous structure of the sella turcica (McKenna et al. 1978) are not available (Thorner 1977; Silvestrini et al. 1978; von Werder and Rjosk 1979). Thus it has been hypothesized that all patients with long-term hyperprolactinemia may harbor a pituitary microprolactinoma (L'Hermite et al. 1978). This is also suggested by the rather uniform responses of PRL secretion to stimulation and suppression in all patients with hyperprolactinemia of the same magnitude, with and without enlarged sella turcica (Tolis 1977; Frantz and Kleinberg 1978).

5.3.1.1.2 Prolactin Secretion

The similarity of PRL secretion patterns in hyperprolactinemic patients with and without radiologic evidence for a prolactinoma has been described above (Sect. 2.1). Though there is no typical response to any of the stimulating or suppressing agents known, the basal PRL level can suggest a pituitary microadenoma in the presence of a normal sella turcica (McKenna et al. 1978). According to Tolis et al. (1974) PRL levels above 100 ng/ml are highly suspicious of a prolactinoma. Levels above 500 ng are always caused by PRL-secreting adenomas (Frantz et al. 1972; Jacobs and Daughaday 1973). Thus out of 133 hyperprolactinemic patients with no radiologic evidence of PRL-secreting adenoma none had PRL levels exceeding 4 000 μU/ml or 200 ng/ml (Fig. 5.7). Similar findings were published by Kleinberg et al. (1977) and Friesen et al. (1973).

Basal PRL levels correlate fairly well with the radiologically determined adenoma size (Malarkey and Johnson 1976; Gomez et al. 1977) and the adenoma size found at operation (Fahlbusch et al. 1978). This could also be demonstrated for 58 microprolactinoma patients (Fig. 5.8) in whom PRL levels ranged from slightly elevated up to 13 000 μU or 650 ng/ml (Fahlbusch et al. 1980).

The overall correlation between adenoma size and PRL levels strongly suggests that many patients with radiologically normal sella turcica and moderately elevated PRL levels, who are now called functionally hyperprolactinemic by many investigators, may in fact harbor a microadenoma which, due to its small size, has not caused deformation of the osseous structure of the sella turcica (Tolis 1977; Thorner 1977; McKenna et al. 1978; von Werder et al. 1980a). The presence of a microadenoma in a radiologically completely normal sella has been repeatedly documented in patients with Cushing's disease (Tyrrell et al. 1978) and in postmortem studies (Costello 1936; Burrow et al. 1981). In addition, follow-up studies of hyperprolactinemic patients with initially a radiologically normal sella turcica have demonstrated that some of these patients develop radiologic evidence documenting a microadenoma (Gomez et al. 1977; Rjosk et al. 1979b, Fig. 5.10).

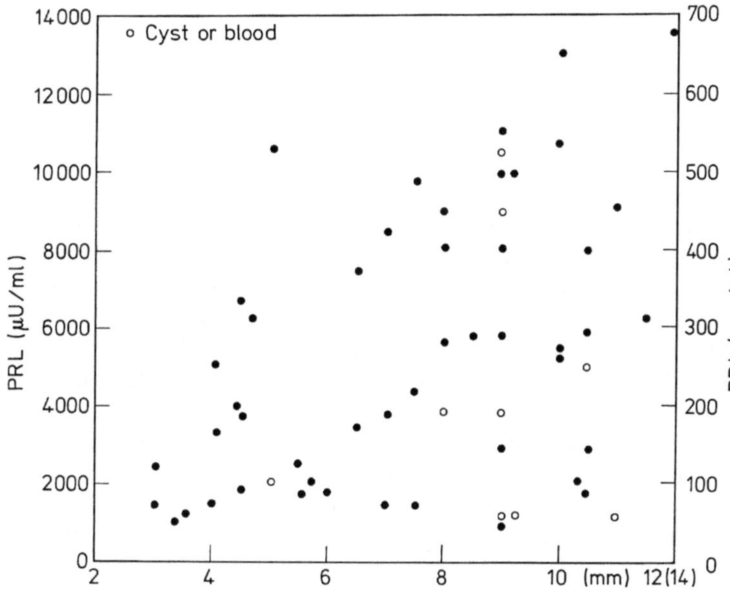

Fig. 5.8. Correlation of prolactin levels with intraoperatively confirmed adenoma size in 58 patients with microadenomas. (Fahlbusch et al. 1980)

5.3.1.1.3 Clinical Signs

The clinical features of females with hyperprolactinemia and radiologic evidence for a microadenoma do not differ from those with a radiologically normal sella turcica (see Chap. 4). Both groups present with symptoms typical of elevated PRL levels without any local tumor symptoms (Friesen et al. 1973; Franks et al. 1975b; Rjosk et al. 1976, 1979b; Bergh et al. 1977; Gomez et al. 1977; L'Hermite et al. 1978; von Werder et al. 1978a). These patients are initially seen almost exclusively by a gynecologist, to whom they may present with primary or secondary amenorrhea, menstrual disturbances, primary or secondary sterility and/or galactorrhea. The incidence of galactorrhea which generally occurs after the establishment of menstrual irregularities (von Werder 1975) varies from 30% to 80% in this group depending on the investigator (del Pozo et al. 1974; Franks et al. 1975b; Jacobs et al. 1976; Thorner and Besser 1977; von Werder et al. 1978a). The mean duration of amenorrhea in 50 hyperprolactinemic females before the discovery of elevated prolactin levels was 8 years, the mean duration of galactorrhea 5 years (Brendel 1978). However, the presence of galactorrhea does not necessarily mean hyperprolactinemia (Seppälä et al. 1977). Thus Kleinberg et al. (1977) found normal PRL levels in 86% of women with galactorrhea. This finding is not surprising since several hormones are involved in the process of lactation (Cowie et al. 1980). Whereas normoprolactinemic females with galactorrhea usually have no menstrual disturbances (Kleinberg et al. 1977), hyperprolactinemic galactorrhea is usually associated with menstrual irregularities (Thorner et al. 1974; Seppälä et al. 1975, 1977; Franks et al. 1975b; Jacobs et al. 1976; Asfour et al. 1977; Kleinberg et al. 1977). In addition mild signs of virilism may be encountered (Forbes et al. 1954; Thorner et al. 1974; see also Chap. 4).

5.3.1.2 Macroprolactinomas

5.3.1.2.1 Incidence

Females with macroprolactinomas usually have higher PRL levels than those with microprolactinomas. However there is considerable overlap between both groups (Fig. 5.7). This is partly explained by cystic tumors, which appear as large macroadenomas on the lateral skull X-ray, but have, due to cystic degeneration, less lactotrophic tissue than suggested by the X-ray findings (Fahlbusch et al. 1978; Fig. 5.9). As has been mentioned above, the percentage of hyperprolactinemic females found to have macroprolactinomas varies depending on the speciality of the investigator. For example, gynecologists usually see fewer patients with macroprolactinoma (Varga et al. 1973; Gomez et al. 1977; Rolland et al. 1974; Tolis et al. 1974; del Pozo et al. 1974; Seppälä et al. 1975); females with these tumors being more frequently encountered in endocrine clinics (Jacobs et al. 1976; Nader et al. 1976; von Werder et al. 1977). In the department of neurosurgery where only patients with radiologic evidence of a prolactinemia are seen, more than 75% may harbor macroprolactinomas (Derome et al. 1979). In a group of 258 hyperprolactinemic females who where first seen either by a gynecologist, a neurosurgeon or an endocrinologist, 57 (26%) had radiologic evidence of a macroprolactinoma (von Werder et al. 1980a).

5.3.1.2.2 Clinical Signs

In addition to the biologic effect of elevated PRL levels, local tumor symptoms in patients with macroprolactinomas can occur. Whereas it is difficult to attribute headaches, which can be observed in 40% of prolactinoma patients, to tumoral origin (Brendel 1978; Kelly et al. 1978; Derome et al. 1979), visual field defects are

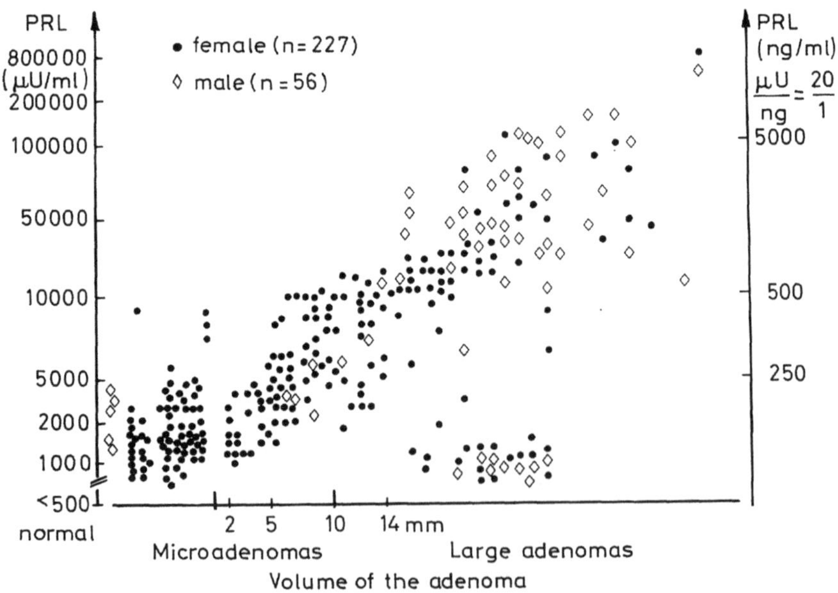

Fig. 5.9. Correlation of prolactin levels and the adenoma size and sex distribution in 283 patients with hyperprolactinemia. The large adenomas with low PRL-levels were found to be cystic. (Fahlbusch et al. 1978, supplemented by Fahlbusch 1981)

clear-cut evidence of suprasellar macroprolactinoma extension. Though the latter is quite frequently observed in male prolactinoma patients, the incidence of visual field defects in female prolactinoma cases does not exceed 25% (Nader et al. 1976; Gomez et al. 1977; Kelly et al. 1978; von Werder et al. 1978a; Derome et al. 1979).

Pituitary insufficiency due to compression of residual anterior pituitary tissue is seen only in patients with very large prolactinomas of the invasive type (Brendel 1978; von Werder 1978a; Peillon et al. 1978; Derome et al. 1979). A decreased response of gonadotropins to stimulation with GnRH is not necessarily due to destruction of the pituitary lactotrophs by tumor compression, but may be caused by long-term hyperprolactinemia (see Sect. 2.2.3).

The history of amenorrhea and galactorrhea is usually longer in patients with large macroprolactinomas (Peillon et al. 1978) suggesting that a larger adenoma takes longer to develop. However, the mean age at the time of discovery of hyperprolactinemia in micro- and macroprolactinoma patients is not significantly different (Eversmann et al. 1980). In addition, the frequency of socalled chromophobe adenomas, before discovery of PRL, leading to local tumor symptoms make it seem unlikely that duration of growth is a determining factor in the development of large macroprolactinomas. If all microprolactinomas which are discovered more frequently every year (Peillon et al. 1978; Reichlin 1979) were to develop into large adenomas there would be a significant increase in pituitary adenomas compared to the preprolactin era. It seems, therefore, appropriate to assume that microprolactinomas or smaller macroprolactinomas are not necessarily precursors of large pituitary adenomas, from which they differ only slightly in histologic appearance but in response to medical therapy (Eversmann et al. 1979; von Werder et al. 1980b). The fact that PRL levels in general remain stable over a long period of observation without treatment (Fig. 5.10) also suggests that most microadenomas remain microadenomas and do not develop into large tumors.

The occasionally longer history of irregular menses, amenorrhea and/or galactorrhea before the discovery of a macroprolactinoma may explain clinical symptoms caused by estrogen deficiency. Thus severe changes in vaginal mucosa are encountered (Rjosk et al. 1979b) which are not observed in patients with microprolactinomas (Gomez et al. 1977). Estrogen deficiency may also explain the fact that females with large macroprolactinomas, who had spontaneous galactorrhea at the onset of menstrual disturbances, do not present with a provocable galactorrhea at the time of diagnosis (von Werder et al. 1978a). Substitution with exogenous estrogens may lead to reinstitution of galactorrhea in these patients (personal observation).

5.3.1.3 Mixed Adenomas

Lactotrophic adenoma growth can occur together with adenomatous growth of other anterior pituitary cells. The mixed somatotrophic and lactotrophic adenoma is the most frequent of all mixed pituitary adenomas leading to hypersecretion of GH and PRL (Zimmerman et al. 1974; von Werder 1975). Thus in up to 40% of all acromegalics prolactin hypersecretion, in addition to GH hypersecretion, is observed (Zimmerman et al. 1974; von Werder 1975; Franks et al. 1975a; Landolt 1978). Though pituitary adenomas in most patients are composed of somatotrophic as well as lactotrophic cells (Zimmerman et al. 1974), a case with two separate adenomas, one producing GH, the other PRL, has also been reported (Tolis et al.

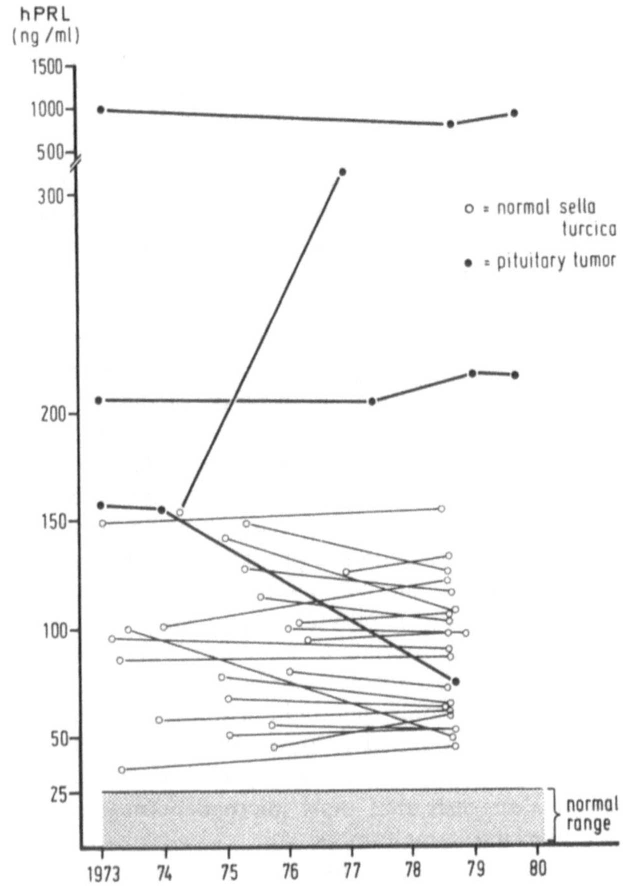

Fig. 5.10. PRL levels in 25 hyperprolactinemic females who have not received any therapy. With the exception of one female, who has developed an abnormal pituitary fossa with an increase in PRL, all other patients had no significant rise of the PRL level suggesting that growth of the pituitary lesion is the exception and not the rule in female hyperprolactinemic patients (modified and supplemented after Rjosk, H. K. et al., Acta Endocrinol, in press)

1978). Hypersecretion of PRL may also be due to the suprasellar extension of somatotrophic adenoma leading to compression of the pituitary stalk and a blockade of PIF transport (MacLeod 1976; Illig et al. 1978; Landolt 1978; Rogol and Eastman 1979). The mixed GH- and PRL-producing adenomas present as macroprolactinomas, sometimes of the invasive type (von Werder et al. 1978a; Fig. 5.14). The clinical picture is dominated by the clinical signs and symptoms of active acromegaly. Most females suffer from amenorrhea and some have galactorrhea (Davidoff 1926; Kleinberg et al. 1977; Tolis et al. 1978; von Werder et al. 1978a). However, galactorrhea in an acromegalic patient (Roth 1918) does not necessarily mean that the patient suffers from hyperprolactinemia as well (Kleinberg et al. 1977), since hGH itself has lactogenic potency comparable to PRL (Friesen et al. 1973).

It is of interest that very high PRL levels are found in those acromegalics in whom the GH secretion pattern is similar to that of PRL in normal individuals (Silvestrini et al. 1978; Lamberts et al. 1979a). Thus GH can be stimulated by TRH and inhibited by dopaminergic drugs (Liuzzi et al. 1974). This has led to the hypothesis of the existence of a mammosomatotrophic cell secreting both GH and PRL (Liuzzi et al. 1976). As has been mentioned above, anatomic proof of such a cell in mixed adenomas has not yet been presented (Zimmerman et al. 1974).

Elevated PRL levels in patients with TSH hypersecretion due to primary hypothyroidism leading to amenorrhea and galactorrhea (Edwards et al. 1971) have been observed. The autonomous hypersecretion of TSH and PRL leading to hyperthyroidism and recurrent goiter and amenorrhea-galactorrhea is extremely rare (Horn et al. 1976; Werner 1979). Parallel inhibition of TSH and PRL after L-dopa and bromocriptine administration (Fig. 5.11) has been demonstrated in such a patient, whereas thyroid hormones had no influence on TSH or PRL levels. Transsphenoidal adenomectomy led to normalization of TSH and PRL levels and separate TSH- and PRL-secreting cells could be demonstrated by electron microscopy and immunofluorescent techniques in the adenomatous tissue (Horn et al. 1976).

Though patients with Cushing's disease sometimes have galactorrhea (Mahesh et al. 1969; Kleinberg et al. 1977), an ACTH and PRL secreting adenoma in a female has not been reported. Also patients with Nelson's syndrome seem to have normal PRL levels (Krieger and Luria 1976). Gonadotropin-producing adenomas are

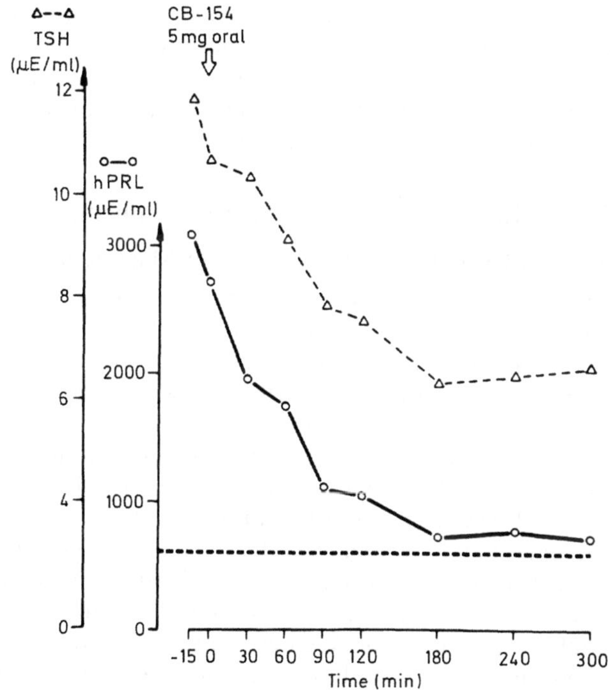

Fig. 5.11. Plasma TSH and PRL levels before and after administration of 5 mg bromocriptine (CB-154) in a patient with a TSH- and PRL-producing adenoma. The *broken line* represents the upper limit of the normal basal value. (Horn et al. 1976)

extremely rare; accordingly concommitant hypersecretion of PRL in patients with such adenomas has not yet been reported.

Endocrine inactive adenomas – so-called true chromophobe adenomas – may cause hyperprolactinemia by PIF transport inhibition (Rogol and Eastman 1979). However, since it is impossible in patients with PRL levels below 200 ng/ml to tell whether or not PRL hypersecretion derives from an adenoma or disinhibited paraadenomatous tissue by preoperative function tests, this cannot be clarified until anatomic evaluation of the adenoma has been performed. Only in cases with clear-cut endocrinologically inactive tumors, such as craniopharyngeomas in the sella region, can it be stated that the source of exaggerated PRL secretion is the extratumoral normal anterior pituitary (Illig et al. 1978). Thus Silvestrini et al. (1978) found 8 out of 12 patients with hypothalamic tumors who had hyperprolactinemia. Typically these patients have only moderately elevated PRL levels which may further increase after TRH administration and will generally not respond to chlorpromazine administration (Jacobs and Daughaday 1973; von Werder et al. 1977; Illig et al. 1978).

Prolactinomas may also be found in patients with multiple endocrine adenomatosis type I (MEA-I). In this syndrome the occurrence of ACTH- and GH-producing adenomas is well known, though most patients were earlier believed to have so-called nonfunctional pituitary adenomas (Ballard et al. 1964). It was therefore not surprising that, after the introduction of the PRL radioimmunoassay, some of these "nonfunctional" adenomas in MEA-I patients were shown to produce PRL (Carlson et al. 1978). In a series presented by Peillon et al. (1978), 3 of 157 cases with prolactinomas occurred as part of the multiple endocrine adenomatosis syndrome. Recently the association of a prolactinoma with multiple neurofibromatosis in a patient with amenorrhea and galactorrhea has been described (Pinnamaneni et al. 1980). Whether this represents a new endocrine tumor association remains to be established.

5.3.2 Male Patients with Hyperprolactinemia

5.3.2.1 Incidence and Clinical Signs of Macroprolactinomas

The prolactinoma is the most common of all pituitary adenomas (Antunes et al. 1977; Franks et al. 1977; von Werder et al. 1977; Fahlbusch et al. 1978). More than two-thirds of all adenomas seen in the hospital at random were found to secrete PRL (Frantz and Kleinberg 1978). Whereas microprolactinomas in the male are extremely rare (Peillon et al. 1975b; 1978; Derome et al. 1979; von Werder et al. 1980a) macroprolactinomas are almost as common in males as in females (von Werder et al. 1980a) and are usually larger in males (Friesen et al. 1973; Peillon et al. 1977; 1978; Carter et al. 1978; Franks et al. 1978; von Werder et al. 1978a; 1980a; Derome et al. 1979, Fig. 5.9). Thus, in a patient group in which only 25% of female patients with macroprolactinomas had visual field defects, 70% of the male patients had suprasellar extension of the tumor documented by visual field defects and confirmed by CT (von Werder et al. 1978a). In most instances the diagnosis of prolactinomas in men has been made because of visual disturbances and severe headaches (Peillon et al. 1977). Derome et al. (1979) report that more than half of their patients have had visual problems. Headaches were found in 70% of 30 of their cases and four of these had epilepsy (Derome et al. 1979). Accordingly none of these patients had tumors with a diameter of less than 10 mm and only 11 out of 30 had purely intrasellar macroprolactinomas. The other 19 adenomas showed suprasellar extension with invasion of the basilar cysterna and the cavernous sinus.

This is in agreement with Franks et al. (1978) who found in 16 out of 20 men with impotence and a prolactinoma suprasellar extension of the tumor. The late recognition of these tumors as a cause of impotence (Franks et al. 1977) may play a role for the advanced stage at the time of diagnosis of these pituitary adenomas in men (Franks et al. 1977). Peillon et al. (1978) found a longer history of prolactinoma-related symptoms in patients with so-called giant adenomas. However, Eversmann et al. (1980) found no difference in the mean age of female and male macroprolactinoma patients, the latter group again having the larger adenomas. Because of the larger size of these adenomas PRL levels are significantly higher compared to females (Friesen et al. 1973; Peillon et al. 1977; Carter et al. 1978; Franks et al. 1978; von Werder et al. 1978a, 1980a; Derome et al. 1979).

Whereas pituitary insufficiency is rarely seen in female macroprolactinoma patients before surgery, this is quite common in male prolactinoma patients. Thus 82% of 26 cases showed lack of GH and 75% lack of gonadotropin secretion after appropriate stimulation (von Werder et al. 1978a). Peillon et al. (1977) also reported no increase of LH and FSH in 35 male prolactinoma patients after GnRH stimulation. This is in contrast to the observation of Franks et al. (1978) and Carter et al. (1978) who found normal gonadotropin responses to GnRH in the presence of low testosterone levels.

In addition to the local tumoral signs, male patients with prolactinomas suffer from clinical hypogonadism with lack of libido and sexual potency and infertility (Thorner et al. 1974; Thorner and Besser 1977; Peillon et al. 1977; von Werder et al. 1977; Carter et al. 1978; Franks et al. 1978; Nagulesparen et al. 1978; Arafah et al. 1981). Galactorrhea is present less often than in women and may be accompanied by gynecomastia (Thorner et al. 1974). Whereas decreased libido and sexual potency are found in almost all male prolactinoma patients, the incidence of gynecomastia and galactorrhea varies from 4% to 30% (Besser et al. 1977; Peillon et al. 1977; Carter et al. 1978; Eversmann et al. 1981). Though mild gynecomastia is relatively common in male prolactinoma patients (Peillon et al. 1977) the majority of patients with gynecomastia have normal PRL levels (Turkington 1972c).

The infertility observed in male prolactinoma patients may be due to complete impotence (Thorner et al. 1974; Besser et al. 1977, Franks et al. 1978), to disturbance of FSH secretion because of tumoral destruction of the gonadotrophs (Peillon et al. 1977; Eversmann et al. 1981), or to a direct effect of PRL on spermatogenesis (Segal et al. 1976). Whereas testosterone levels are usually below the normal range in male prolactinoma patients (Friesen et al. 1973; Carter et al. 1978; Franks et al. 1978) and rise after exogenous hCG stimulation (Carter et al. 1978; Franks et al. 1978), sperm counts may differ considerably. Thus normal sperm counts (Besser et al. 1977; Thorner et al. 1980a), polyspermia (Peillon et al. 1977) and oligospermia or azoospermia (Segal et al. 1976) have been observed. From these contradictory findings it may be concluded that PRL, though implicated in the development of human spermatozoa (Shah et al. 1976), has only an inconsistent indirect effect on spermatogenesis. This is also suggested by the finding that only 2 out of 71 male patients with disturbance of fertility and none of 53 patients with impotentia coeundi had moderate hyperprolactinemia (Rjosk and Schill 1979).

5.3.2.2 Mixed Adenomas

As in female acromegalics male patients with GH hypersecretion may also have concommitant hypersecretion of PRL (Zimmerman et al. 1974; Franks et al. 1975a; von Werder 1975). The clinical picture in these patients is also dominated by acromegaly, hyperprolactinemia being an accidental finding, since loss of libido and sexual impotence are usually present in most normoprolactinemic men with active acromegaly (von Werder 1975). The origin of hyperprolactinemia in these patients has been discussed above (Sect. 5.3.1.3). Müller et al. (1978) recently described a unique male patient who had a triple hormone excess of the anterior pituitary. This patient, who had developed the clinical picture of Cushing's disease due to pituitary ACTH excess and acromegaly, was found also to have hyperprolactinemia with PRL levels above 500 ng/ml. Somatostatin infusion led to inhibition af GH and ACTH secretion in this patient without an effect on basal PRL levels (Müller et al. 1978). However, the administration of the dopamine agonist bromocriptine led to parallel inhibition of all three hormones, suggesting derangement of central dopaminergic control as a primary cause of this triple hormone excess (Fig. 5.15).

5.4 Therapy of Prolactinomas

Patients with pituitary adenomas require treatment either for an expanding lesion which may lead to destruction of the sella turcica and visual disturbances, or for the pituitary hormone excess caused by the endocrine active adenoma. The latter alone would not represent an indication for surgery if adequate medical therapy were available. In patients with Cushing's disease effective medical therapy is still lacking (Orth 1978; von Werder et al. 1980a). However, with the introduction of the dopamine agonistic substances such as bromocriptine, medical therapy of acromegaly has become possible (Thorner et al. 1975b; Liuzzi et al. 1976; Schwinn et al. 1977), although only a certain percentage of acromegalies will respond to bromocriptine treatment (Summers et al. 1975).

This is in contrast to hyperprolactinemic states where dopamine agonists are unequivocally effective and have therefore become a real alternative to surgical and radiotherapeutical procedures. Most of our knowledge derives from our experience with bromocriptine, which has been specifically developed as a prolactin inhibitor (see Chap. 2) and was introduced into clinical research soon after the establishment of hPRL as a separate anterior pituitary hormone. Though the beneficial effect of bromocriptine in hyperprolactinemia associated with a PRL-secreting adenoma has been widely documented, it is still controversial which patients should receive primary medical and which patients primary surgical therapy. Up to now the therapeutical policy in the treatment of prolactinoma patients varies considerably from location to location, depending on such factors as the availability of dopamine agonistic substances for treatment in humans, the availability and the quality of neurosurgery, and the availability of radiotherapeutical equipment, such as generators of heavy particles for irradiation (for review see Kohler and Ross 1973; Fahlbusch and von Werder 1978; Linfoot 1979; Faglia et al. 1980a). In the following pages accepted indications for the various therapeutical procedures in the treatment of prolactinoma patients will be discussed and existing controversies will be pointed out.

5.4.1 Medical Treatment

Hyperprolactinemia, with few exceptions, will respond to medical treatment. It is a well-established fact that dopamine agonists are effective in lowering PRL levels even when patients harbour invasive macroprolactinomas (Thorner 1977; Franks et al. 1978; Frantz and Kleinberg 1978; von Werder et al. 1978a, 1980a; Thorner et al. 1980a). L-dopa has been used to suppress PRL secretion (Turkington 1972a) although its short duration of biologic activity and the need for frequent administration make it unsuitable for long-term treatment (Frantz et al. 1972; Frantz and Kleinberg 1978). Clinically much more useful are the ergot derivatives, which lead to long-lasting dopamine receptor stimulation (see Chap. 1). Though lisuride hydrogen maleate (de Cecco et al. 1978; Liuzzi et al. 1978), lergotrile mesylate (Cleary et al. 1975; Leebaw et al. 1978) and other agents in addition to bromocriptine have been used for the treatment of hyperprolactinemia, most of today's knowledge derives from experience in treating hyperprolactinemic patients with bromocriptine (Lutterbeck et al. 1971; Besser et al. 1972; Varga et al. 1972, 1973; del Pozo et al. 1974; Fossati et al. 1976; Thorner 1977; Frantz and Kleinberg 1978; Thorner et al. 1980a). Therefore, unless stated differently, all results of medical therapy reported in this chapter are based on the treatment with bromocriptine.

5.4.1.1 Microprolactinomas

In patients with microprolactinomas PRL levels can be normalized to the same extent as in so-called functional hyperprolactinemia (Besser et al. 1972; Thorner et al. 1974; Fossati et al. 1976; Mornex et al. 1978; von Werder et al. 1978a). In female microprolactinomas moderate doses such as 2.5 mg bromocriptine twice daily will lead to normalization of PRL secretion (Fig.5.12) as well as to normalization of hypogonadism (del Pozo et al. 1974; Rolland et al. 1974; Thorner et al. 1974; Rjosk et al. 1976, 1979b; L'Hermite et al. 1978; Jewelewicz and Zimmermann 1978; von Werder et al. 1978a). The PRL-lowering efficacy of bromocriptine is more pronounced in patients with higher PRL levels (Fossati et al. 1976). However, in patients with extremely elevated PRL (above 1000 ng/ml), doses up to 40 mg per day are not always sufficient to normalize PRL secretion (von Werder et al. 1978a). This failure of bromocriptine to normalize PRL levels is observed more frequently in females (Fossati et al. 1976; von Werder et al. 1978a, 1980a).

Bromocriptine must be taken with meals in order to avoid side effects such as postural hypotension, nausea and vomiting (see Chap. 2). Treatment usually begins with an evening dose of 1.25 mg which is increased to 1.25 mg three times a day the following day. Every 4 days the dose can be increased up to the amount needed for normalization of PRL levels.

Larger doses must sometimes be administered to patients with only moderate hyperprolactinemia (von Werder et al. 1980a). Most patients respond with restoration of menses within 4 weeks (Thorner 1977), though considerably longer treatment periods are sometimes needed (von Werder et al. 1980b, Fig 5.25). If normalization of PRL secretion has been achieved but amenorrhea persists, a pregnancy test is indicated, since patients may become pregnant at the first ovulation without ever having had a menstrual period (von Werder et al. 1978a).

In patients with microprolactinomas without overall enlargement of the sella turcica and without suprasellar extension and pituitary insufficiency due to

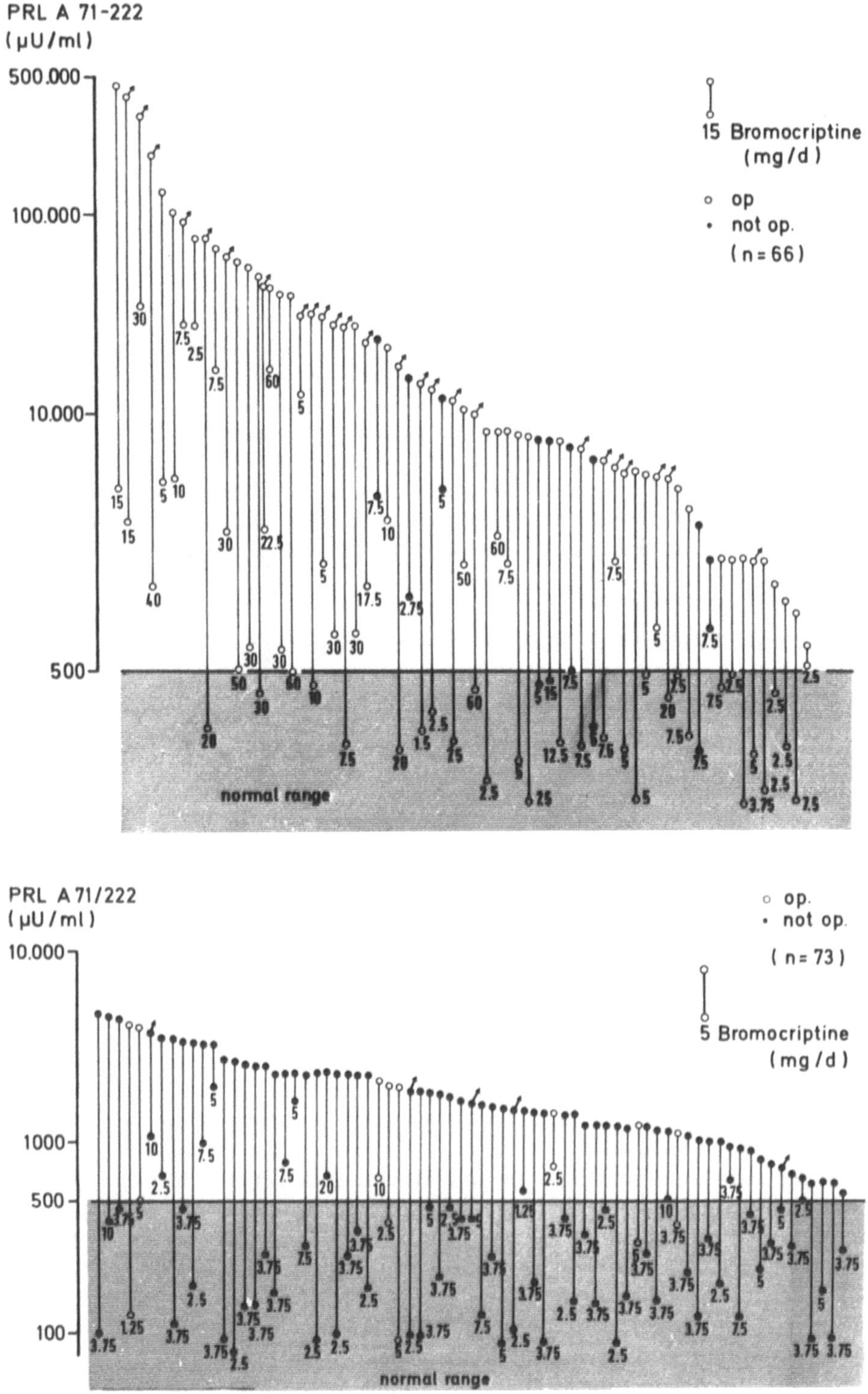
PRL A 71-222
(µU/ml)
500.000
100.000
10.000
500
15 Bromocriptine
(mg/d)
o op
• not op.
(n = 66)
normal range
PRL A 71/222
(µU/ml)
10.000
1000
500
100
o op.
• not op.
(n = 73)
5 Bromocriptine
(mg/d)
normal range

Fig. 5.12

compression of anterior pituitary tissue (see Sect. 2.4) bromocriptine as a primary therapy is generally accepted (Rolland et al. 1974; Rjosk et al. 1976, 1979b; Bergh et al. 1978; Mornex et al. 1978; von Werder et al. 1978a, 1980a). However, some authors have recommended primary radiotherapy before bromocriptine is started in patients who want to become pregnant (Besser et al. 1972; Thorner et al. 1974; Thorner 1977). Other investigators prefer selective removal of the microprolactinoma by the transsphenoidal approach (Hardy 1969, 1973, 1980; Fahlbusch et al. 1978; Guiot 1978; Cowden et al. 1979; Derome et al. 1979).

In the rare male cases with microprolactinomas, bromocriptine therapy may be indicated. Such treatment leads to normalization of hypogonadism with a rise in basal testosterone levels and restoration of normal libido and sexual potency (Thorner et al. 1980a). Since microprolactinomas in men may be more often precursors of macroprolactinomas (von Werder et al. 1980a), surgical treatment can be considered while the microadenoma is completely resectable (Hardy 1969, 1973, 1980; Fahlbusch et al. 1978, 1980; Cowden et al. 1979; Derome et al. 1979; Pont et al. 1979; Arafah et al. 1981). Spontaneous regression of sellar enlargement may also be encountered (Vogt 1977) and frequent radiologic examination of the sella turcica may be indicated before surgical intervention for selective adenoma removal is undertaken.

5.4.1.2 Macroprolactinomas

As has been discussed above, hyperprolactinemia is susceptible to medical treatment regardless of its origin. Thus even patients with large, invasive macroprolactinomas who remain hyperprolactinemic after surgery will respond with a lowering of PRL levels (Besser et al. 1972; Thorner et al. 1974; Fossati et al. 1976; von Werder et al. 1978a, 1980a). However, since PRL levels in these patients are sometimes extremely high, complete normalization, even after the administration of large dosages of bromocriptine is not always achieved (Fossati et al. 1976; Franks et al. 1978; Nagulesparen et al. 1978; von Werder et al. 1978a, 1980a) (Fig.5.12).

Since reduction of tumor growth is an important therapeutic aim, most investigators agree that patients with macroprolactinomas should be operated on (von Werder et al. 1978a; Fahlbusch et al. 1978; Guiot 1978; Derome et al. 1979; Tucker et al. 1981), irradiated (Antunes et al. 1977; Kleinberg et al. 1977), or both (Guiot 1978; Derome et al. 1979). However, primary medical therapy has recently been performed in patients with macroprolactinomas and extensive suprasellar extension (Thorner et al. 1980b) and in female patients who wish to become pregnant (Bergh et al. 1978). For the tumorostatic effect of bromocriptine and the effect of pregnancies on prolactinoma growth see Sect. 5.4.1.3 and Sect. 5.5.

Though there is also controversy about primary treatment of macroprolactinomas, it is generally agreed that patients with macroprolactinomas with persistent hyperprolactinemia after surgery or radiotherapy should be treated with bromocriptine (Besser et al. 1972; Thorner et al. 1974; Fossati et al. 1976; Thorner

◀ **Fig. 5.12.** Effect of long-term bromocriptine treatment on hPRL levels in 66 hyperprolactinemic patients with macroprolactinomas and 73 patients with microprolactinomas with and without prior surgery. The rate of normalization of hPRL secretion decreases with the increase in the basal hPRL levels. The mean dosage of bromocriptine in the macroprolactinoma patients was 15 mg per day, and in the microprolactinoma patients 5 mg per day

1977; Carter et al. 1978; Nagulesparen et al. 1978; Franks et al. 1978; von Werder et al. 1978a, 1980a). Postoperative therapy leads to normalization of PRL levels and resumption of ovarian function if gonadotrophic tissue has not been destroyed by the surgical procedure. Male patients with persistent hyperprolactinemia and low basal testosterone levels with loss of libido and sexual potency generally show a rise in plasma testosterone following pharmacologic normalization of PRL levels (Fossati et al. 1976; Carter et al. 1978; Franks et al. 1978; Eversmann et al. 1981) and normalization of libido and sexual potency (Thorner et al. 1974, 1980a; von Werder 1975; Thorner 1977; von Werder et al. 1978a). In patients with long-standing hyperprolactinemic hypogonadism additional testosterone therapy together with bromocriptine may be required (Nagulesparen et al. 1978). In addition to the biochemical and clinical reversal of hypogonadism in males and females, unspecific symptoms such as headaches and depression may improve following normalization of PRL levels with bromocriptine (Fig. 5.13).

In mixed adenomas bromocriptine may be specifically indicated. The effectiveness of this drug in the treatment of acromegaly has been established (Thorner et al. 1975b; Schwinn et al. 1977). Those acromegalics who also have hyperprolactinemia seem in particular, to benefit from this treatment (Sivelstrini et al. 1978; Lamberts et al. 1979a; Fig.5.14). In patients with Cushing's disease due to pituitary ACTH hypersecretion bromocriptine does not seem to be very potent in lowering ACTH

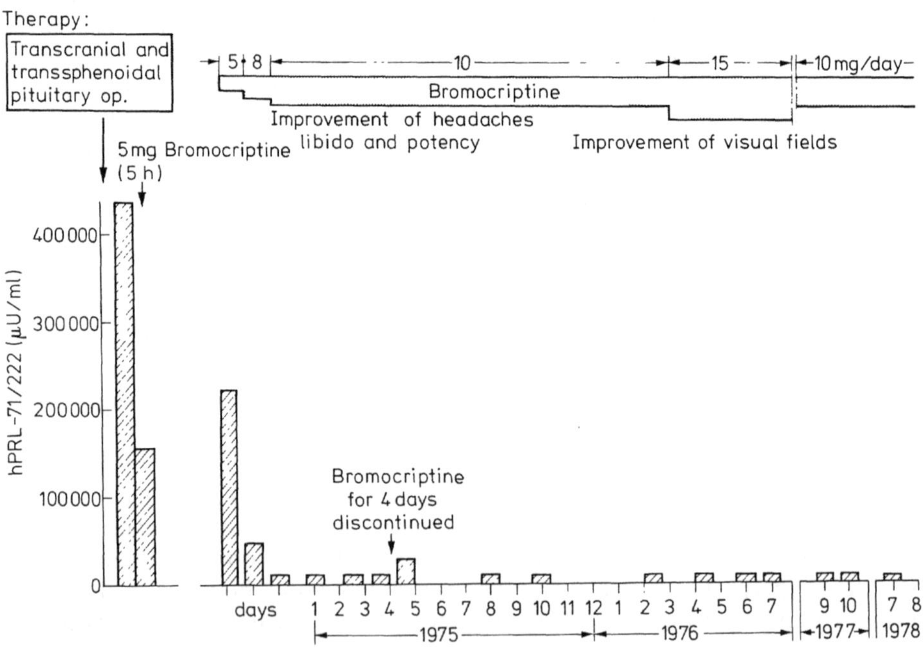

Fig. 5.13. Male patient with an invasively growing prolactinoma and extremely elevated prolactin levels after several transcranial and transsphenoidal pituitary operations. This patient had recurrent liquorrhea and despite adequate testosterone substitutional therapy, lack of libido and potency. Long-term bromocriptine therapy led to improvement of headaches, libido, and sexual potency as well as improvement of visual fields. His recurrent liquorrhea disappeared completely. This has been regarded as clinical evidence for an antiproliferative effect of bromocriptine. (von Werder et al. 1978a)

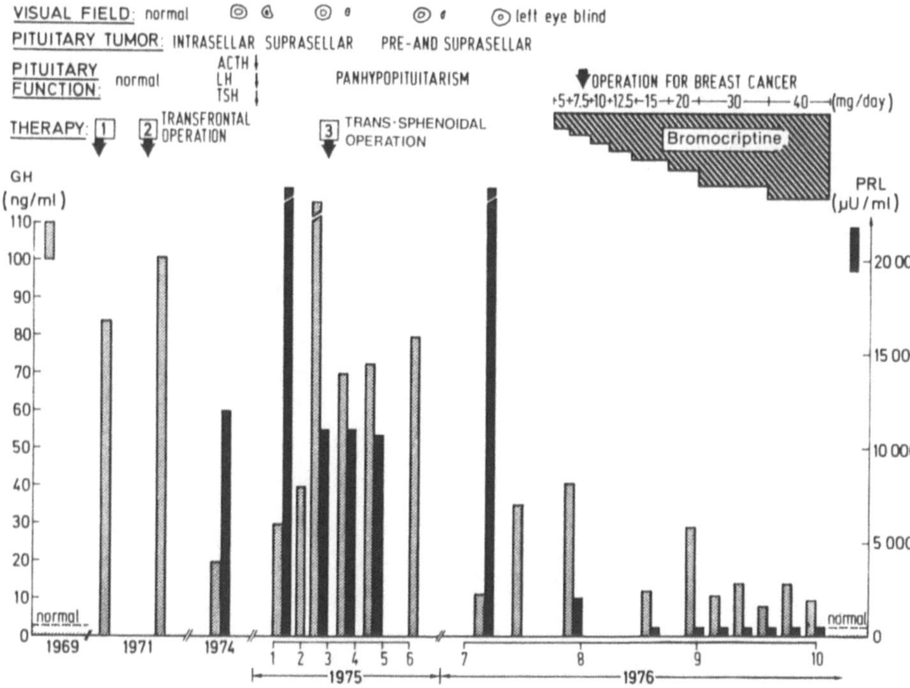

Fig. 5.14. Effect of multiple surgery and bromocriptine on prolactin and growth hormone levels in a 40-year-old patient with acromegaly and hyperprolactinemia. (von Werder et al. 1978 a)

levels (Lamberts et al. 1977b; von Werder et al., 1980a). However in patients with a pituitary adenoma producing PRL, GH and ACTH (O. A. Müller et al. 1978) bromocriptine therapy led to normalization of all three hormone levels (Fig.5.15).

5.4.1.3 Antiproliferative Effect of Ergot Alkaloid Derivatives on Prolactinoma Growth

Evidence based on clinical improvement of visual fields under bromocriptine (Friesen and Tolis 1977) and on long-term follow-up of patients with invasively growing adenomas (Fig. 5.13) suggests that bromocriptine might stop adenoma growth or lead to actual tumor shrinkage. Thus, ergot derivatives were shown to cause reduction of tumor size in transplanted pituitary adenomas (Quadri et al. 1972; MacLeod and Lehmeyer 1973). Several mechanisms may explain this antiproliferative effect. Increased PRL concentration in the lactotroph after inhibition of its secretion may cause reduced PRL synthesis by an intracellular negative feedback mechanism leading to inhibition of DNA synthesis (Davies et al. 1974; Lloyd et al. 1975). In addition, inhibition of enzymatic activity after bromocriptine was found (MacLeod and Nagy 1979). Since 3′-5′-cyclic-AMP is also implicated in stimulating PRL secretion (Lloyd et al. 1975) the generation of the cyclic nucleotide may be influenced by bromocriptine. Recent studies, however, seem to indicate that bromocriptine does not exert its intracellular effect through cyclic AMP. The direct effect of dopamine agonistic drugs on the anterior pituitary

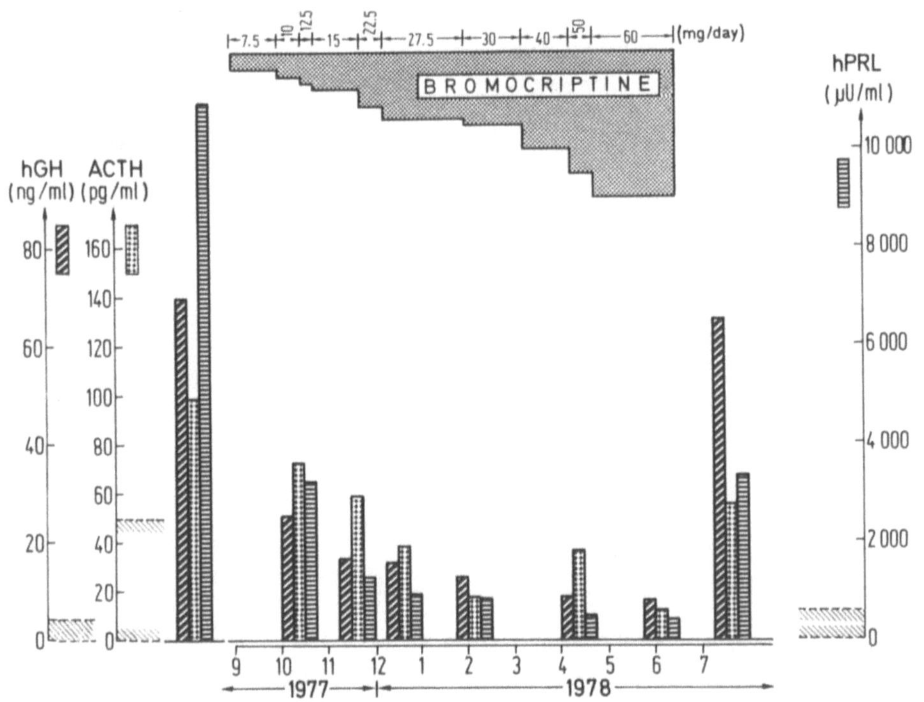

Fig. 5.15. Effect of bromocriptine on growth hormone, ACTH, and prolactin secretion in one patient with a pituitary triple hormone excess. This patient had been treated with bilateral adrenalectomy for Cushing's disease and by transfrontal pituitary surgery for acromegaly 9 years ago. He now has recurrence of acromegaly and hyperprolactinemia whereas his ACTH levels are only slightly elevated. Bromocriptine leads to a suppression of all three hormones, which rise to pretreatment level after bromocriptine withdrawal, with the exception of prolactin which again remains significantly lower compared to the pretreatment level. (von Werder et al. 1980a)

gland has also been documented by specific binding of dopamine agonists to pituitary cell membranes (Caron et al. 1978).

Recently, several case reports describing regression in size of macroprolactinomas under bromocriptine therapy have been published. In some patients reduction of adenoma size after bromocriptine therapy were documented by conventional X-ray techniques (Corenblum et al. 1975; Vaidya et al. 1977; Mühlenstedt et al. 1978; Sobrinho et al. 1978, 1981; George et al. 1979). In another case with a macroprolactinoma and PRL levels of 620 ng/ml, an operation was performed after 26 months of bromocriptine therapy. At the time of surgery no adenoma could be identified, suggesting regression of the prolactinoma during prolonged treatment with bromocriptine (Nillius et al. 1978). Regression of suprasellar extension of a pituitary macroadenoma has also been documented by computerized tomography (Fig. 5.16) and metrizamide cisternography after bromocriptine therapy (Landolt et al. 1979; McGregor et al. 1979a, b; Thorner et al. 1980b; von Werder et al. 1980a) and lisuride administration (Chiodini et al. 1980). Recently Thorner et al. (1980b) presented evidence that bromocriptine-induced reduction of tumor size may occur very rapidly. Thus, improvement of visual fields

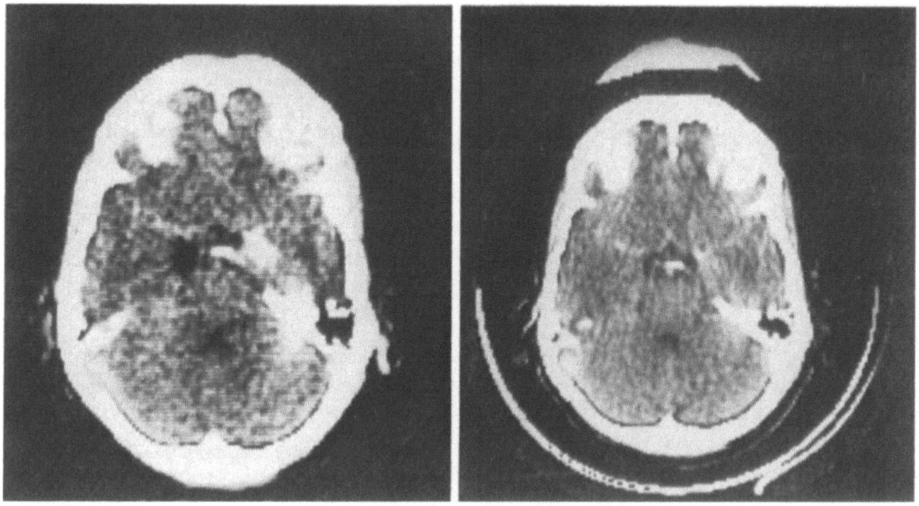

Fig. 5.16. Effect of long-term bromocriptine therapy on a macroprolactinoma with parasellar extension in male patient (M. H., 37 years). Whereas the computerized tomography with contrast enhancement shows a right parasellar adenoma (*left side*) before therapy, the computerized tomography after 2 years bromocriptine therapy (7.5 mg/day) shows no evidence of a pituitary adenoma extending into the parasellar area. The prolactin level decreased at the same time from 13082 μU/ml to 71 μU/ml. (von Werder et al. 1980a)

could be documented within 3 days and metrizamide cisternography showed objective tumor reduction within 6 weeks (Fig. 5.17).

Shrinking of adenomas can be accompanied by permanent suppression of PRL secretion. Von Werder et al. (1980b) measured PRL levels in 14 female and 15 male patients with macroprolactinomas before, during and after withdrawal of bromocriptine treatment. PRL levels after bromocriptine withdrawal were found to be significantly lower compared to pretreatment values (Fig. 5.18). This effect was less pronounced in seven females with microprolactinomas who showed no significant suppression after 1 month of bromocriptine withdrawal (von Werder et al. 1980a). Eversmann et al. (1979) investigating 37 patients with prolactinomas found persistent suppression of PRL secretion after long-term withdrawal of bromocriptine therapy in those patients with large PRL-secreting adenomas and excessively high PRL levels. In contrast, in patients with macroprolactinomas and only moderately elevated PRL levels or in patients with microprolactinomas, significant long-lasting suppression of PRL secretion was not observed.

Since basal PRL levels correlate well with the lactotrophic adenoma mass (Malarkey and Johnson 1976; Fahlbusch et al. 1978; Kelly et al. 1978; Derome et al. 1979; Klijn et al. 1980, Fig. 5.9), a persistent reduction in PRL levels after long-term withdrawal may be indicative of shrinkage of the larger prolactinomas. The efficacy of bromocriptine predominantly in large prolactinomas may be due to the increased mitotic rates and signs of enhanced proliferation, which are observed in theses adenomas (Peillon et al. 1978). The degree of persisting suppression correlated not only with the height of PRL levels before treatment but also with the duration of bromocriptine therapy whereas no correlation could be shown between PRL levels after bromocriptine withdrawal and the length of withdrawal time (Eversmann et al. 1979).

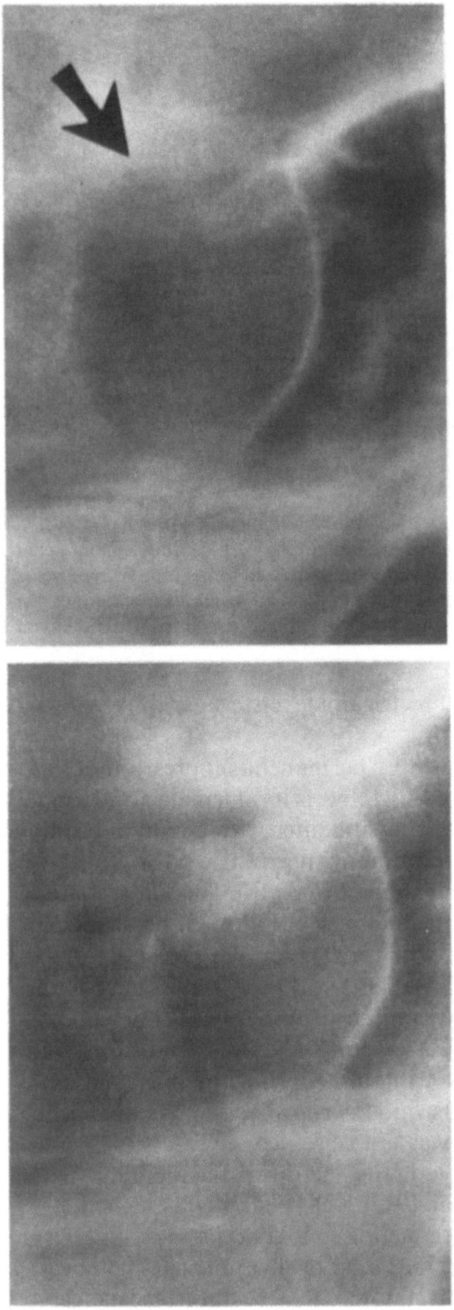

Fig. 5.17. Lateral tomographic sections after intrathecal metrizamide injection in a patient with prolactinoma. Before commencement of bromocriptine (*upper part*) and 6 weeks after starting bromocriptine (7.5 mg/day) (*lower part*). Before therapy there is upward bulging of the diaphragma sellae into the chiasmatic cistern (*arrow, on the upper part*). Six weeks after starting bromocriptine shows metrizamide within the pituitary fossa, indicating cisternal herniation due to shrinkage of the pituitary mass (*lower part*). (Thorner et al. 1980b)

It may be concluded that bromocriptine is effective in inhibiting PRL secretion and pituitary adenoma growth in most patients with large macroprolactinomas and high basal PRL levels. In contrast, no such effect may be anticipated in patients with microprolactinomas or macroprolactinomas with only moderately elevated PRL levels. The difference in behavior of these tumors may be explained by the different biology of these adenomas (see Sect. 5.1.2 and Sect. 5.3.1). However, all patients with

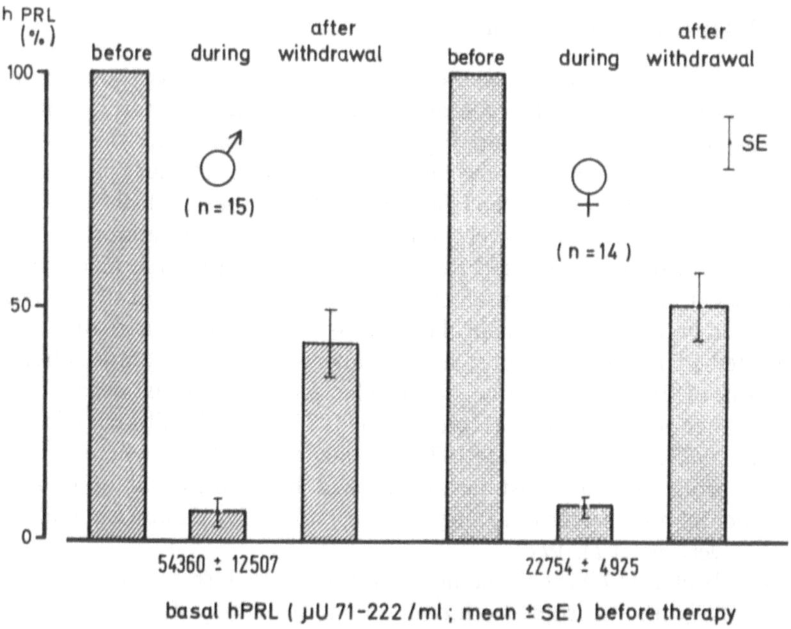

Fig. 5.18. Prolactin levels before, during and after withdrawal of bromocriptine therapy in 15 male and 14 female patients with macroprolactinomas. The decrease in prolactin levels under and after withdrawal of bromocriptine therapy is shown as a percent of the basal level before commencement of treatment. Despite the fact that bromocriptine was withdrawn for at least 1 month, PRL levels remained significantly lower, compared to the pretreatment level. (von Werder et al. 1980b)

prolactinomas and persistent hyperprolactinemia after surgery should be treated with bromocriptine. Though only macroprolactinomas with extremely elevated PRL levels will show reduction of adenoma size, all patients will benefit from clinical improvement of sexual function after normalization or lowering of PRL levels during treatment.

Clinical trials are now mandatory to find out whether patients with invasive macroprolactinomas should receive bromocriptine as a primary therapy in order to transform an invasive macroprolactinoma into a surgically resectable tumor, since the limitations of operative procedures in patients with invasive macroprolactinoma are obvious (Thorner et al. 1980a; von Werder et al. 1980a).

Though bromocriptine treatment has become an established mode of therapy in acromegaly (Thorner et al. 1975b), there is little evidence that bromocriptine also leads to shrinkage of somatotrophic adenomas. Thus, GH levels returned to the pretreatment range when bromocriptine was withdrawn (Eversmann et al. 1979). This is also observed in patients with mixed adenomas, in whom PRL levels remain persistently suppressed while GH rises to the pretreatment level (Fig. 5.15). In agreement with these findings an increase in tumor size has been observed in acromegalic patients while receiving bromocriptine therapy in GH suppressive doses (Salti and Istfan 1979). This is in contrast, however, with the observation of shrinkage of the pituitary adenoma in two acromegalics who were treated with bromocriptine only (Wass et al. 1979). In addition pituitary apoplexy under bromocriptine therapy in 2 acromegalics has been reported (Yamayi et al. 1981).

5.4.2 Operative Treatment

There is still little doubt about the mode of treatment of large prolactinomas leading to marked changes in the osseous structure of the sella turcica and suprasellar extension with rapidly developing chiasma syndrome. Since the treatment should lead to diminution of adenoma mass as soon as possible, surgical procedures seem to be the therapy of choice (Jefferson 1978; Fahlbusch et al. 1978; Guiot 1978; Derome et al. 1979). However, neurosurgical procedures almost inevitably fail to normalize extremely elevated PRL levels in cases with large macroprolactinomas (Franks et al. 1977; Fahlbusch et al. 1978; von Werder et al.1978a; Derome et al. 1979). In patients with microprolactinomas or intrasellar macroadenomas with less elevated PRL concentrations, surgical intervention results more often in normalization of PRL levels (Fig. 5.19). The latter group is usually operated by transsphenoidal approach (Hardy 1969, 1973, 1980; Fahlbusch et al. 1978, 1980; Guiot 1978; Derome et al. 1979; Arafah et al. 1981), whereas large invasive macroprolactinomas must be approached by transfrontal or combined transsphenoidal and transfrontal surgery (Jefferson 1978; Fahlbusch et al. 1978; Guiot 1978; Derome et al. 1979).

Hardy et al. (1978) achieved normalization of PRL secretion in 91% of a group of patients with microprolactinomas and in 78% of those with intrasellar macroadenomas.

Fahlbusch et al. (1980) reported recently results of a joint study of the departments of neurosurgery of the universities of Milan and Munich. Sixty percent of 58 patients with microprolactinomas in whom selective adenomectomy by

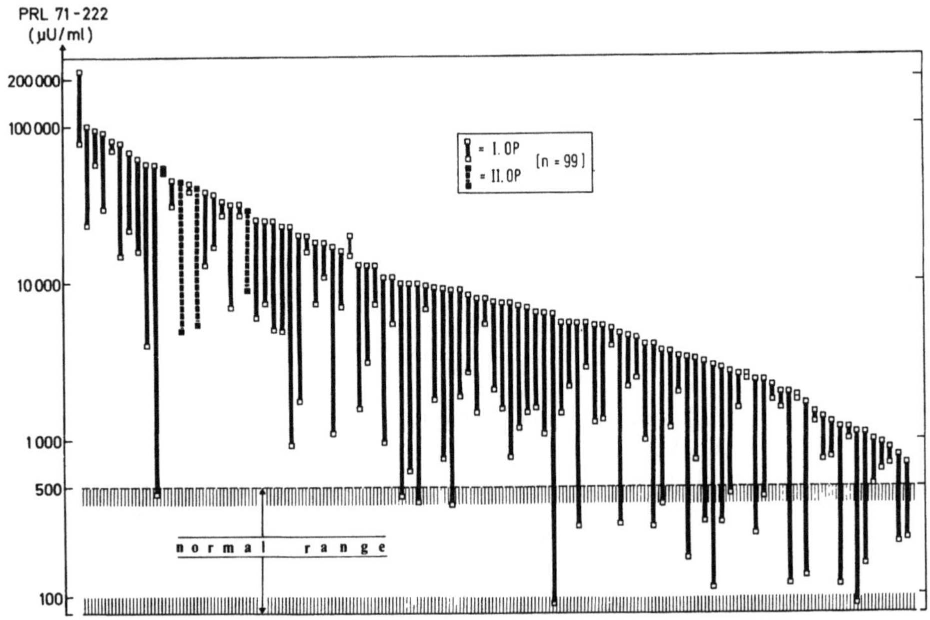

Fig. 5.19. Prolactin levels before and after pituitary surgery in 99 patients with prolactinomas. Postoperative PRL levels were measured six weeks after the operation. It becomes obvious that patients with PRL levels above 10000 μU/ml are rarely normalized by the surgical procedure

transsphenoidal route was undertaken had normal PRL levels after the operation. In the series of Derome et al. (1979) 62% of the patients with microadenomas had normal PRL levels after operation. Similar results have been reported by other investigators (Chang et al. 1977; Guitelman et al. 1977; Jaquet et al. 1978). The existing discrepancies in neurosurgical results may be due not only to the different techniques but also to the different classification of these adenomas (Derome et al. 1979; Fahlbusch et al. 1980). There is general agreement that very small microadenomas with a diameter of 3–6 mm accompanied by only moderately elevated PRL levels are almost always normalized if they are operated by transspenoidal approach (Hardy et al. 1978; Fahlbusch et al. 1978, 1980; Derome et al. 1979). Since in some centers patients with such small microadenomas are not subjected to transsphenoidal surgery but receive bromocriptine as primary therapy (von Werder et al. 1978a; Editorial 1980), the patients who come to operation at the various departments of neurosurgery present with differing PRL levels and adenoma size (Hardy et al. 1978; Fahlbusch et al. 1978; Derome et al. 1979).

As has been pointed out, in patients with macroprolactinomas and higher PRL levels, surgical results are less favorable. Whereas in one study the normalization rate of PRL levels was 61% in female microadenoma patients, it was only 40% in those with macroprolactinomas (Derome et al. 1979). The same surgeon achieved normalization in 5 out of 30 male patients, all harboring smaller adenomas. Normalization of PRL secretion in patients with macroprolactinomas may be achieved by combining transsphenoidal surgery with cryosurgical procedures (Fahlbusch et al. 1977). With this approach normalization of PRL levels has been observed (von Werder et al. 1978a), though insufficiency of anterior pituitary function is frequently encountered postoperatively (Levin et al. 1974; von Werder 1975).

PRL levels return to normal within 24 h after adenoma removal. As has been shown in 30 patients with microprolactinomas, acutely normalized PRL levels have remained normal over a follow-up period ranging from 2 to 5 years (Fahlbusch et al. 1977). However, in a patient with a macroprolactinoma with suprasellar extension, in whom PRL concentrations had been normalized by transsphenoidal surgery with additional cryosurgery, PRL levels began to increase 1 year after surgery (Fig. 5.20). This suggests that normal PRL levels after surgery in this patient with panhypopituitarism derived from residual adenoma tissue which started growing again under necessary estrogen substitutional therapy. In patients with invasive macroprolactinomas and extremely elevated PRL levels, PRL secretion will not revert to normal even if transsphenoidal and transfrontal surgery are combined (Chang et al. 1977; Guitelman 1977; Fahlbusch et al. 1978; Guiot 1978; Derome et al. 1979). In these patients additional postoperative radiotherapy to reduce residual adenoma mass has been advocated (Guiot 1978; Derome et al. 1979). So far no clear-cut data about the efficacy of postoperative radiotherapy in macroprolactinomas with persistent hyperprolactinemia are available. However, there is general agreement that further lowering or normalization of postoperative PRL secretion with bromocriptine leads to clinical improvement of persistent hypogonadism and in a large number of patients to further reduction of residual adenoma mass (see Sect. 5.4.1.3).

However, even if normalization of PRL secretion in macroprolactinomas is not achieved by surgical intervention, patients benefit from the improvement of local

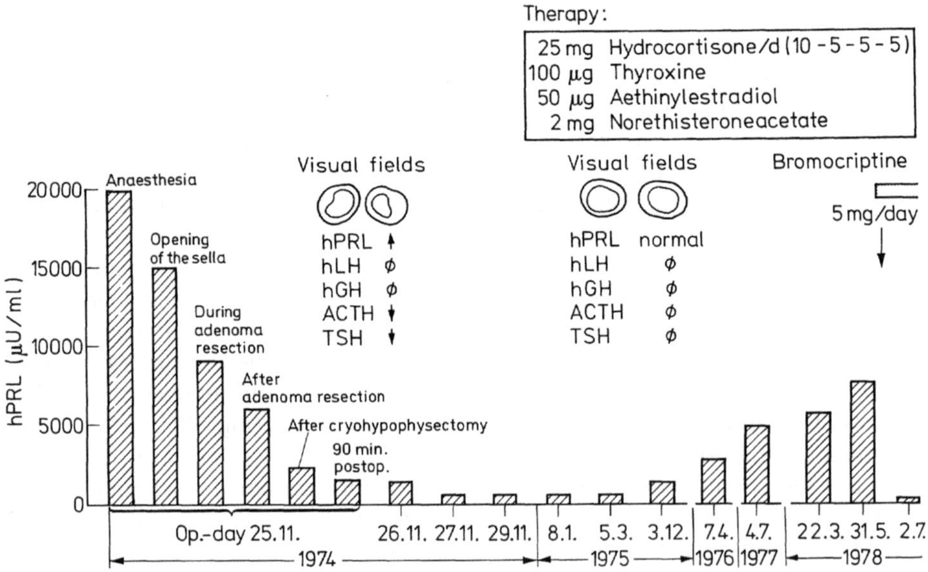

Fig. 5.20. Intra- and postoperative prolactin levels (μU A 71/222) in a 28-year-old female with a macroprolactinoma with suprasellar extension. The late rise of PRL after initial normalization signals adenoma recurrence. (von Werder et al. 1978a)

adenoma symptoms. Thus, headaches and visual field defects disappear immediately after surgery (Derome et al. 1979), whereas visual acuity improves more slowly or not at all if atrophy of the optic nerves existed before the surgical intervention (Derome et al. 1979).

An alternative procedure for the treatment of endocrine active adenomas is stereotactic Yttrium^{-90} implantation (Wright et al. 1970; Burke et al. 1972; Joplin et al. 1978; Kelly et al. 1978). In 21 female patients Yttrium^{-90} implantation of 20 000 rads was well tolerated and led to a mean fall of PRL levels by 60% of baseline values. Anterior pituitary function remained normal after the stereotactic operation. However, as with transsphenoidal surgery Yttrium^{-90} implantation fails to normalize PRL levels if these are above 200 ng/ml. Thus only 3 out of 13 females with prolactinomas desiring pregnancy conceived after Yttrium^{-90} implantation without additional therapy. Further six patients became pregnant after postimplantation bromocriptine treatment (Kelly et al. 1978).

5.4.3 Radiotherapy

External radiation of the pituitary in the treatment of endocrine active pituitary adenomas has been advocated (Roth et al. 1970). Thus primary treatment with 4 500 rads from 15-MeV linear accelerator or a radiocobalt source in females with microprolactinomas who wish to become pregnant has been recommended (Thorner et al. 1975a; Besser et al. 1977; Gomez et al. 1977; Jewelewicz and Zimmerman 1978).

Radiation has been administered as a prophylactic procedure to inhibit pituitary enlargement during pregnancy (Thorner et al. 1975a). However, the efficacy of this treatment in lowering PRL levels has not been documented, since these patients received bromocriptine 3 months after completion of external radiation (Thorner et al. 1975a; Besser et al. 1977). Therefore the overall efficacy of external radiation as primary therapy in patients with microprolactinomas is still not clear. In four patients with presumed microprolactinomas, significantly lower PRL levels were observed 2–3 months after conventional radiation. In three other patients with only moderately elevated PRL levels this effect was less pronounced (Gomez et al. 1977). Heavy particle treatment seems to be superior to conventional radiotherapy (Kliman and Kjellberg 1977) reaching the efficacy of transsphenoidal surgery in the management of microprolactinomas. This kind of treatment, however, can only be administered where the necessary equipment for the generation of accelerated heavy particles is available. In addition the long-term effects of this form of therapy on pituitary function are not yet known.

In patients with macroprolactinomas Kleinberg et al. (1977) reported substantial reduction of PRL levels following conventional external radiation. In two further patients receiving proton beam treatment, PRL levels fell to 41% of pretreatment levels (Kleinberg et al. 1977). In contrast to the rapid reduction of PRL concentrations after surgery (Fahlbusch et al. 1977) PRL levels fall slowly after pituitary irradiation (Gomez et al. 1977). The slow decrease of secretory activity of the tumorous lactotrophs after radiotherapy has also been shown in 12 patients with persistent hyperprolactinemia receiving radiotherapy after surgery (Fig. 5.21). The gradual fall of PRL levels within the first 6 months after radiotherapy is similar to the situation in acromegalic patients in whom the decrease in the GH levels proceeds slowly over several years (Roth et al. 1970). The efficacy of radiotherapy in lowering postoperative persistent hyperprolactinemia varies in individual patients. Whereas some authors find no beneficial effect of radiotherapy (Samaan et al. 1977), others have observed a positive response in one-third of the patients treated with postoperative radiotherapy (von Werder et al. 1978c). However, radiotherapy rarely leads to normalization of PRL secretion in patients with macroprolactinomas and postoperative residual elevated PRL levels (Samaan et al. 1977; Carter et al. 1978; von Werder et al. 1978c). Occasionally, further enlargement of the prolactinoma several months after radiotherapy with deterioration of visual function has been encountered (Fig. 5.22).

Landolt et al. (1979) have suggested that postoperative radiotherapy together with bromocriptine treatment might have a synergistic effect on further adenoma regression and normalization of PRL secretion. However, Eversmann et al. (1979) found no difference in the PRL levels in patients with macroprolactinomas and high postoperative PRL levels following either radiotherapy and bromocriptine or bromocriptine alone.

Because of the rather inconsistent and sometimes divergent results of conventional external radiation in the treatment of prolactinomas no definite conclusions can be drawn. However, since in individual patients beneficial responses with inhibition of PRL secretion and improvement of local tumor symptoms have been reported (Kleinberg et al. 1977; Gomez et al. 1977; Carter et al. 1978; von Werder et al. 1978c), postoperative radiotherapy of macroprolactinomas – particularly those with invasive features – together with bromocriptine treatment, seems justified (Carter et al. 1978; von Werder et al. 1978a; Derome et al. 1979).

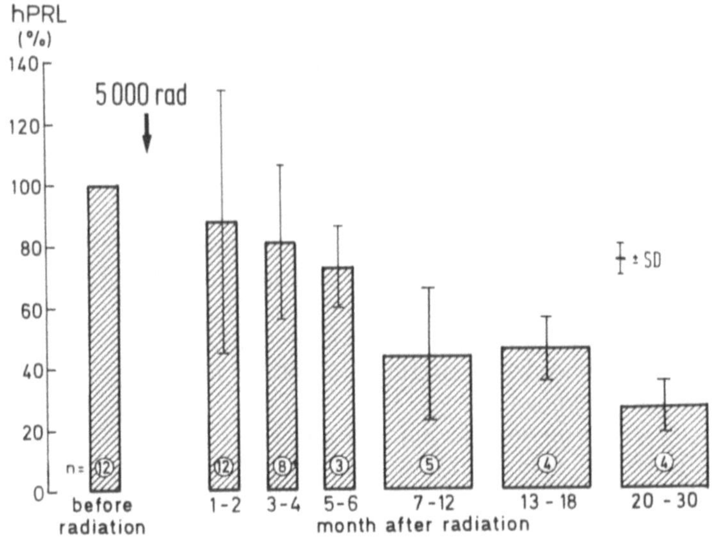

Fig. 5.21. Effect of radiotherapy (18 MeV from a supervoltage source) on basal prolactin levels (100%). All 12 patients had macroprolactinomas with postoperatively persisting hyperprolactinemia (range 4000–120000 μU/ml). (von Werder et al. 1978c)

With the exception of individual cases (Kleinberg et al. 1977), no large patient series has been reported documenting the efficacy of treatment of macroprolactinomas with heavy particles. This is in contrast to the well documented results of this treatment in a great number of patients with acromegaly and Cushings's disease (Kjellberg and Kliman 1973; Linfoot 1979).

5.5 Pregnancy and Prolactinoma

Due to amenorrhea and/or anovulation caused by hyperprolactinemia, pregnancies seldom occur spontaneously in patients with prolactinomas (Gemzell and Wang 1979). Thus, before it became possible to measure hPRL and before the introduction of transsphenoidal surgery and dopaminergic agents in the treatment of intrasellar adenomas, these patients required treatment with human gonadotropins in order to become pregnant (Gemzell 1975; Gemzell and Wang 1979). However, it is well established that the normal pituitary enlarges during pregnancy (Erdheim and Stumme, 1909; Goluboff and Ezrin 1969; Pasteels 1972). This is due to the increased estrogen levels during pregnancy which stimulate lactotrophic cell growth. However, not only the normal anterior pituitary, but also the adenomatous tissue can grow under the influence of estrogens (Peillon et al. 1970). Thus, the enlargement of pituitary tumors during pregnancy, leading to development of clinical signs such as visual field defects and headaches, have been reported (Child et al. 1975a; Falconer and Stafford-Bell 1975; Gemzell 1975; Lamberts et al. 1977a, 1979b; Gemzell and Wang 1979).

Before bromocriptine became available gonadotropins were used to induce ovulatory cycles in hyperprolactinemic females, since these patients rarely respond

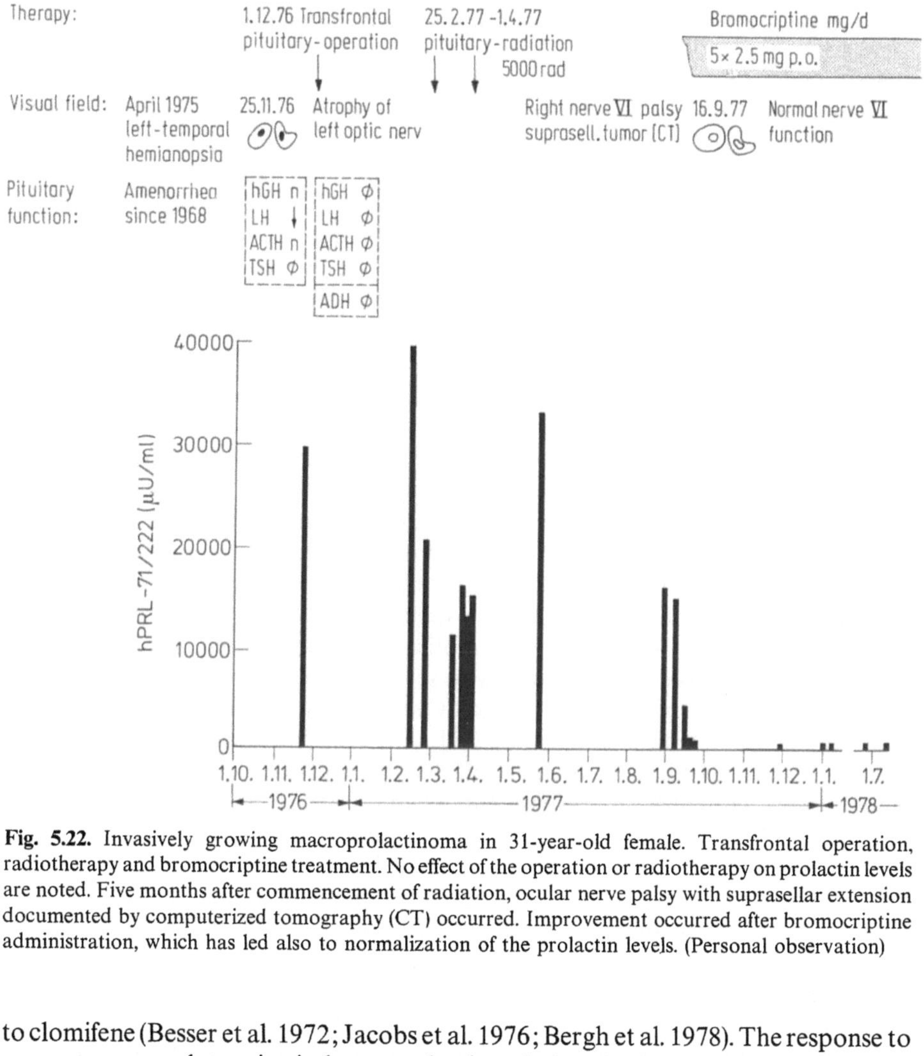

Fig. 5.22. Invasively growing macroprolactinoma in 31-year-old female. Transfrontal operation, radiotherapy and bromocriptine treatment. No effect of the operation or radiotherapy on prolactin levels are noted. Five months after commencement of radiation, ocular nerve palsy with suprasellar extension documented by computerized tomography (CT) occurred. Improvement occurred after bromocriptine administration, which has led also to normalization of the prolactin levels. (Personal observation)

to clomifene (Besser et al. 1972; Jacobs et al. 1976; Bergh et al. 1978). The response to exogenous gonadotropins in hyperprolactinemic females is not different from that of normoprolactinemia (Bergh et al. 1978). However, this treatment is not suitable for a large number of hyperprolactinemic females desiring fertility, since it requires highly specialized hospitals with adequate laboratory facilities for daily estrogen monitoring (Bergh et al. 1978). Therefore, the introduction of bromocriptine, which has been shown to normalize ovarian function in females with micro- and macroprolactinomas (Besser et al. 1972; Thorner et al. 1974; Rolland et al. 1974; Rjosk et al. 1976; von Werder et al. 1978a) dramatically increased the rate of pregnancies in hyperprolactinemic females. This has enabled many previously sterile women, some of whom harbor large macroprolactinomas, to now become pregnant (Magyar and Marshall 1978; Gemzell and Wang 1979; Griffith et al. 1979).

5.5.1 Clinical Course of Prolactinomas During Pregnancy

Women with pituitary adenomas may develop visual complications during pregnancy, which appear mostly in the second and third trimester, rarely at the beginning of pregnancy (Child et al. 1975a; Falconer and Stafford-Bell 1975; Gemzell 1975; Lamberts et al. 1977a, 1979b; Gemzell and Wang 1979; Rjosk et al. 1979a). In addition hemorrhagic phenomena inside the adenoma may lead to a sudden increase in adenoma size (Peillon et al. 1970) leading to permanent blindness (Winter 1931), cranial nerve palsy (Lamberts et al. 1979b) or fatal intracranial hemorrhage (Christensen and Larsson 1954; Rand 1957). However, the latter are scattered case reports which give no information as to the incidence of these complications. Recently reviews about the natural history of pituitary tumors in nonpregnant and pregnant females have appeared, based on data drawn from published reports and from responses to a questionnaire (Magyar and Marshall 1978; Gemzell and Wang 1979). Magyar and Marshall (1978) reported about 91 pregnancies occuring in 73 women harboring an untreated pituitary adenoma. During pregnancy headaches occurred in 26% of the women and visual field disturbances were reported in 25%, 61% remaining without complications. The mean time for the development of visual disturbances during pregnancy was 14 weeks. The development of headaches usually preceeded the occurrence of visual changes. The chance of developing complications during pregnancy was independent of parity. However, symptoms were severe enough in 14 patients to warrant surgery or radiotherapy during pregnancy. One additional patient was operated on purely on the basis of rapidly increasing PRL levels. Seven patients received surgery or radiotherapy postpartum because radiologic evidence suggested progressive tumor growth. Thus antitumor treatment due to pregnancy was required in 30% of the patients. Of all 73 patients who became pregnant without prior therapy for the pituitary adenoma, 69 had no permanent sequelae. Only four patients who were operated on and received radiotherapy developed permanent sequelae such as mild visual field defects. In contrast 78 pregnancies were observed in 73 females whose prolactinomas had been treated previously. In these patients only 4% developed headaches and 5% visual disturbances, only one of whom required treatment. The abortion rate or perinatal mortality were not increased by the treatment of the pituitary adenoma during pregnancy though prematurity rates were found to be increased.

Gemzell and Wang (1979) reviewed the outcome of 217 pregnancies in 187 patients with pituitary adenomas. In patients with microadenomas 94.5% of all pregnancies were uneventful. However in patients with previously untreated macroprolactinomas the complication rate during pregnancy was considerably higher. Thus in 25% visual disturbances and headaches, and in 8.9% headaches were experienced during pregnancy. In this group the majority already had some symptoms in the first trimester of pregnancy.

Griffith et al. (1979) reported about 137 patients, most of them harboring prolactinomas. Of the 116 completed pregnancies in these patients, eight were accompanied by visual disturbances and in nine pregnancies severe headaches were described. With the exception of craniotomy performed in two cases and bromocriptine reinstitution in one during pregnancy, no further intervention was undertaken. However, the results of these large series have to be interpreted with

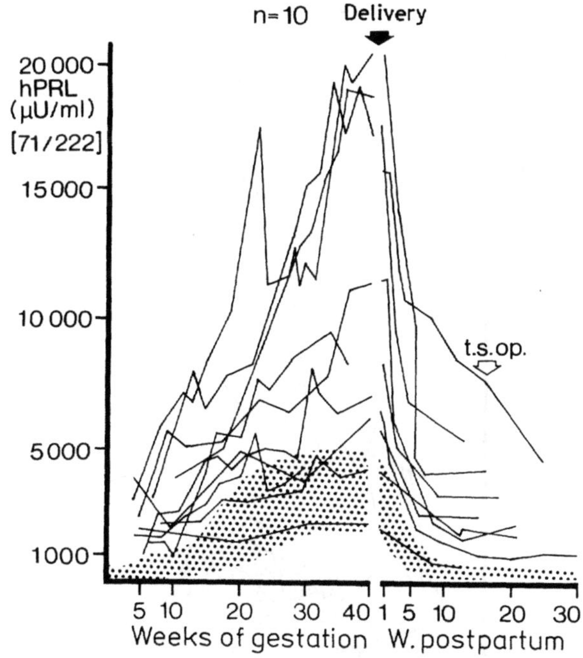

Fig. 5.23. Prolactin levels (µU A 71/222) during pregnancy in ten patients with prolactinomas, who were not operated on before induction of ovulation with bromocriptine. The *shaded area* represents the normal range of prolactin levels during pregnancy. (Rjosk et al. 1980)

caution since they derive from a heterogenous group with varying elevations of PRL levels and adenoma size and different forms of treatment before pregnancy.

From the data of various investigators reporting the outcome of pregnancies in patients with comparable adenoma sizes, it can be deduced that complications are found predominantly in those patients, who harbor a large adenoma which was documented by X-ray and high PRL levels before pregnancy (Child et al. 1975a; Bergh et al. 1978; Rjosk et al. 1978). Thus, patients with intrasellar adenomas and no evidence of suprasellar extension before pregnancy experienced no adverse symptoms during pregnancy if pregnancy was achieved by bromocriptine treatment only (Rjosk et al. 1976, 1979b; Bergh et al. 1978; Mornex et al. 1978). PRL levels may differ widely in individual patients during pregnancy (see Figs. 5.23–5.25). Thus, excessive increases (Fig. 5.25), no change in PRL levels during pregnancy, and resolution of hyperprolactinemia after pregnancy (Cowden and Thomson 1979) have been reported. In general PRL levels are in the same range as before pregnancy within 3 weeks post partum, regardless of the PRL increase during pregnancy (Figs. 5.23–5.25).

5.5.2 Treatment of Females with Prolactinomas Seeking Fertility

5.5.2.1 Surgery

There is general agreement that patients with macroprolactinomas extending into the sphenoid sinus and into the suprasellar area should be operated on before pregnancy is induced (Hardy 1973; Chang et al. 1977; Fahlbusch et al. 1978; L'Hermite et al. 1978; Derome et al. 1979; Gemzell and Wang 1979).

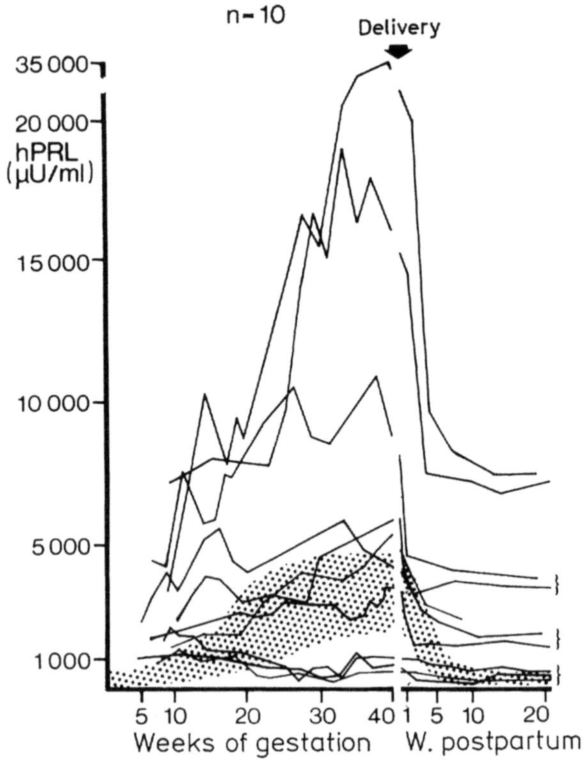

Fig. 5.24. Prolactin levels (μU A 71/222) during pregnancy in ten females who had been operated on by transsphenoidal approach. The *shaded area* represents the normal range of prolactin levels during pregnancy. (Rjosk et al. 1980)

A high normalization rate has been described by Hardy (1973) and Gomez et al. (1977). Chang et al. (1977) reported on 17 out of 20 hyperprolactinemic patients with prolactinomas who experienced normal menstruation after transsphenoidal microsurgery. This is in contrast to a report by Kleinberg et al. (1977) in which only 2 out of 15 prolactinoma patients resumed menstruation and became pregnant after transsphenoidal surgery. The possible explanation for the discrepancy in the operative results has been discussed elsewhere (pages 194, 195). Since in most patients with large prolactinomas, postoperative persistent hyperprolactinemia can be normalized by bromocriptine treatment, pregnancy rates are considerably higher, if gonadotropic function has remained intact after surgery (Rjosk et al. 1978, 1979b; von Werder et al. 1978a). Sometimes large doses of bromocriptine given over a long period of time are necessary to normalize high postoperative PRL levels (Fig. 5.25).

As an alternative to transsphenoidal surgery Yttrium^{-90} implantation has been proposed (Child et al. 1975a; Kelly et al. 1978). Additional postimplantation bromocriptine therapy increases the chances of becoming pregnant (Kelly et al. 1978).

In patients with large macroprolactinomas transsphenoidal surgery or Yttrium^{-90} implantation is not performed with the aim of normalizing PRL secretion but reducing lactotrophic growth potential (Rjosk et al. 1976; Kelly et al. 1978; Fahlbusch et al. 1978; von Werder et al. 1978a). If normalization of PRL secretion was intended radical surgery would be necessary. However, this procedure

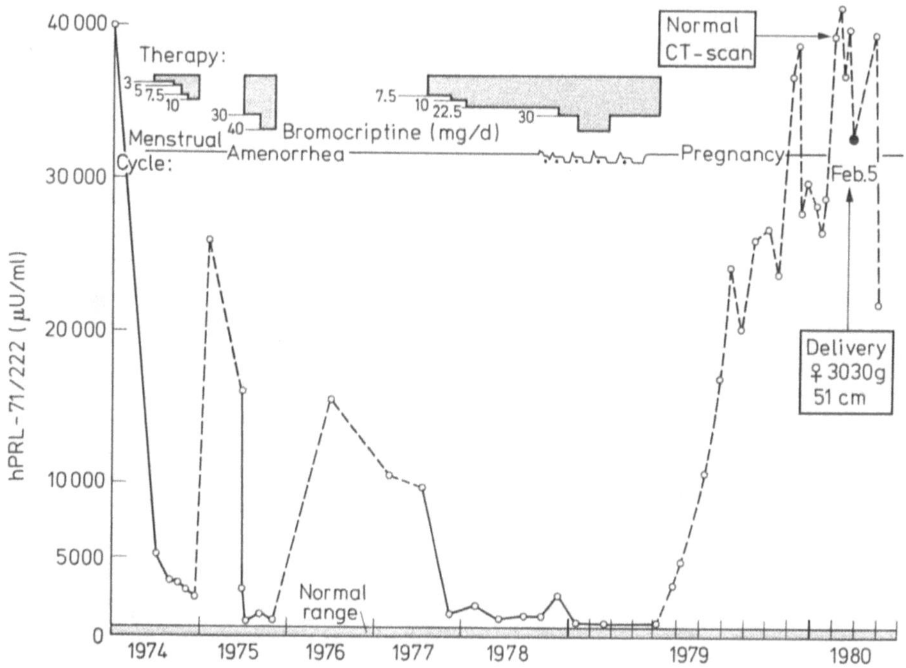

Fig. 5.25. Twenty-six-year-old female patient with macroprolactinoma and persistent hyperprolactinemia after transsphenoidal surgery. This patient had been treated for several months with bromocriptine in a dosage up to 40 mg per day without achieving normalization of postoperatively persistent hyperprolactinemia. However, PRL levels after withdrawal of the second bromocriptine treatment did not increase up to pretreatment level. A third course of bromocriptine treatment led to normalization of PRL secretion and a normal ovulatory menstrual cycle, documented by a clinically uneventful pregnancy with, however, an excessive increase in PRL levels, which fell into the pretreatment range four weeks after delivery. (Modified after von Werder et al. 1980b)

could endanger gonadotropic function making postoperative pregnancies impossible despite normal PRL levels (von Werder et al. 1978a; Fahlbusch et al. 1978, 1980; Kelly et al. 1978; Derome et al. 1979). However, if gonadotropic function is restored after surgery, normalization of PRL levels then leads to normal ovulatory cycles and pregnancies.

5.5.2.2 Medical Therapy

Dopaminergic agents can restore PRL secretion to normal in hyperprolactinemic females, even in those with large prolactinomas. Though pregnancies in these patients may be feasible with medical therapy alone, the chance of developing complications during pregnancy (Child et al. 1975a; Bergh et al. 1978; Magyar and Marshall 1978; Gemzell and Wang 1979; Rjosk et al. 1979a) demands combining medical therapy with prior treatment directed against the adenoma. Thus in addition to surgery, radiotherapy as a primary treatment has been advocated (Thorner et al. 1975a; Besser et al. 1977). However, prior radiotherapy does not exclude development of visual field defects (Thorner et al. 1975a; Lamberts et al. 1977a, 1979b). The latter complication is more likely to occur in females with large macroprolactinomas, who have received bromocriptine only in order to induce ovulation (Bergh et al. 1978; L'Hermite et al. 1978; Magyar and Marshall 1978). Most patients will recover after termination of pregnancy without permanent sequelae (Child et al. 1975a; Magyar and Marshall 1978; Bergh et al. 1978; Gemzell and Wang 1979), though long-term complications cannot be excluded (Magyar and Marshall 1978; Griffith et al. 1979). Fatal outcome is extremely rare (Magyar and Marshall 1978).

In contrast, in patients with microprolactinomas medical treatment with dopamine agonistic substances seems to be a safe procedure leading to a high pregnancy rate and no tumor-associated complications (Rjosk et al. 1976, 1978, 1979b; Bergh et al. 1978; Mornex et al. 1978; Griffith et al. 1979). However, even in these patients no general agreement concerning their management has been reached (Jewelewicz and Zimmerman 1978). Whereas neurosurgeons generally advocate that all females including those with minute changes of the sella turcica suggesting the presence of a small microadenoma, should be operated on (Chang et al. 1977; Derome et al. 1979; Hardy 1973; Hardy et al. 1978) other physicians recommend primary radiotherapy (Thorner et al. 1975a; Besser et al. 1977; Kleinberg et al. 1977). Endocrinologists with experience with Yttrium^{-90} implantation favor the stereotactic procedure before allowing pregnancy to occur (Kelly et al. 1978), whereas those groups who have access to proton beam radiotherapy (Kliman and Kjellberg 1977; Linfoot 1979) use heavy particle treatment to reduce the adenomatous growth potential. The considerable differences in therapeutic approach can be explained by the different therapeutic tools available in institutions where prolactinoma patients are seen.

At the University of Munich the following policy has been adopted (Rjosk et al. 1976, 1978, 1979b; von Werder et al. 1977, 1978a, 1980a, b; Fahlbusch et al. 1978, 1980): All patients with hyperprolactinemia who desire pregnancy are seen by the gynecologist, neurosurgeon and endocrinologist together. If radiologic evidence of a macroprolactinoma is found, computerized tomography and visual field examination is performed to clarify if suprasellar extension is present. In addition, evaluation of pituitary function is performed in all patients with radiologic evidence of a prolactinoma. All those patients who have PRL levels below 4000 μU or 200 ng/ml, an intrasellar adenoma without suprasellar extension, and no impairment of anterior pituitary function, are treated with bromocriptine only. In patients with higher PRL levels, evidence for destruction of the sella turcica, or suprasellar extension, or partial anterior pituitary insufficiency, transsphenoidal surgery is advocated. The main aim is reduction of adenomatous growth potential and preservation of gonadotropic function and not necessarily normalization of PRL levels, since these can be normalized by the postoperative bromocriptine therapy. Following this policy 82 pregnancies without any complications associated with the pituitary tumor (Table 5.1) have been followed.

After the problem arising from pituitary adenoma growth during pregnancy, the possible risks of teratogenecity of bromocriptine has to be considered. According to the available documentation on 1410 pregnancies in women who conceived while taking bromocriptine the incidence of fetal abnormalities recorded was not different from that of the untreated population (Turkalj et al. to be published). Up to now follow-up data on 206 children born to these women have revealed normal development 2 years after birth. Thus, it may be assumed that bromocriptine has no teratogenic effect although evidence is still being accumulated.

Table 5.1. Eighty-two pregnancies in 62 patients with hyperprolactinemia. (Rjosk et al., Acta Endocrinol, in press)

Diagnosis (pregnancies/patients)	Has not come to term	Aborted	Children	Complications associated with the pituitary tumor
Normal sella turcica (34/28)	2	6	26	0
Prolactinoma (not operated) (25/20)	1	3	21	0
Prolactinoma (operated) (22/13)	3	1 1 mole	17	0
Suprasellar tumor (1/1)	0	0	1	0
Total (82/62)	6	11	65	0

One patient is represented in two groups.

5.5.2.3 Treatment of Local Tumor Complications During Pregnancy

As was mentioned above, pituitary adenoma growth during pregnancy may be anticipated if patients with macroprolactinomas become pregnant without prior treatment of the prolactinoma. The clinical signs caused by the enlargement of the prolactinoma are headaches and visual field disturbances (Thorner et al. 1975a; Child et al. 1975a; Lamberts et al. 1977a, 1979b; Magyar and Marshall 1978; Bergh et al. 1978; Griffith et al. 1979). In order to detect these symptoms as early as possible, all hyperprolactinemic females with evidence of a prolactinoma must be monitored carefully during pregnancy. Thus, monthly PRL measurements during pregnancy (von Werder et al. 1977; Rjosk et al 1979b) and visual field examinations (Rjosk et al. 1976, 1979b; von Werder et al. 1977; Jewelewicz and Zimmerman 1978) at monthly intervals starting in the second trimester are recommended. If headaches or visual symptoms appear the patients should be admitted to the hospital and CT should be performed. Emergency surgery to preserve visual function may be necessary (Kajtar and Tomkin 1971; Emperaire et al. 1972; Falconer and Stafford-Bell 1975). Since the pregnancy is not dependent on an intact pituitary, the outcome of this procedure is usually favorable for mother and child. If visual problems are less severe, large doses of glucocorticoids which lower estrogen levels and may reduce swelling and pressure on the optic chiasm, may be administered with therapeutic success (Jewelewicz et al. 1977). Implantation with Yttrium^{-90} during pregnancy has also been performed without adverse effects on the pregnancy (Burke et al. 1972; Child et al 1975a).

Recently, beneficial effects of bromocriptine treatment of prolactinoma complications during pregnancy have been reported. Thus Bergh et al. (1978) observed rapid regression of visual field defects during bromocriptine treatment in one hyperprolactinemic pregnant patient. However, HoYuen (1978) reported one hyperprolactinemic female who developed bitemporal visual field defects in the second half of pregnancy, though she had been treated during the entire pregnancy with bromocriptine. This is supported by an observation of Rjosk et al. (1979a), who observed further deterioration of vision under bromocriptine therapy. In their patient rapid restitution of the visual fields occurred only after premature caesarean section. That induction of labor leading to a rapid fall in circulating estrogens may reverse visual field defects has also been shown by Thorner et al. (1975a).

Though adenoma-related complications occur rarely during pregnany in hyperprolactinemic females harboring prolactinomas and are susceptible to surgical and sometimes medical treatment, adequate planning and interdisciplinary management of each individual patient should make prolactinoma-associated complications an extremely rare event during pregnancy.

5.6 References

Ahumada JC, del Castillo EB (1932) Sobre un caso de galactorrea y amenorrea. Bol Soc Ginecol Obstet 11:64–66

Annegers JF, Coulam CB, Abbond CF, Laws ER, Kurland LT (1978) Pituitary adenoma in Olmsted county, Minnesota, 1935–1977. Mayo Clin Proc 53:641–643

Antunes JL, Housepian EM, Frantz AG, Holub DA, Hai RM, Carmel PW, Quest DO (1977) Prolactin secreting pituitary tumors. Ann Neurol 2:148–158

Arafah BM, Manni A, Brodkey JS, Kaufman B, Velasco M, Pearson OH (1981) Cure of hypogonadism after removal of prolactin-secreting adenomas in men. J Clin Endocrinol Metab 52:91–94

Argonz J, del Castillo EB (1953) A syndrome characterized by estrogenic insufficiency, galactorrhea and decreased urinary gonadotropin. J Clin Endocrinol 13:79–87

Asfour M, L'Hermite M, Hedouin-Quincampoix M, Fossati P (1977) Hypogonadism, galactorrhea and hyperprolactinemia: Evaluation of pituitary gonadotrophins reserve before and under treatment. Acta Endocrinol (Copenh) 84:738–749

Assies J, Schellekens APM, Touber JL (1978) Prolactin in human cerebrospinal fluid. J Clin Endocrinol Metab 46:576–586

Assies J, Schellekens APM, Touber JL (1978) Protein hormones in cerebrospinal fluid: Evidence for retrograde transport of prolactin from the pituitary to the brain in man. Clin Endocrinol 8:487–491

Aubert ML, Grumbach MM, Kaplan SL (1975) The ontogenesis of human fetal hormones. III. Prolactin. J Clin Invest 56:155–164

Ballard HS, Frame B, Hartsock RJ (1964) Familial multiple endocrine adenoma: peptic ulcer complex. Medicine (Baltimore) 43:481–516

Bailey P (1932) Tumors of the hypophysis cerebri. In: Penfield W (ed) Cytology and pathology of the nervous system. Hueber, New York, pp 905–951

Banna M, Nicholas W, McLachelin M (1978) The borderline pituitary fossa in patients with amenorrhea and galactorrhea. Neuroendocrinology 16:440–442

Bar RS, Mazzaferri EL, Malarkey WB (1975) Primary empty sella, galactorrhea, hyperprolactinemia and renal tubular acidosis. Am J Med 59:863–866

Baumann G, Loriaux DL (1976) Failure of endogenous prolactin to alter renal salt and water excretion and adrenal function in man. J Clin Endocrinol 43:643–649

Bergh T, Nillius SJ, Wide L (1977) Hyperprolactinemia in amenorrhea, incidence and clinical significance. Acta Endocrinol (Copenh) 86:683–694

Bergh T, Nillius SJ, Wide L (1978) Clinical course and outcome of pregnancies in amenorrhoeic women with hyperprolactinemia and pituitary tumors. Br Med J 2:875–880

Besser GM (1976) The pituitary fossa – normal or abnormal? Br J Radiol 49:652–653

Besser GM, Parke L, Edwards CRW, Forsyth IA, McNeilly AS (1972) Galactorrhea: Successful treatment with reduction of plasma prolactin levels by Brom-ergocryptine. Br Med J 3:669–672

Besser GM, Thorner MO, Wass JAH (1977) Hyperprolactinemia-hypogonadism syndrome – medical treatment. In: James VHT (ed) Endocrinology, vol II. Excerpta Medica, Amsterdam Oxford, pp 353–357

Bloodworth JMB, Kovacs K, Horvath E (1979) Light and electron microscopy of pituitary tumors. In: Linfoot JA (ed) Recent advances in the diagnosis and treatment of pituitary tumors. Raven, New York, pp 141–159

Bohnet HG, Dahlen HG, Wuttke W, Schneider HPG (1976) Hyperprolactinemic anovulatory syndrome. J Clin Endocrinol 42:132–143

Boyar RM, Kapen S, Finkelstein JW, Perlow M, Sassin JF, Fukushima DK, Weitzmann ED, Hellman L (1974) Hypothalamic-pituitary function in diverse hyperprolactinemic states. J Clin Invest 53:1588–1598

Boyar RM, Kapen S, Weitzmann ED, Hellman L (1976) Pituitary microadenoma and hyperprolactinemia. N Engl J Med 294:263–265

Brendel C (1978) Erfolge in der Behandlung von Prolactin produzierenden Tumoren. Medical dissertation, University of Munich

Bruneton JN, Drouillard JP, Sabatier JC, Elie GP, Tavernier JF (1979) Normal variants of the sella turcica. Radiology 131:99–104

Buchborn E, Irmscher K (1969) Diabetes insipidus bei Tumoren der Hypophyse. In: Kracht J (ed) Oestrogene, Hypophysentumoren. Springer, Berlin Heidelberg New York, pp 256–262

Burke CW, Joplin GF, Fraser TR (1972) Pituitary tumor treated by pituitary implantation of Yttrium-90 during or before pregnancy (two cases). Proc R Soc Med 65:486–488

Burrow GN, Wortzman G, Rewcastle NB, Holgate RC, Kovacs K (1981) Microadenomas of the pituitary and anormal sellar tomograms in an unselected autopsy series. New Engl J Med 304: 156–158

Buvat J, Thomas K, Racadot A, Blacker C, Buvat-Herbant M, Ferin F, Linquette M (1978) Changes in pituitary gonadotropins during amenorrhea-galactorrhea syndrome due to sulpiride. Clin Endocrinol (Oxf) 9:499–504

Carlson HE, Levine GA, Goldberg NJ, Hershman JM (1978) Hyperprolactinemia in multiple endocrine adenomatosis. type 1. Arch Intern Med 138:1807–1808

Caron MC, Beaulieu M, Raymond V, Gagné B, Drouin J, Lefkowitz RJ, Labrie F (1978) Dopaminergic receptors in the anterior pituitary gland. J Biol Chem 252:2244–2253

Carter JN, Tyson JE, Tolis G, vanVliet S, Faiman C, Friesen HG (1978) Prolactin-secreting tumors and hypogonadism in 22 men. N Engl J Med 299:847–852

Catania A, Cantalamessa L, Reschini E (1976) Plasma prolactin response to luteinizing hormone releasing hormone in acromegalic patients. J Clin Endocrinol 43:689–691

Chang RJ, Keye WR, Young JR, Wilson CB, Jaffe RB (1977) Detection, evaluation, and treatment of pituitary microadenomas in patients with galactorrhea and amenorrhea. Am J Obstet Gynecol 128:356–363

Chiari J, Braun C, Spaeth J (1855) Klinik der Geburtshilfe und Gynäkologie. Enke, Erlangen, pp 371–372

Child DF, Gordon H, Mashiter K, Joplin GF (1975 a) Pregnancy, prolactin and pituitary tumors. Br Med J 4:87–89

Child DF, Nader S, Mashiter K, Kjeld M, Banks L, Fraser TR (1975 b) Prolactin studies in functionless pituitary tumors. Br Med J 1:604–606

Chiodini PG, Liuzzi A, Verde G, Cozzi R, Silvestrini F, Marsili MT, Horowski R, Passerini F, Luccarelli G, Borghi PG (1980) Size reduction of a prolactin secreting adenoma during long term treatment with the dopamine agonist lisuride. Clin Endocrinol (Oxf) 12:47–51

Christensen E, Larson H (1954) Fatal subarachnoid hemorrhages in pregnant women with intracranial and intramedullary vascular malformations. Acta Psychiatr Neurol Scand 29:441–451

Cleary RE, Crabtree R, Lemberger L (1975) The effect of lergotrile on galactorrhea and gonadotropin secretion. J Clin Endocrinol 40:830–833

Clemens JA, Shaar CJ (1980) Control of prolactin secretion in mammals. Fed Proc 39:2588–2592

Corenblum B, Webster BR, Mortimer CB, Ezrin C (1975) Possible anti-tumor effects of 2 bromoergocryptine (CB-154, Sandoz) in 2 patients with large prolactin-secreting pituitary adenomas. Clin Res 23:614A

Costello RJ, (1936) Subclinical adenoma of the pituitary gland. Am J Pathol 12:205–215

Cowden EA, Thomson JA (1979) Resolution of hyperprolactinemia after bromocriptine-induced pregnancy. Lancet 1:613

Cowden EA, Thomson JA, Doyle D, Ratcliffe JG, Macpherson P, Teasdale GM (1979) Tests of prolactin secretion in diagnosis of prolactinomas. Lancet 1:1155–1158

Cowie AT, Forsyth IA, Hart IC (1980) Hormonal control of lactation. Springer, Berlin Heidelberg New York (Monographs on Endocrinology, Vol. 15)

D'Agata R, Scapagnini U (1979) Effect of bromocriptine (CB-154) on estrogen induced prolactin release. Acta Endocrinol (Copenh) 90:193–197

Danoff BR, Pripstein S, Croce N, Kramer S, Lee KF (1978) The value of computerized tomography in delineating suprasellar extension of pituitary adenoma for radiotherapeutic management. Cancer 42:1066–1072

Davidoff LM (1926) Studies in acromegaly. III. The anamnesis and symptomatology in one hundred cases. Endocrinology 10:461–463

Davies C, Jacobi J, Lloyd HM, Meares JD (1974) DNA synthesis and the secretion of prolactin and growth hormone by the pituitary gland of the male rat: Effects of diethylstilbeostrol and 2 bromo-alpha-ergocryptine methane-sulphonate. J Endocrinol 61:411–417

De Cecco L, Foglia G, Ragni N, Rossato P, Venturini PL (1978) The effect of lisuride hydrogen maleate in the hyperprolactinemia-amenorrhea-syndrome: Clinical and hormonal responses. Clin Endocrinol (Oxf) 9:491–498

Deck MDF (1973) Radiographic and radioisotopic techniques in diagnosis of pituitary tumors. In: Kohler PO, Ross GT (eds) Diagnosis and treatment of pituitary tumors. Excerpta Medica, Amsterdam, pp 71–85

Dellmann H-D, Johnson JA, Klachko DM (eds) (1977) Comparative endocrinology of prolactin. Plenum, New York London

Derome PJ, Peillon F, Bard RH, Jedynak CP, Racadot J, Guiot G (1979) Adénomas à prolactine: résultats du traitement chirurgical. Nouv Presse Med 8:577–583

Domingue JN, Wing SD, Wilson CB (1978) Coexisting pituitary adenomas and partially empty sellas. J Neurosurg 48:23–28

Doyle F, McLachlan M (1977) Radiological aspects of pituitary-hypothalamic disease. Clin Endocrinol (Oxf) 6:53–81

Dubois PJ, Orr DP, Hoy RJ, Herbert DL, Heinz ER (1979) Normal sellar variations in frontal tomograms. Radiology 131:105–110

Du Boulay G (1976) The pituitary fossa – normal or abnormal. Br J Radiol 49:653

Editorial (1980) Hyperprolactinemia: Pituitary tumor or not? Lancet 2:517–519

Edwards CRW, Forsyth IA, Besser GM (1971) Amenorrhea, galactorrhea and primary hypothyroidism with high circulating levels of prolactin. Br Med J 3:462–464

Edwards CRW, Thorner MO, Miall PA, Al-Dujaili EAS, Hanker JP, Besser GM (1975) Inhibition of the plasma-aldosterone response to frusemide by bromocriptine. Lancet 2:903–904

El-Etreby MF, Günzel P (1973) Prolaktinzell-Tumoren im Tierexperiment und beim Menschen. Arzneim Forsch 23:1768–1790

Emperaire JC, Riemens V, Dubecq JJ, Palmade J, Leuret JP (1972) Hypophysectomie d'urgence à deux mois de grossesse après induction de l'ovulation. Bordeaux Med 15:1901–1904

Erdheim J, Stumme E (1909) Über die Schwangerschaftsveränderung in der Hypophyse. Beitr Pathol Anat 46:1–132

Eversmann T, Fahlbusch R, Rjosk HK, von Werder K (1979): Persisting suppression of prolactin secretion after longterm treatment with bromocriptine in patients with prolactinomas. Acta Endocrinol (Copenh) 92:413–427

Eversmann T, Fahlbusch R, Rjosk HK, von Werder K (1980) Hyperprolactinemic infertility. In: Salvadori B, Semm K, Vadora E (eds) Fertility and sterility. Proceedings of the Vth European congress on sterility and fertility. C.I.C. Edizioni Internationali Gruppo Editoriale Medico, Rome, pp 140–142

Eversmann T, Eichinger R, Fahlbusch R, Rjosk HK, von Werder K (1981) Die Hyperprolaktinämie beim Mann: Klinik und Therapie. Schweiz Med Wschr 111 (in press)

Eversmann T, Gottsmann M, Uhlich E, Ulbrecht G, von Werder K, Scriba PC (1978) Increased secretion of growth hormone, prolactin, antidiuretic hormone and cortisol induced by the stress of motion sickness. Aviat Space Environ Med 49:53–57

Faglia G, Beck-Peccoz P, Ferrari C, Travaglini P, Ambrosi B, Spada A (1973) Plasma growth hormone response to thyreotropin releasing hormone in patients with active acromegaly. J Clin Endocrinol 36:1259–1262

Faglia G, Giovanelli MA, MacLeod RM (eds) (1980a) Pituitary microadenomas. Academic Press, London

Faglia G, Moriondo P, Beck-Peccoz P, Travaglini P, Ambrosi B, Spada A, Nissim M (1980b) Use of neuroactive drugs and hypothalamic regulatory hormones in the diagnosis of hyperprolactinemic states. In: Müller EE (ed) Neuroactive drugs in endocrinology: Physiologic, diagnostic and therapeutic applications. Elsevier/North-Holland, Amsterdam, pp 263–278

Fahlbusch R (1978) Endokrine Funktionsstörungen bei cerebralen Prozessen. Thieme, Stuttgart

Fahlbusch R, von Werder K (eds) (1978) Treatment of pituitary adenomas. Thieme, Stuttgart

Fahlbusch R, Grumme T, Aulich A, Wende S, Steinhoff H, Lanksch W, Kazner E (1976) Suprasellar tumors in the CT scan. In: Lanksch W, Kazner E (eds) Cranial computerized tomography. Springer, Berlin Heidelberg New York, pp 114–127

Fahlbusch R, Rjosk HK, von Werder K (1977) Perioperative prolactin levels in patients with prolactinomas. Acta Endocrinol [Suppl] (Copenh) 208:46–47

Fahlbusch R, Rjosk HK, von Werder K (1978) Operative treatment of prolactin producing adenomas. In: Fahlbusch R, von Werder K (eds) Treatment of pituitary adenomas. Thieme, Stuttgart, pp 225–237

Fahlbusch R, Giovanelli MA, Crosignani PG, Faglia G, Rjosk HK, von Werder K (1980) Differentiated therapy of microprolactinomas; significance of transsphenoidal adenomectomy. In: Faglia G, Giovanelli MA, MacLeod RM (eds) Pituitary microadenomas. Academic Press, London, 443–456

Falconer MA, Stafford-Bell MA (1975) Visual failure from pituitary and parasellar tumors occuring with favourable outcome in pregnant women. J Neurol Neurosurg Psychiatry 38:919–930

Farquhar MG, Rinehart JF (1954 a) Electron microscopy studies of the anterior pituitary gland of castrate rats. Endocrinology 54:516–541

Farquhar MG, Rinehart JF (1954 b) Cytologic alterations in the anterior pituitary gland following thyroidectomy. An electronmicroscope study. Endocrinology 55:857–876

Fine SA, Frohman LA (1978) Loss of central nervous system component of dopaminergic inhibition of prolactin secretion in patients with prolactin-secreting pituitary tumors. J Clin Invest 61:973–980

Foncin JF (1966) Etudes sur l'hypophyse humaine au microscope électronique. Pathol Biol (Paris) 14:893–902

Forbes AP, Henneman PH, Griswold GC, Albright F (1954) Syndrome characterized by galactorrhea, amenorrhea and low urinary FSH: comparison with acromegaly and normal lactation. J Clin Endocrinol 14:265–271

Fossati P, L'Hermite M, Derrien G, Golstein J, Vanhaelst L, Robyn C, Linquette M (1975) La prolactinémie chez les myxoedémateuses. Etude statique et dynamique, son rôle dans le syndrome hypothyroidie-galactorrhée. Ann Endocrinol (Paris) 36:145–51

Fossati P, Strauch G, Tourniaire J (1976) Etude de l'activité de la bromocriptine dans les états d'hyperprolactinémie. Nouv Presse Med 5:1687–1690

Franks S, Jacobs HS, Nabarro JDN (1975a) Prolactin concentrations in patients with acromegaly: Relation of growth hormone levels before and after surgery. J Endocrinol 65:25–26

Franks S, Murray MAF, Jequier AM, Steele SJ, Nabarro JDN, Jacobs HS (1975 b) Incidence and significance of hyperprolactinemia in women with amenorrhea. Clin Endocrinol (Oxf) 4:597–607

Franks S, Nabarro JDN, Jacobs HS (1977) Prevalence and presentation of hyperprolactinemia in patients with "functionless" pituitary tumors. Lancet 1:778–780

Franks S, Jacobs HS, Martin N, Nabarro JDN (1978) Hyperprolactinemia and impotence. Clin Endocrinol 8:277–287

Frantz AG (1978) Prolactin. N Engl J Med 298:201–207

Frantz AG, Kleinberg DL (1978) The pathophysiology of hyperprolactinemic states and the role of newer ergot compounds in their treatment. Fed Proc 37:2192–2196

Frantz AG, Kleinberg DL, Noel GL (1972) Studies on prolactin in man. Recent Prog Horm Res 28:527–590

Friesen HG, Tolis G (1977) The use of bromocriptine in the galactorrhea-amenorrhea syndromes: The Canadian Cooperative Study. Clin Endocrinol (Oxf) 6:91s–99s

Friesen H, Webster BR, Hwang P, Guyda H, Munro RE, Read L (1972) Prolactin synthesis and secretion in a patient with the Forbes-Albright syndrome. J Clin Endocrinol 34:192–199

Friesen H, Tolis G, Shiu R, Hwang P, Hardy J (1973) Studies on human prolactin: Chemistry, radioreceptor assay and clinical significance. In: Pasteels JL, Robyn C (eds) Human prolactin. Excerpta Medica, Amsterdam, pp 11–23

Frommel R (1882) Über puerperale Atrophie des Uterus. Geburtshilfe Gynaekol 7:305–313

Gemzell C (1975) Induction of ovulation in infertile women with pituitary tumors. Am J Obstet Gynecol 121:311–315

Gemzell C, Wang CF (1979) Outcome of pregnancy in women with pituitary adenoma. Fertil Steril 31:363–372

Genazzani AR, Camanni F, Massara F, Picciolini E, Cocchi D, Belforte L, Müller EE (1980) A new pharmacologicyl approach to the diagnosis of hyperprolactinemic states. The nomifensine test. Acta Endocrinol (Copenh) 93:139–148
George SR, Burrow GN, Zinman B, Ezrin C (1979) Regression of pituitary tumors, a possible effect of bromoergocryptine. Am J Med 66:697–702
Goluboff LG, Ezrin C (1969) Effect of pregnancy on the somatotroph and the prolactin cell of the human adenohypophysis. J Clin Endocrinol 29:1533–1538
Gomez F, Reyes FI, Faiman C (1977) Nonpuerperal galactorrhea and hyperprolactinemia. Am J Med 62:648–660
Griffith RW, Turkalj I, Braun P (1979) Pituitary tumors during pregnancy in mothers treated with bromocriptine. Br J Clin Pharmacol 7:393–396
Guillemin R, Vargo T, Rossier J, Minick S, Ling N, Rivier C, Vale W, Bloom F (1977) β-Endorphin and adrenocorticotropin are secreted concomittantly by the pituitary gland. Sience 197:1367–1369
Guiot G (1978) Considerations on the surgical treatment of pituitary adenomas. In: Fahlbusch R, von Werder K (eds) Treatment of pituitary adenomas. Thieme, Stuttgart, pp 203–218
Guitelman A, Aparicio NJ, Mancini A, Encinas MT, Levalle O, Schally AV (1977) Correlation of serum prolactin and LH response to stimulation with LH-RH before and after exstirpation of pituitary adenomas with clinical outcome. J Clin Endocrinol 45:810–813
Guyda H, Robert F, Colle E, Hardy J (1973) Histologic, ultrastructural and hormonal characterization of a pituitary tumor secreting both hGH and prolactin. J Clin Endocrinol 36:531–547
Hachmeister U, Fahlbusch R, von Werder K (1972) Ultrastructural identity of pituitary adenoma cells in Forbes-Albright syndrome and of adeno hypophyseal pregnancy cells. Acta Endocrinol [Suppl] (Copenh) 159:42
Haenel H (1928) Ein Fall von dauernder Milchsekretion beim Manne. MMW 75:261–263
Hagen C, Lindholm J, Suenson E, Riishede J, Hummer L, Jacobsen H-H (1979) Relationship between plasma prolactin concentration and pituitary function in patients with pituitary adenoma. Clin Endocrinol (Oxf) 11:671–679
Hardy J (1969) Transsphenoidal microsurgery of the normal and pathological pituitary. Clin Neurosurg 16:185–217
Hardy J (1973) Transsphenoidal surgery of hypersecreting pituitary tumors. In: Kohler PO, Ross GT (eds) Diagnosis and treatment of pituitary tumors. Excerpta Medica, Amsterdam, pp 179–194
Hardy J (1980) Ten years after recognition of pituitary microadenoma. In: Faglia G, Giovanelli MA, MacLeod RM (eds) Pituitary microadenomas. Academic Press, London, pp 7–14
Hardy J, Beauregard H, Robert F (1978) Prolactin secreting pituitary adenomas: Transphenoidal microsurgical treatment. In: Robyn C, Harter M (eds) Progress in prolactin physiology and pathology. Elsevier/North-Holland Biomedical Press, Amsterdam, pp 361–70
Herbert DC, Ishikawa H, Rennels EG (1979) Evidence for the autoregulation of hormone secretion by prolactin. Endocrinology 104:97–100
Herlant M (1943) Recherches sur la localisation histologique des hormons gonadotropes feinelles au niveau de l'hypophyse antérieure. Arch Biol (Liege) 54:225–357
Herlant M (1960) Etude critique de deux techniques nouvelles destinées à mettre en évidence les différentes catégories cellulaires présentes dans la glande pituitaire. Bull Microsc Appl 10:37–44
Herlant M, Pasteels JL (1967) Histophysiology of human anterior pituitary. In: Bajusz E, Jasmin G, Basel S (eds) Methods and achievements in experimental pathology, vol 3. Karger, Basel, pp 250–305
Herlant M, Laine E, Fossati P, Linquette M (1965) Syndrome aménorrhée-galactorrhée par adénome hypophysaire à cellules à prolactine. Ann Endocrinol (Paris) 26:65–71
Herlant M, Linquette M, Laine E, Fossati P, May JP, Lefevre J, (1966) Adenome hypophysaire avec cellules thyréotropes avec syndrome aménorrhée galactorrhée chez une malade porteuse d'un myxoedeme congenital par ectopie thyroidienne. Ann Endocrinol (Paris) 27:181–198
Hollenhorst RW, Younge BR (1973) Ocular manifestations produced by adenomas of the pituitary gland: Analysis of 1000 cases. In: Kohler PO, Ross GT (eds) Diagnosis and treatment of pituitary tumors. Excerpta Medica, Amsterdam, pp 53–64
Horn K, Erhardt F, Fahlbusch R, Pickardt CR, von Werder K, Scriba PC (1976) Recurrent thyrotoxic goiter and galactorrhea amenorrhea syndrome due to an autonomous thyrotropin and prolactin producing pituitary adenoma. J Clin Endocrinol 43:137–143
Horowski R, Wendt H, Gräf K-J (1978) Prolactin-lowering effect of low doses of lisuride in man. Acta Endocrinol (Copenh) 87:234–240

Horvath E, Kovacs K (1974) Misplaced exocytosis: Distinct ultrastructural feature in some pituitary adenomas. Arch Pathol 97:221–224

Ho Yuen B (1978) Bromocriptine, pituitary tumors and pregnancy. Lancet 2:1314

Hwang P, Guyda H, Friesen H (1971) A radioimmunoassay for human prolactin. Proc Natl Acad Sci USA 68:1902–1906

Hwang P, Robertson M, Guyda H, Friesen H (1973) The purification of human prolactin from frozen pituitary glands. J Clin Endocrinol 36:1110–1118

Illig R, Torresani T, Exner GU (1978) Plasma prolactin before and after TRH in 24 children with craniopharyngeoma. In: Fahlbusch R, von Werder K (eds) Treatment of pituitary adenomas. Thieme, Stuttgart, pp 193–196

Jacobs LS, Daughaday WH, (1973) Pathophysiology and control of prolactin secretion in patients with pituitary and hypothalamic disease. In: Pasteels JL, Robyn C (eds) Human prolactin. Excerpta Medica, Amsterdam pp 189–205

Jacobs HS, Franks S, Murphy MAF, Hull MGR, Steele SJ, Nabarro JDN (1976) Clinical and endocrine features of hyperprolactinemic amenorrhea. Clin Endocrinol (Oxf) 5:439–454

Jaquet P, Grijoli F, Guibout M, Lissitzky JC, Carayon P (1978) Prolactin secreting tumors. Endocrine status before and after surgery in 33 women. J Clin Endocrinol 46:459–466

Jefferson A (1978) The treatment of chromophobe pituitary adenomas by means of transfrontal surgery, radiation therapy and supportive hormon therapy. In: Fahlbusch R, von Werder K (eds) Treatment of pituitary adenomas. Thieme, Stuttgart, pp 237–252

Jelden GL, Chernak ES, Rodriguez-Antunez A, Haaga JR, Lavik PS, Dhaliwal PS (1976) Further progress in CT-scanning and computerized radiation therapy treatment planning. AJR 127: 179–185

Jewelewicz R, Zimmerman EA (1978) Current management of the amenorrhea-galactorrhea syndrome. Fertil Steril 29:597–603

Jewelewicz R, Zimmerman EA, Carmel PW (1977) Conservative management of a pituitary tumor during pregnancy following induction of ovulation with gonadotropins. Fertil Steril 28:35–40

Joplin GF, Banks L, Child DF, Diamant L, Doyle FH, Fraser TR Jadresic A, Koochek MH (1978) Treatment of acromegaly by pituitary implantation of 90Y. In: Fahlbusch R, von Werder K (eds) Treatment of pituitary adenomas.Thieme, Stuttgart, pp 261–266

Jordan RM, Kendall JW, Search JL (1976) Cerebrospinal fluid hormone concentration in evaluation of pituitary tumors. Ann Intern Med 85:49–55

Jordan RM, Kendall JW, Kerber CW (1977) The primary empty sella syndrome. Am J Med 62:569–580

Kajtar T, Tomkin GH (1971) Emergency hypophysectomy in pregnancy after induction of ovulation. Br Med J 4:88–90

Kaufman B, Pearson OA, Chamberlin WB (1973) Radiographic features of intrasellar masses and progressive asymmetrical non-tumorous enlargements of the sella turcica, the empty sella. In: Kohler PO, Ross GT (eds) Diagnosis and treatment of pituitary tumors. Excerpta Medica, Amsterdam, pp 100–129

Kazner E, Fahlbusch R, Lanksch W, Rothe R, Scherer U, Steinhoff H, Grumme T, Lange S, Meese W, Aulich A, Wende S (1978) Computerized tomography in diagnosis and follow up of pituitary adenomas. In: Fahlbusch R, von Werder K (eds) Treatment of pituitary adenomas. Thieme, Stuttgart, pp 101–114

Kelly WF, Mashiter K, Doyle FH, Banks LM, Joplin GF (1978) Treatment of prolactin-secreting pituitary tumors in young women by needle implantation of radioactive yttrium. Q J Med 188: 473–493

Kjellberg RN, Kliman B (1973) A system for therapy of pituitary tumors. In: Kohler PO, Ross GT (eds) Diagnosis and treatment of pituitary tumors. Excerpta Medica, Amsterdam, pp 234–252

Kleinberg DL, Noel GL, Frantz AG (1971) Chlorpromazine stimulation and L-dopa suppression of plasma prolactin in man. J Clin Endocrinol 33:873–876

Kleinberg DL, Noel GL, Frantz AG (1977) Galactorrhea: Study of 235 cases including 48 with pituitary tumors. N Engl J Med 296:589–600

Klijn JGM, Lamberts SWJ, de Jong FH, Docter R, van Dongen KJ, Birkenhäger JC (1980) The importance of pituitary tumour size in patients with hyperprolactinemia in relation to hormonal variables and extrasellar extension of tumour. Clin Endocrinol (Oxf) 12:341–355

Kliman B, Kjellberg RN (1977) Proton beam therapy of pituitary tumors in women with galactorrhea-amenorrhea syndrome. 59th Annual Meeting of the Endocrine Society, Chicago, Abstr no 330

Knazek RA, Skyler JS (1976) Secretion of human prolactin in vitro. Proc Soc Exp Biol Med 151:561–564

Kohler PO, Ross GT (eds) (1973) Diagnosis and treatment of pituitary tumors. Excerpta Medica, Amsterdam
Kovacs K (1977) Morphology of prolactin producing adenomas. Clin Endocrinol (Oxf) 6:71–79
Kovacs K, Horvath E, Corenblum B, Sirek AMT, Penz G, Ezrin C (1975) Pituitary chromophobe adenomas consisting of prolactin cells. A histologic immuno-cytological and electron microscopic study. Virchows Arch [Pathol Anat] 366:113–123
Kovacs K, Ryan N, Horvath E, Ezrin C, Penz G (1978) Prolactin cell adenomas of the human pituitary. Morphologic features of prolactin cells in the nontumorous portions of the anterior lobe. Horm Metab Res 10:409–412
Krieger DT, Luria M (1976) Effectiveness of cyproheptadine in decreasing plasma ACTH concentrations in Nelson's syndrome. J Clin Endocrinol 43:1179–1182
Krieger DT, Howanitz PJ, Frantz AG (1976) Absence of nocturnal elevation of plasma prolactin concentrations in Cushing's disease. J Clin Endocrinol 42:260–272
Lachelin GCL, Abu-Fadil S, Yen SSC (1977) Functional delineation of hyperprolactinemic amenorrhea. J Clin Endocrinol 44: 1163–1174
Lamberts SWJ, MacLeod RM (1979) The inability of bromocriptine to inhibit prolactin secretion by transplantable rat pituitary tumors. Observations on the mechanism and dynamics of the autofeedback regulation of prolactin secretion. Endocrinology 104:65–70
Lamberts SWJ, Birkenhäger JC, Kwa HG (1976) Basal and TRH-stimulated prolactin in patients with pituitary tumors. Clin Endocrinol (Oxf) 5:709–711
Lamberts SWJ, Seldenrath HJ, Kwa HG, Birkenhäger JC (1977a) Transient bitemporal hemianopsia during pregnancy after treatment of galactorrhea-amenorrhea syndrome with bromocriptine. J Clin Endocrinol 44:180–184
Lamberts SWJ, Timmermans HAT, de Jong FH, Birkenhäger JC (1977 b) The role of dopaminergic depletion in the pathogenesis of Cushing's disease and the possible consequences for medical therapy. Clin Endocrinol (Oxf) 7:185–193
Lamberts SWJ, Klijn JGM, KWA GH, Birkenhäger JC (1979 a) The dynamics of growth hormone and prolactin secretion in acromegalic patients with "mixed" pituitary tumors Acta Endocrinol (Copenh) 90:198–210
Lamberts SWJ, Klijn JGM, de Lange SA, Singh R, Stefanko SZ, Birkenhäger JC (1979 b) The incidence of complications during pregnancy after treatment of hyperprolactinemia with bromocriptine in patients with radiologically evident pituitary tumors. Fertil Steril 31:614–619
Landolt AM (1975) Ultrastructure of human sella tumors. Acta Neurochir [Suppl] (Wien) 22
Landolt AM (1978) Progress in pituitary adenoma biology. Adv tech Standards Neurosurg 5:3–49
Landolt AM, Rothenbühler V, Kistler GS (1978) Morphology of chromophobe adenoma. In: Fahlbusch R, von Werder K (eds) Treatment of pituitary adenomas. Thieme, Stuttgart, pp 154–171
Landolt AM, Wüthrich R, Fellmann H (1979) Regression of pituitary prolactinoma after treatment with bromocriptine. Lancet 1:1082–1083
Leebaw W, Lee L, Woolf P (1978) The effect of the dopamine agonist, lergotryle mesylate, on prolactin secretion in women. Acta Endocrinol (Copenh) 87:12–18
Leonardi M, Bertolissi F, Penco T, Fabris G, de Nardi F (1980) CT evidence of functioning microadenomas of the pituitary. In: Faglia G, Giovanelli MA, MacLeod RM (eds) Pituitary microadenomas. Academic Press, London, pp 329–335
Levin ME, Daughaday WH, Levy J (1959) Persistant lactation associated with pituitary tumor and hyperadrenal corticism. Am J Med 27:172–175
Levin SR, Hofeldt FD, Schneider V, Becker N, Karam JH, Seymour RJ, Adams JE, Forsham PH (1974) Cryohypophysectomy for acromegaly: Factors associated with altered endocrine function and carbohydrate metabolism. Am J Med 57:526–535
Lewis UJ, Singh RNP (1973) Recovery of prolactin from human pituitary glands. In: Pasteels JL, Robyn C (eds) Human prolactin. Excerpta Medica, Amsterdam, pp 1–10
Lewis UJ, Singh RNP, Seavey BK (1971) Human prolactin: isolation and some properties. Biochem Biophys Res Commun 44:1169–1176
L'Hermite M, Delvoye P, Nokin J, Vekemans M, Robyn C (1972) Human prolactin secretion as studied by radio-immunoassay: Some aspects of its regulation. In: Boyns AR, Griffiths K (eds) Prolactin and carcinogenesis. Alpha Omega Alpha, Cardiff, pp 81–97
L'Hermite M, Robyn C, Golstein J, Rothenbucher G, Birk J, Loos U, Bonnyns M, Vanhealst L (1974) Prolactin and thyrotropin in thyroid diseases: Lack of evidence for a physiological role of thyrotropin-releasing hormone in the regulation of prolactin secretion. Horm Metab Res 6:190–5

L'Hermite M, Caufriez A, Virasoro E, Stevenaert A, Copinschi G, Robyn C (1978) Endocrinology of prolactin producing tumors. In: Fahlbusch R, von Werder K (eds) Treatment of pituitary adenomas. Thieme, Stuttgart, pp 60–77

Linfoot JA (ed) (1979) Recent advances in the diagnosis and treatment of pituitary tumors. Raven, New York

Linquette M, Fossati P, Dupont-Lecompte J, Gasnault JP, Montois R (1970) Adenomes hypophysaires et gravido puerpéralité. Rev Fr Endocrinol Clin 11:223–232

Liuzzi A, Chiodini PG, Botalla L, Silvestrini F, Müller EE (1974) Growth hormone (GH)-releasing activity of TRH and GH-lowering effect of dopaminergic drugs in acromegaly: Homogeneity in the two responses. J Clin Endocrinol 39:871–876

Liuzzi A, Panerai AE, Chiodini PG, Secchi C, Cocchi D, Botalla L, Silvestrini F, Müller EE (1976) Neuroendocrine control of growth hormone secretion: Experimental and clinical studies. In: Pecile E, Müller EE (eds) Growth hormone and related peptides. Excerpta Medica, Amsterdam, pp 236–251

Liuzzi A, Chiodini PG, Oppizzi G, Botalla L, Verde G, De Stefano L, Colussi G, Gräf KJ, Horowski R (1978) Lisuride hydrogen maleate: Evidence for a long lasting dopaminergic activity in humans. J Clin Endocrinol 46:196–202

Lloyd HM, Meares JD, Jacobi J (1975) Effects of oestrogen and bromocriptine on in vivo secretion and mitosis in prolactin cells. Nature 255:497–498

Loepp W, Lorenz R (1971) Röntgendiagnostik des Schädels. Thieme, Stuttgart, p 181

Lutterbeck PM, Pryor S, Varga L, Wenner R (1971) Treatment of non-puerperal galactorrha with an ergot alkaloid. Br Med J 3:228–229

MacLeod RM (1976) Regulation of prolactin secretion. In: Martini L, Ganong WF (eds) Frontiers in neuroendocrinology, vol 4. Raven, New York, pp 169–194

MacLeod RM, Lehmeyer JE (1973) Suppression of pituitary tumor growth and function by ergot alkaloids. Cancer Res 33:849–855

MacLeod RM, Nagy I (1979) The activity of pituitary enzymes during the stimulation and inhibition of prolactin secretion. Fed Proc 38:1025 A

Magyar DM, Marshall JR (1978) Pituitary tumors and pregnancy. Am J Obstet Gynecol 132:739–751

Mahesh VB, Dalla Pria S, Greenblatt RB (1969) Abnormal lactation with Cushing's syndrome. A case report. J Clin Endocrinol 29:978–981

Malarkey WB, Johnson JC (1976) Pituitary tumors and hyperprolactinemia. Arch Intern Med 136: 40–44

Mashiter K, van Norden S, de Marco L, Adams E, Joplin GF (1979) Hormone secretion by human somatotrophic lactotrophic and mixed pituitary adenomas in culture. J Clin Endocrinol 48:108–113

McComb DJ, Kovacs K (1978) Ultrastructural morphometry of sparsely granulated prolactin cell adenomas of the human pituitary. Acta Endocrinol (Copenh) 89:521–529

McGregor AM, Scanlon MF, Hall R, Hall K (1979 a) Effects of bromocriptine on pituitary tumour size. Br Med J 2:700–703

McGregor AM, Scanlon MF, Hall K, Cook DB, Hall R (1979 b) Reduction in size of a pituitary tumor by bromocriptine therapy N Engl J Med 300:291–293

McKenna TJ, Glick AD, Cobb CA, Jacobs LS (1978) Galactorrhea and hypogonadism associated with a radiologically inapparent prolactin-secreting pituitary tumor. Acta Endocrinol (Copenh) 87:225–233

Moore DM, Buckingham MS, Singh MM, Milligan MP, Elstein M (1978) Serum prolactin in female infertility. Lancet 2:1243–1247

Mornex R, Orgiazzi J, Hugues B, Gagnaire J-C, Claustrat B (1978) Normal pregnancies after treatment of hyperprolactinemia with bromoergocryptine, despite suspected pituitary tumors. J Clin Endocrinol 47:290–295

Mühlenstedt D, Osmers F, Schneider HPG (1978) Regression eines Hypophysenadenoms unter Bromocriptin. Arch Gynecol 226:341–346

Müller EE, Genazzani AR, Murru S (1978) Nomifensine: Diagnostic test in hyperprolactinemic states. J Clin Endocrinol 47:1352–1357

Müller OA, von Werder K (1977) Diagnostik bei Hypophysentumoren. Med Klin 72:1563–1577

Müller OA, Fink R, von Werder K, Scriba PC (1978) Hypersecretion of ACTH, growth hormone and prolactin in a patient with pituitary adenoma. Acta Endocrinol [Suppl] (Copenh) 215:4–5

Nader S, Mashiter K, Doyle FH, Joplin GF (1976) Galactorrhoea, hyperprolactinaemia and pituitary tumours in the female. Clin Endocrinol (Oxf) 5:245–251

Nagulesparen M, Ang V, Jenkins JS (1978) Bromocriptine treatment of males with pituitary tumours hyperprolactinemia and hypogonadism. Clin Endocrinol (Oxf) 9:73–79

Newton TH, Richmond I (1980) Radiologic evaluation of pituitary microadenomas. In: Faglia G, Giovanelli MA, MacLeod RM (eds) Pituitary microadenomas. Academic Press, London, pp 313–322

Nillius SJ, Bergh T, Lundberg PO, Stahle J, Wide L (1978) Regression of a prolactin-secreting pituitary tumor during long-term treatment with bromocriptine. Fertil Steril 30:710–712

Noel GL, Suh HK, Stone JG, Frantz AG (1972) Human prolactin and growth hormone release during surgery and other conditions of stress. J Clin Endocrinol 35:840–851

Nokin J, Vekemans M, L'Hermite M, Robyn C (1972) Circadian periodicity of serum prolactin concentration in man. Br Med J 3:561–562

Olivier L, Vila-Porcile E, Racadot O, Peillon F, Racadot J (1975) Ultrastructure of pituitary tumor cells: A critical study. In: Tixier-Vidal A, Farquhar MG (eds) The anterior pituitary. Academic Press, San Francisco London New York, pp 231–276

Orth D (1978) Metyrapone is useful only as adjunctive therapy in Cushing's disease. Ann Intern Med 89:128–130

Osamura RY, Watanabe K, Teramoto A, Hirakawa K, Kawano N, Morii S (1978) Male prolactin secreting pituitary adenomas in humans studied by peroxidase-labelled antibody method. Acta Endocrinol (Copenh) 88:643–652

Paiz C, Hennigar GR (1970) Electron microscopy and histochemical correlation of human anterior pituitary cells. Am J Pathol 59:43–74

Parker DC, Rossman LG, Vanderlaan EF (1973) Sleep-related, nyctohemeral and briefly episodic variation in human plasma prolactin concentrations. J Clin Endocrinol 36:1119–1124

Pasteels JL (1963) Recherches morphologiques et expérimentales sur la sécrétion de prolactine. Arch Biol (Liege) 74:439–533

Pasteels JL (1972) Morphology of prolactin secretion. In: Wolstenholme GEW, Knight J (eds) Lactogenic hormones. Livingstone; Edinburgh London, pp 241–255

Pasteels JL, Gausset P, Danguy A, Ectors F (1972 a) Immunofluorescent studies on prolactin and the pituitary. In: Boyns AR, Griffiths K (eds) Prolactin and carcinogenesis. Alpha Omega Alpha, Cardiff, pp 128–136

Pasteels JL, Gausset P, Danguy A, Ectors F, Nicoll CS, Varavudhi P (1972 b) Morphology of the lactotropes and somatotropes of man and rhesus monkeys. J Clin Endocrinol 34:959–967

Peillon F, Vila-Porcile E, Olivier L, Racadot J (1970) L'action des oestrogenes sur les adénomes hypophysaires chez l'homme. Documents histopathologiques en microscopie optique et électronique et apport de l'expérimentation. Ann Endocrinol (Paris) 31:259–270

Peillon F, Boyet F, Andre C, Dreyfus G (1975 a) La fonction gonadotrope dans les adénomes a prolactine. Ann Endocrinol (Paris) 36:331–332

Peillon F, Racadot J, Olivier L, Racadot O, Vila-Porcile E (1975 b) Les adénomes a prolactine. Ann Endocrinol (Paris) 36:277–278

Peillon F, Bard H, Boyet F (1977) Les adénomes à prolactine chez l'homme. Gynecologie 28: 433–435

Peillon F, Racadot J, Moussy D, Vila-Porcile E, Olivier L, Racadot O (1978) Prolactin-secreting adenomas. A correlative study of morphological and clinical data. In: Fahlbusch R, von Werder K (eds) Treatment of pituitary adenomas. Thieme, Stuttgart, pp 114–122

Pineda G, Valdes M, Castro M, Stevenson C (1979) Pituitary microadenoma: Diagnostic studies. Clin Endocrinol (Oxf) 10:583–594

Pinnamaneni K, Birge SJ, Avioli LV (1980) Prolactin-secreting pituitary tumor associated with von Recklinghausen's disease. Arch Intern Med 140:397–399

Pont A, Shelton R, Odell WD, Wilson CB (1979) Prolactin secreting tumors in men: Surgical cure. Ann Intern Med 91:211–213

Pozo E del, Friesen H, Burmeister P (1973) Endocrine profile of a specific prolactin inhibitor: Br-ergocryptine (CB-154). A preliminary report. Schweiz Med Wochenschr 103:847–848

Pozo E del, Varga L, Wyss H, Tolis G, Friesen H, Wenner R, Vetter L, Uettwiler A (1974) Clinical and hormonal response to bromocriptine (CB 154) in the galactorrhea syndromes. J Clin Endocrinol 39:18–25

Quadri, SK, Lu KH, Meites J (1972) Ergot induced inhibition of pituitary tumor growth in rats. Science 176:417–418

Quigley ME, Judd SJ, Gilliland GB, Yen SSC (1979) Effects of a dopamine antagonist on the release of

gonadotropin and prolactin in normal women and women with hyperprolactinemic anovulation. J Clin Endocrinol 48:718–720
Racadot J, Vila-Porcile E, Peillon F, Olivier L (1971) Adénomes hypophysaires à cellules à prolactine: étude structurale et ultrastructurale corrélations anatomo-cliniques. Ann Endocrinol (Paris) 32:298–305
Racadot J, Vila-Porcile E, Olivier L, Peillon F (1975) Electron microscopy of pituitary tumours. Prog Neurol Surg 6:95–141
Rand CW (1957) Two cerebral complications of pregnancy: Brain tumor and spontaneous subarachnoid hemorrhage. Clin Neurosurg 3:104–141
Reichlin S (1979) The prolactinoma problem. N Engl J Med 300:313–315
Reschini E, Guistina G, Cantalamessa L, Peracchi H (1976) Hyperthyroidism with elevated plasma TSH – levels and pituitary tumor: Study with somatostatin. J Clin Endocrinol 43:924–927
Rinehart JF, Farquhar MG (1953) Electron microscopic studies of the anterior pituitary gland. J Histochem Cytochem 1:93–113
Rjosk HK, Schill WB (1979) Serum prolactin in male infertility. Andrologia 11:297–304
Rjosk HK, von Werder K, Fahlbusch R (1976) Hyperprolaktinämische Amenorrhoe. Geburtshilfe Frauenheilkd 36:575–587
Rjosk HK, Fahlbusch R, Huber H, von Werder K (1978) Growth of prolactin-producing pituitary adenomas during pregnancy. In: Fahlbusch R, von Werder K (eds) Treatment of pituitary adenomas. Thieme, Stuttgart, pp 395–400
Rjosk HK, Fahlbusch R, Greite JH, von Werder K (1979 a) Prevention and therapy of visual complications during pregnancy in patients with prolactinomas (Abstr). In: Derome PJ, Jedynak CP, Peillon F (eds) Pituitary adenomas, biology, physiopathology and treatment. Asclepios Publishers, Paris, p 340
Rjosk HK, Fahlbusch R, von Werder K (1979 b) Hyperprolactinämie und Sterilität. Arch Gynecol 665–667
Rjosk HK, Fahlbusch R, Huber H, von Werder K (1980) Growth of prolactinomas during pregnancy. In: Faglia G, Giovanelli MA, MacLeod RM (eds) Pituitary microadenomas. Academic Press, London, pp 535–541
Robert F, Hardy J (1975) Prolactin secreting adenomas. A light and electron microscopical study. Arch Pathol 99:625–633
Robertson WD (1978) Radiologic assessment of pituitary microadenomas. A J R 131:489–492
Robyn C, Delvoye P, Nokin J, Vekemans M, Badawi M, Perez-Lopez FR, L'Hermite M (1973) Prolactin and human reproduction. In: Pasteels JL, Robyn C (eds) Human prolactin. Excerpta Medica, Amsterdam, pp 167–188
Rogol AD, Eastman RC (1979) Prolactin and pituitary tumors. Am J Med 66:547–548
Rogol AD, Rosen SW (1974) Prolactin of apparent large molecular size: The major immunoactive prolactin component in plasma of a patient with a pituitary tumor. J Clin Endocrinol 38:714–716
Rolland R, Schellekens LA, Lequin RM (1974) Successful treatment of galactorrhea and amenorrhea and subsequent restoration of ovarian function by a ergot alkaloid 2-bromo-α-ergocryptine. Clin Endocrinol (Oxf) 3:155–166
Romeis B (1940) Hypophyse. In: Möllendorff W (ed) Innersekretorische Drüsen II: Hypophyse. Springer, Berlin (Handbuch der mikroskopischen Anatomie des Menschen, vol 6/3, pp 79–80)
Rosen SW, Weintraub BD (1977) Ectopic prolactin and placental lactogen. In: James VHT (ed) Endocrinology, vol 2. Excerpta Medica, Amsterdam Oxford, pp 425–429
Roth J, Gordon P, Brace K (1970) Efficacy of conventional pituitary irradiation in acromegaly. N Engl J Med 282:1385–1391
Roth O (1918) Auftreten von Milchsekretion bei einem an Akromegalie leidenden Patienten. Berl Klin Wochenschr 55:305–307
Rubin AL, Levin R, Bernstein RI, Tyrrell JB, Noacco C, Forsham PH (1973) Stimulation of growth hormone by luteinizing hormone releasing hormone in active acromegaly. J Clin Endocrinol 37:160–162
Säger W (1975) Hypophysenadenome bei Galactorrhoe. Virchows Arch Pathol Anat 368:123–139
Säger W (1977) Die Morphologie der paraadenomatösen Adenohypophyse. Virchows Arch [Pathol Anat] 372:299–314
Salti IS, Istfan N (1979) Bromocriptine fails to stop growth of eosinophilic adenomas in acromegaly. N Engl J Med 301:386
Samaan NA, Leavens ME, Jesse RH jr (1974) Serum growth hormone and prolactin response to

thyrotropin releasing hormone in patients with acromegaly before and after surgery. J Clin Endocrinol 38:957–963
Samaan NA, Leavens ME, Jesse JH (1977) Serum prolactin in patients with "functionless" chromophobe adenomas before and after therapy. Acta Endocrinol (Copenh) 84:449–460
Sassin JF, Frantz AG, Weitzman ED, Kapen S (1972) Human prolactin: 24 hour pattern with increased release during sleep. Science 177:1205–1207
Sassin JF, Frantz AG, Kapen S, Weitzman ED (1973) The nocturnal rise of human prolactin is dependent on sleep. J Clin Endocrinol 37:436–440
Scanlon MF, Rodriguez-Arnao MD, McGregor AM, Weightman D, Lewis M, Cook DB, Gomez-Pan A, Hall R (1981) Altered dopaminergic regulation of thyrotropin release in patients with prolactinomas: comparison with other tests of hypothalamic pituitary function. Clin Endocrinol 14:133–143
Schaison G, Metzger J (1976) The primary empty sella. An endocrine study on 12 cases. Acta Endocrinol (Copenh) 83:483–492
Schneider AB, Kowalski K, Sherwood LM (1975) Big human placental lactogen: disulfide peptide chains. Biochem Biophys Res Commun 64:717–724
Schroeder LL, Johnson JC, Malarkey WB (1976) Cerebrospinal fluid prolactin: A reflection of abnormal prolactin secretion in patients with pituitary tumors. J Clin Endocrinol 43:1255–1260
Schürmann K, Reulen HJ, Beyer T (1978) A dramatic bleeding during transsphenoidal operation on an apparent pituitary adenoma, caused by an intrasellar aneurysm. In: Fahlbusch R, von Werder K (eds) Treatment of pituitary adenomas. Thieme, Stuttgart, pp 316–323
Schwarz K (1969) Pathophysiologie und Klinik der Hypophysentumoren. In: Kracht J (ed) Oestrogene, Hypophysentumoren. Springer, Berlin Heidelberg New York, pp 223–235
Schwinn G, Dirks H, McIntosh C, Köbberling J (1977) Metabolic and clinical studies on patients with acromegaly treated with bromocriptine over 22 months. Eur J Clin Invest 7:101–107
Segal S, Polishuk WZ, Ben-David M (1976) Hyperprolactinemic male infertility. Fertil Steril 27:1425–1427
Seppälä M, Hirvonen E, Ranta T, Virkkunen P, Leppäluoto J (1975) Raised serum prolactin levels in amenorrhoea. Br Med J 2:305–306
Seppälä M, Lehtovirta P, Ranta T (1977) Discordant patterns of hyperprolactinemia and galactorrhea in secondary amenorrhoea. Acta Endocrinol (Copenh) 86:457–463
Shah GV, Desai RB, Sheth AR (1976) Effect of prolactin on metabolism of human spermatozoa. Fertil Steril 27:1292–1294
Sherman BM, Schlechte J, Halmi NS, Chapler FK, Harris EC, Duello TM, van Gilder J, Granner DK (1978) Pathogenesis of prolactin secreting pituitary adenomas. Lancet 2: 1019–1021
Siler TM, Yen SSC, Guillemin R (1974) Inhibition by somatostatin on the release of TSH induced in man by thyrotropin releasing factor. J Clin Endocrinol 38:742–745
Silvestrini F, Liuzzi A, Chiodini PG (1978) Prolactin and pituitary tumors. In: Martini L, James VHT (eds) Current topics in experimental endocrinology, vol 3. Academic Press, New York, pp 131–172
Sinha YN, Selby FW, Vanderlaan W (1973) Radioimmunoassay of prolactin in the urine of mouse and man. J Clin Endocrinol 36:1039–1042
Sobrinho LG, Nunes MCP, Calhaz-Jorge C, Mauricio JC, Santos MA (1981) Effect of treatment with bromocriptine on the size of prolactin producing pituitary tumours. Acta Endocrinol 96:24–29
Sobrinho LG, Nunes MCP, Santos MA, Mauricio JC (1978) Radiological evidence for regression of prolactinoma after treatment with bromocriptine. Lancet 2:257–258
Solbach HG, Wiegelmann W, Kley HK, Rudorff KH, Krüskemper HL (1978) Endocrine evaluation of pituitary insufficiency. In: Fahlbusch R, von Werder K (eds) Treatment of pituitary adenomas. Thieme, Stuttgart, pp 38–47
Summers VK, Hipkin LJ, Diver MJ, Davis JV (1975) Treatment of acromegaly with bromocriptine. J Clin Endocrinol 40:904–906
Thorner MO (1977) Prolactin: clinical physiology and the significance and management of hyperprolactinemia. In: Martini L, Besser GM (eds) Clinical neuroendocrinology. New York San Francisco London, Academic Press, pp 319–361
Thorner MO, Besser GM (1977) Hyperprolactinemia and gonadal function: Results of bromocriptine treatment. In: Crosignani PC, Robyn C (eds) Prolactin and human reproduction. Academic Press, London, pp 285–301
Thorner MO, McNeilly AS, Hagan C, Besser GM (1974) Longterm treatment of galactorrhea and hypogonadism with bromocriptine. Br Med J 2:419–422

Thorner MO, Besser GM, Jones A, Dacie J, Jones AE (1975a) Bromocriptine treatment of female infertility: report of 13 pregnancies. Br Med J 4:694–697

Thorner MO, Chait A, Aitken M, Benker G, Bloom SR, Mortimer CH, Sanders P, Stuart Mason A, Besser GM (1975b) Bromocriptine treatment of acromegaly. Br Med J 1:299–303

Thorner MO, Evans WS, MacLeod RM, Nunley WC Jr, Rogol AD, Morris JC, Besser GM (1980a) Hyperprolactinemia: Current concepts of management including medical therapy with bromocriptine. In: Thorner MO, Calne DB, Lieberman A, Goldstein G (eds) Ergot compounds and brain function – Neuroendocrine and neuropsychiatric aspects. Raven, New York, pp 165–189

Thorner MO, Martin WH, Rogol AD, Morris JL, Perryman RL, Conway BP, Howards S-S, Wolfman MG, MacLeod RM (1980b) Rapid regression of pituitary prolactinomas during bromocriptine treatment. J Clin Endocrinol 51:438–445

Tolis G (1977) Galactorrhea-amenorrhea and hyperprolactinemia. Pathophysiological aspects and diagnostic tests. Clin Endocrinol 6:815s–895s

Tolis G, Goldstein M, Friesen HG (1973) Functional evaluation of prolactin secretion in patients with hypothalamic pituitary disorders. J Clin Invest 53:783–788

Tolis G, Somma G, van Campenhout J, Friesen H (1974) Prolactin secretion in sixty-five patients with galactorrhea. Am J Obstet Gynecol 118:91–101

Tolis G, Bertrand G, Carpenter S, McKenzie JM (1978) Acromegaly and galactorrhea-amenorrhea with two pituitary adenomas secreting growth hormone or prolactin. Ann Intern Med 89:345–348

Tucker HS, Grubb SR, Wigand JP, Taylor A, Lankford HV, Blackard WG, Becker DP (1981) Galactorrhea-amenorrhea syndrome: follow-up of forty-five patients after pituitary tumor removal. Am. Intern. Med. 94: 303–307

Türkalj I, Braun P, Krupp P (to be published) Surveillance of bromocriptine during pregnancy.

Turkington RW (1971) Ectopic production of prolactin. N Engl J Med 285:1455–1458

Turkington RW (1972a) Inhibition of prolactin secretion and successful therapy of the Forbes-Albright syndrome with L-Dopa. J Clin Endocrinol 34:306–311

Turkington RW (1972b) Secretion of prolactin by patients with pituitary and hypothalamic tumors. J Clin Endocrinol 34:159–164

Turkington RW (1972c) Serum prolactin levels in patients with gynecomastia. J Clin Endocrinol 34:62–66

Turkington RW, Underwood LE, van Wyk JJ (1971) Elevated serum prolactin levels after pituitary-stalk section in man. N Engl J Med 285:707–710

Tyrrell JB, Brooks RM, Fitzgerald PA, Cofoid PB, Forsham PH, Wilson CB (1978) Cushing's disease. Selective trans-sphenoidal resection of pituitary microadenomas. N Engl J Med 298:753–758

Tyson JE, Friesen HG, Anderson MS (1972) Human lactational and ovarian response to endogenous prolactin release. Science 177:897–899

Vaidya R, Aloorkar S, Sheth A (1977) Therapeutic regression of putative pituitary hyperplasia and/or microadenoma with CB-154. Fertil Steril 18:363

Varga L, Lutterbeck PM, Pryor JS, Wenner R, Erb H (1972) Suppression of puerperal lactation with an ergot alkaloid: A double-blind study. Br Med J 2:743–744

Varga L, Wenner R, del Pozo E (1973) Treatment of galactorrhea-amenorrhea syndrome with Br-ergocryptine (CB-154): restoration of ovulatory function and fertility. Am J Obstet Gynecol 17:75–79

Vezina JL, Sutton TJ (1974) Prolactin secreting pituitary microadenomas. AJR 120:46–54

Vogt JH (1977) Spontaneous regression of enlargement of the sella turcica and associated panhypopituitary symptoms. Acta Endocrinol (Copenh) 85:684–691

Vorherr H (1979) Renal and vascular activity of prolactin preparations. Klin Wochenschr 57:101–109

Wass JAH, Thorner MO, Charlesworth M, Moult PJA, Dacie JE, Jones AE, Besser GM (1979) Reduction of pituitary tumour size in patients with prolactinomas and acromegaly treated with bromocriptine with or without radiotherapy. Lancet 2:66–69

Weber E, Voigt KH, Martin R (1978) Concomitant storage of ACTH and endorphin-like immunoreactivity in the secretory granules of anterior pituitary corticotrophs. Brain Res 157:385–390

Weisberg LA (1977) Pituitary apoplexy. Am J Med 63:109–115

Werder K von (1975) Wachstumshormone und Prolaktinsekretion des Menschen. Urban & Schwarzenberg, München Berlin Vienna

Werder K von, Clemm C (1974) Evidence for "big" and "little" components of circulating immunoreactive prolactin in humans. FEBS Lett 47:181–184

Werder K von, Fahlbusch R (1978) GnRH and TRH stimulated hGH-secretion in active and inactive

acromegaly. In: Gupta D, Voelter W (eds) Hypothalamic hormones. Verlag Chemie, Weinheim New York, pp 677–683
Werder K von, Rjosk HK (1979) Menschliches Prolaktin. Klin Wochenschr 57:1–12
Werder K von, Müller OA, Scriba PC, Fahlbusch R, Hachmeister U, Goebel R, L'Hermite M, Robyn C (1973) Function and ultrastructure of the pituitary in Forbes-Albright syndrome. In: Pasteels JL, Robyn C (eds) Human prolactin. Excerpta Medica, Amsterdam, pp 226–230
Werder K von, Fahlbusch R, Rjosk HK (1977) Hyperprolaktinaemie. Internist 18:520–528
Werder K von, Fahlbusch R, Landgraf R Pickardt CR, Rjosk HK, Scriba PC (1978 a) Treatment of patients with prolactinomas. J Endocrinol Invest 1:47–58
Werder K von, Felixberger F, Gottsmann M, Kerner W, Glöckner B (1978 b) A homologous human prolactin (hPRL) radioimmunoassay with an antibody against "little" hPRL. In: Radioimmunoassay and related procedures in medicine. International Atomic Energy Agency, Vienna, pp 43–56
Werder K von, Gottsmann M, Brendel C, Landgraf R, von Lieven H, Rjosk HK, Fahlbusch R (1978 c) Treatment of prolactinomas: efficacy of radiotherapy. Acta Endocrinol [Suppl] (Copenh) 215:1
Werder K von, Brendel C, Eversmann T, Fahlbusch R, Müller OA, Rjosk HK (1980 a) Medical therapy of hyperprolactinemia and Cushing's disease associated with pituitary adenomas. In: Faglia G, Giovanelli MA, MacLeod RM (eds) Pituitary microadenomas. Academic Press, London, pp 383–397
Werder K von, Eversmann T, Fahlbusch R, Rjosk HK (1980 b) Treatment of prolactinomas. In: Müller EE (ed) Neuroactive drugs in endocrinology: physiologic, diagnostic and therapeutic applications. Elsevier/North-Holland, Amsterdam, pp 347–360
Werner S (1979) Human pituitary adenomas with hypersecretion of TSH and prolactin. Horm Metab Res 11:452–453
White MC, Rosenstock J, Anapliotou M, Mashiter K, Joplin GF (1981) Heterogeneity of prolactin responses to oestradiol benzoate in women with prolactinomas. Lancet I: 1394–1396
Wiegelmann W, Solbach HG, Kley HK, Rudorff KH, Hermann J, Zimmermann H, Krüskemper HL (1975) Die Wirkung von synthetischem Somatostatin bei männlichen Normalpersonen und Akromegalen. Dtsch Med Wochenschr 100:331–334
Winter EW (1931) Hypophysentumor und Schwangerschaft. Arch Gynaekol 147:95–112
Wright AD, Hartog M, Palter H, Tevaarwerk GG, Doyle FH, Arnot R, Joplin GF, Fraser TR (1970) The use of yttrium-90 implantation in the treatment of acromegaly. Proc R Soc Med 63:221–223
Yamada T, Tshukui T, Ikejiri K, Yukimura Y, Kotani M (1976) Volume of the sella turcica in normal subjects and in patients with primary hypothyroidism and hyperthyroidism. J Clin Endocrinol 42:817–822
Yamayi T, Ishibashi M, Kosaka K, Fukushima T, Hori T, Manaka S, Sano K (1981) Pituitary apoplexy in acromegaly during bromocriptine therapy. Acta Endocrinol 98: 171–177
Yen SSC (1978) Chronic anovulation. In: Yen SSC, Jaffe RB (eds) Reproductive endocrinology. Saunders, Philadelphia London Toronto, pp 341–372
Yen SSC, Siler TM, De Vane GW (1974) Effect of somatostatin in patients with acromegaly. N Engl J Med 290:935–938
Zarate A, Canales ES, Villalobos H, Soria J, Jacobs S, Kastin AJ, Schally AV (1975) Pituitary hormonal reserve in patients presenting hyperprolactinemia, intrasellar masses and amenorrhea without galactorrhea. J Clin Endocrinol 40:1034–1037
Zimmerman EA, Defendini R, Frantz AG (1974). Prolactin and growth hormone in patients with pituitary adenomas: a correlative study of hormone in tumor and plasma by immunoperoxidase technique and radioimmunoassay. J Clin Endocrinol Metab 38, 57–85

Subject Index

Other Volumes in This Series:

Volume 22: D. T. Krieger
Cushing's Syndrome
1981. 27 figures in 42 separate illustrations (3 in color), Approx. 170 pages
ISBN 3-540-10811-4

Volume 21: A. E. Schindler
Hormones in Human Amniotic Fluid
1982. 23 figures, 133 tables. Approx. 250 pages
ISBN 3-540-10810-6

Volume 20: R. Volpé
Auto-immunity in the Endocrine System
1981. 32 figures, 15 tables. X, 188 pages
ISBN 3-540-10677-4

Volume 19: P. Mauvais-Jarvis, F. Kuttenn, I. Mowszowicz
Hirsutism
1981. 32 figures, 10 tables. XI, 110 pages
ISBN 3-540-10509-3

Volume 18: I.J. Chopra
Triiodothyronines in Health and Disease
With a contribution by V. Cody
1981. 76 figures, 18 tables. IX, 145 pages
ISBN 3-540-10400-3

Volume 17: J. Chayen
The Cytochemical Bioassay of Polypeptide Hormones
1980. 72 figures, 7 tables. XIV, 208 pages
ISBN 3-540-10040-7

Volume 16: J.E.A. McIntosh, R.P. McIntosh
Mathematical Modelling and Computers in Endocrinology
1980. 73 figures, 56 tables. XII, 337 pages
ISBN 3-540-09693-0

Volume 15: A.T. Cowie, I.A. Forsyth, I.C. Hart
Hormonal Control of Lactation
1980. 64 figures, 7 tables. XIV, 275 pages
ISBN 3-540-09680-9

Volume 14: J.H. Clark, E.J. Peck, Jr.
Female Sex Steroids
Receptors and Function
1979. 116 figures, 18 tables. XII, 245 pages
ISBN 3-540-09375-3

Volume 13: H.F. De Luca
Vitamin D – Metabolism and Function
1979. 14 figures. VIII, 80 pages
ISBN 3-540-09182-3

Volume 12
Glucocorticoid Hormone Action
Editors: J.D. Baxter, G.G. Rousseau
1979. 176 figures, 58 tables. XIX, 638 pages
ISBN 3-540-08973-X

Springer-Verlag
Berlin
Heidelberg
New York

Volume 11: S. Ohne
Major Sex-Determining Genes
1979. 34 figures, 6 tables. XIII, 140 pages
ISBN 3-540-08965-9

Volume 10: W.I.P. Mainwaring
The Mechanism of Action of Androgens
1977. 12 figures, 17 tables. XI, 178 pages
ISBN 3-540-07941-6

Volume 9: R.E. Mancini
Immunologic Aspects of Testicular Functions
1976. 36 figures, 8 tables. IX, 114 pages
ISBN 3-540-07496-1

Volume 8: E. Gurpide
Tracer Methods in Hormone Research
1975. 35 figures. XI, 188 pages
ISBN 3-540-07039-7

Volume 7: E.W. Horton
Prostaglandins
1972. 97 figures. XI, 197 pages
ISBN 3-540-05571-1

Volume 6: K. Federlin
Immunopathology of Insulin
Clinical and Experimental Studies
1971. 53 figures. XIII, 185 pages
ISBN 3-540-05408-1

Volume 5: J. Müller
Regulation of Aldosterone Biosynthesis
1971. 19 figures. VII, 137 pages
ISBN 3-540-05213-5

Volume 4: U. Westphal
Steroid-Protein Interactions
1971. 144 figures. XIII, 567 pages
ISBN 3-540-05313-3

Volume 3: F.G. Sulman
Hypothalamic Control of Lactation
In collaboration with numerous experts.
1970. 58 figures. XII, 235 pages
ISBN 3-540-04973-8

Volume 2: K.B. Eik-Nes, E.C. Horning
Gas Phase Chromatography of Steroids
1968. 85 figures. XV, 382 pages
ISBN 3-540-04277-6

Volume 1: S. Ohno
Sex Chromosomes and Sex-linked Genes
1967. 33 figures. X, 192 pages
ISBN 3-540-03934-1

Springer-Verlag
Berlin
Heidelberg
New York

MIX
Papier aus verantwortungsvollen Quellen
Paper from responsible sources
FSC® C105338

If you have any concerns about our products,
you can contact us on
ProductSafety@springernature.com

In case Publisher is established outside the EU,
the EU authorized representative is:
Springer Nature Customer Service Center GmbH
Europaplatz 3, 69115 Heidelberg, Germany

Printed by Libri Plureos GmbH
in Hamburg, Germany